Dramatic Effect of Cross-Correlations in Random Vibrations of Discrete Systems, Beams, Plates, and Shells

Isaac Elishakoff

Dramatic Effect of Cross-Correlations in Random Vibrations of Discrete Systems, Beams, Plates, and Shells

Springer

Isaac Elishakoff
Boca Raton, FL, USA

ISBN 978-3-030-40393-5 ISBN 978-3-030-40394-2 (eBook)
https://doi.org/10.1007/978-3-030-40394-2

This Springer imprint is published by the registered company Springer Nature Switzerland AG
The registered company address is: Gewerbestrasse 11, 6330 Cham, Switzerland

Preface

The idea of this book occurred when I browsed through the book *This Idea Must Die: Scientific Theories That Are Blocking Progress* edited by John Brockman (2015), a famous publication of the online salon Edge.org. He edited numerous thought-provoking and mind boggling books, like *What Is Your Dangerous Idea? What Have You Changed Your Mind About? Is Internet Changing the Way You Think? This Explains Everything*, and many more, and I am sure, many more to come. Numerous answers of prominent thinkers of our time (he forgot to ask very many others) attempted to reply to Brockman's inquiry of naming the idea that must die. One of the first answers that interested me was given by Bart Kosko (Brockman, 2015), author of several definitive books such as Fuzzy Sets.

Kosko (Brockman, 2015) starts his entry by boldly stating that "It's time for science to retire the fiction of statistical independence". Why do people use this assumption? Kosko (Brockman, 2015) tells us: "An easy justification for this is that almost everyone else does it and it's in the textbooks. This assumption has to be one of the most widespread instances of groupthink in all science". Moreover, "we assume statistical independence because of its armchair appeal: It makes the math easy. It often makes the intractable tractable".

It appears instructive to review, in some detail, the case that vividly illustrates the misuse of the concept of independence. The Florida newspaper *Tallahassee Democrat* reported on May 7, 1983 (Hollander and Proshan, 1984, p. 60). "An Eastern Airlines L-lolljet carrying 172 people between Miami and Nassau lost its engine oil, its power and 12,000 feet of altitude over the Atlantic Ocean on Thursday before one of its three engines could be restarted and a safe landing made. Each engine was afflicted by a mysterious and nearby simultaneous loss of oil pressure and supply". Late in the afternoon, Eastern announced the discovery that an oil seal was missing from each engine. Traub said the crew was warned of low oil pressure and supply by indicator lights like those on the dashboard of a car. At first, they thought something was wrong with the lights. "They considered the possibility of a malfunction in the indication system because it's such an unusual thing to see all three with low pressure indications", Traub said. "The odds are so great that you won't get three indications like this. The odds are way out of sight, so the first thing you'd suspect is a problem with the indication system". The very next day, the same newspaper reported additional detail: "Mistakes by two mechanics caused all three engines of an Eastern Airlines jumbo jet to stall, dropping the aircraft perilously close to the Atlantic Ocean, federal investigators said Friday. One mechanic failed to install six tiny rubber seals on the engine's oil plugs during routine maintenance on the L-1011, leaving a gap for all of the oil to leak out when the engines were fired up, National Transportation Safety Board spokesman Ira Furman said". The pertinent question arises on how such a blunder could occur. On May 8, 1983 (Hollander and Proschan, 1984, pp. 60–61), *Tallahassee Democrat* provided answer to this inquiry: "The spokesman gave this account on the maintenance lapse: In the back shop Wednesday night, a mechanic pulled out of magnetized oil plugs for a routine check done after 25 h of flying. The magnet collects metal from the oil, and mechanics analyze the chips to see if any engine parts are wearing abnormally. The supervisor was out of the replacement plugs, so that mechanic went to the stock room to get one for each of the three Rolls Royce RB-211 engines. He and a partner screwed the plugs in, neither of them noticing that the sealer rings, called "O rings", were missing. The mechanic might have forgotten, Furlow said, because plugs kept at the supervisor's desk are equipped with rings. Stock room plugs are not".

Hollander and Proschan (1984, p. 65) provide the following comment on this incident: "One error that can lead to inflated estimates of reliability is the often unjustified assumption of *independence* between subsystems. The assumed independence may fail to be valid because the subsystems operate in a *common environment*. As an example, consider the L-1011 jet whose three engines stalled… The engines were subject to a common environment, one which allowed for the same O-ring error to be made on each engine during routine maintenance. This error caused *dependent* failures. Yet Traub said "…the odds are way out of sight…" because he was using an incorrect independence-based intuition."

Since my first paper with analogous ideas in the context of random vibrations of structures was exposed some 40 years ago (specifically in my 1977 paper, in several ones that followed, the latest one dated 2015), I thought it would be instructive to devote entire monograph to this topic.

Second reason for my excitement with this topic is a series of papers by Stephen Harry Crandall, MIT's late Ford Professor of Mechanical Engineering. Specifically, in 1976, I attended a symposium in the wonderful English city of Southampton organized by the talented researchers of ISVR, Institute for Sound and Vibration Research. There, Crandall gave a lecture on the surprising patterns observed in random vibrations of beams and plates. He wrote (1977): "When lightly damped uniform continuous structures are excited by wide-band stationary random forces applied at isolated points, the spatial distributions of the mean square velocity exhibit **surprisingly regular patterns** (by present writer—I.E.) if the number of modes responding is sufficiently large. This was demonstrated for taught strings and rectangular plates (Crandall and Wittig, 1972) and for circular plates (Crandall, 1974). Similar results have also been obtained for rectangular membranes (Young,

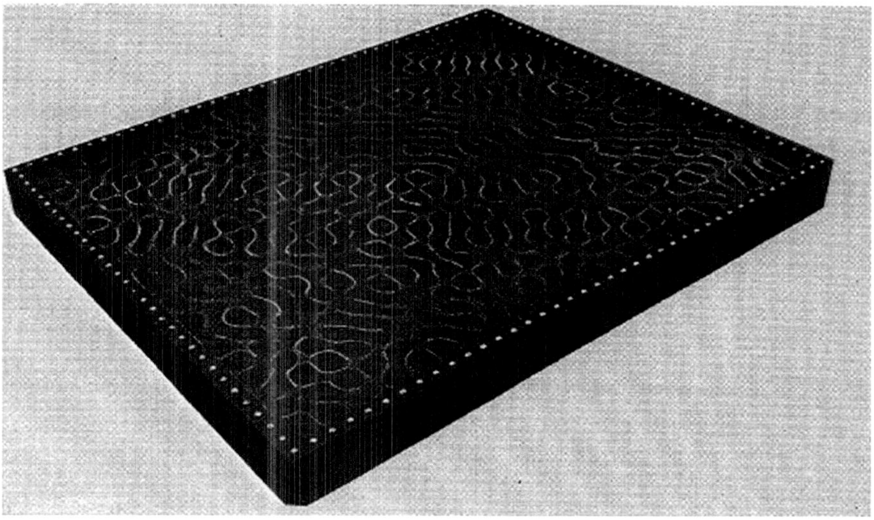

Fig. 1 Chladni pattern arising in rectangular plate due to deterministic harmonic excitation

1975) and contain trapezoidal plates (Lee, 1975)." Figures 1 and 2 show the patterns described by Crandall (1977). (See also Figs. 3–6.)

During his lecture at Southampton, Prof. Crandall was asked by Prof. K.V. Frolov some questions. Fortunately, there was no need for good memory or detailed files to recall the questions. Professor Brian Clarkson, organizer of the Symposium in Southampton, and the Director of the ISVR ably organized the recordings of the

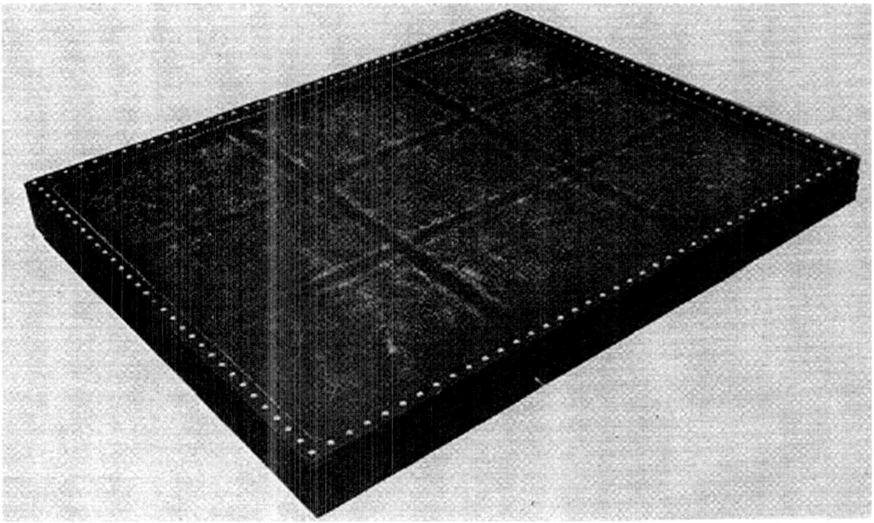

Fig. 2 Chladni pattern arising in rectangular plate due to random wide-band excitation

question/answer sessions which have been recorded in the proceedings (Clarkson, Hammond, Holmes, and Kistner, 1977). Thus, the paper by Crandall (1977, pp. 388–389) is followed by Discussion, in which "Prof. K.V. Frolov commented on the development of the special random theory from one dimension to two dimensions and asked if it is possible to extend these ideas for application to shell structures. Professor Crandall replied that there are great experimental difficulties with shells… In principle, the analytical methods work well. Bolotin and Elishakoff (1971) have done work on shells but they were not looking for order in patterns of surface motion".

I asked Prof. Crandall if the cross-correlation effect was important in the problems considered by him. He replied that the cross-correlations were not included. I suggested that we cooperate on this topic to study the effect of the neglected terms in the analysis. In the summer of 1976, I was spending at some German universities through the support of the DAAD (Deutsche Akademische Austauschdienst, German Academic Exchange Foundation). Specifically, in August 1976, I stayed at the University of Stuttgart, with Prof. John Argyris. His co-worker, Dr. Anton van Zanten got interested in cooperating on the effect of cross-correlations and the possibility of Crandall-like patterns in cylindrical shells. Thus our three-some cooperation ensued, culminating by publishing our paper (Elishakoff, van Zanten, and Crandall, 1979).

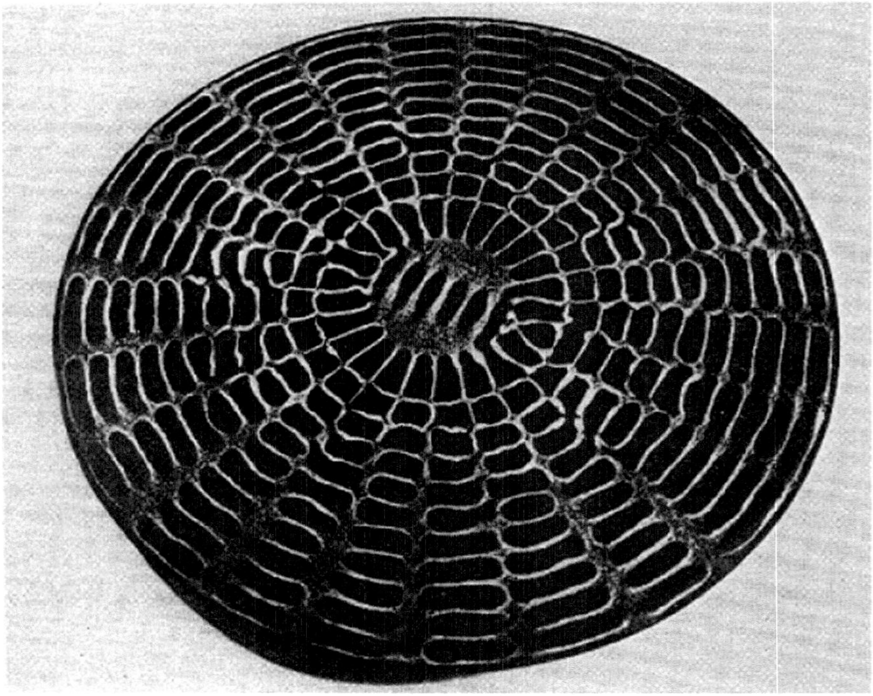

Fig. 3 Chladni pattern arising in a circular plate due to deterministic harmonic excitation

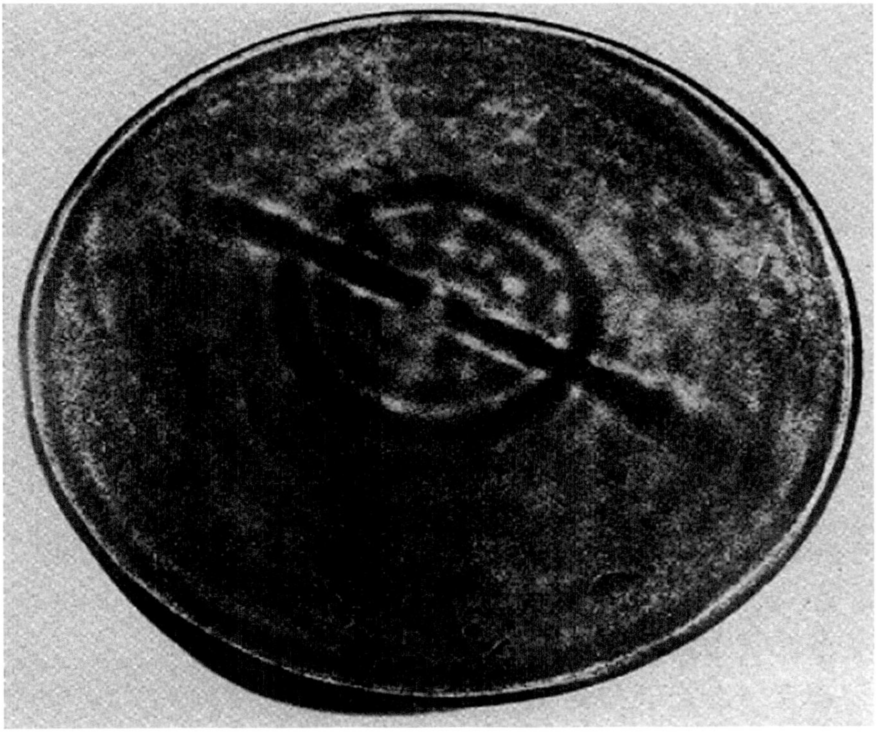

Fig. 4 Chladni pattern arising in a circular plate due to random wide-band excitation

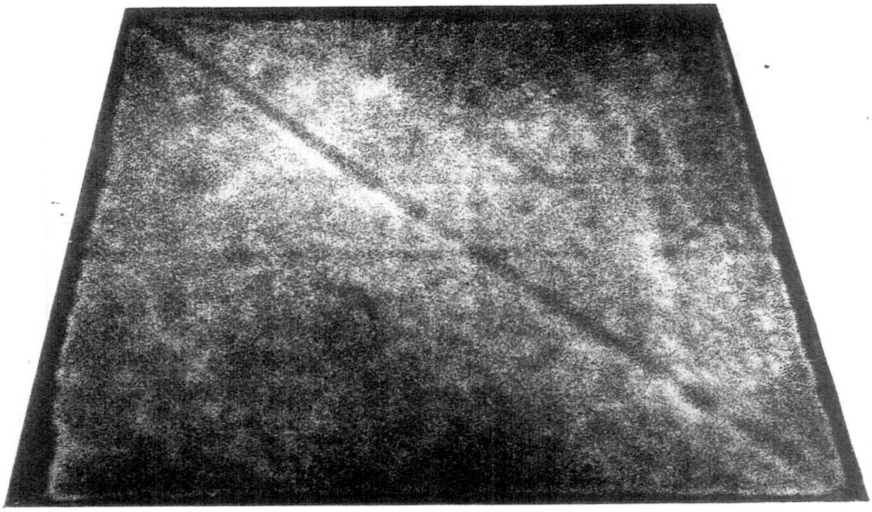

Fig. 5 Random vibration pattern of the plate under two concentrated random loads

Fig. 6 Some of the salt patterns created during wide-band random vibrations remind Union Jack, the national flag of Great Britain

It turned out that the probabilistic response at driving cross-section does not equal to response at the cross-section in the symmetric location, contrary to Gandall's research. His theoretical results about equality of response were paradoxal; they were the artifact of the neglect of cross-correlations; in strings and beams there were small, but not zero.

Twenty years later, Crandall (1996) addressed the topic of selection of successful research subjects. He argued: "Significant research results in Mechanics are produced by hard work, a little ingenuity, and a lot of good luck. The most important factor in the last of these is the choice of problem. Many young researchers pick problems, which are natural extensions of their doctoral thesis. My first efforts applied to determination of nonlinear vibration problems produced only dull and unremarkable papers. Essentially all of my important papers are based on early work on problems that were new to me. Where did these problems come from? I can identify three main sources: consulting assignments, anomalous experimental results, and the preparation of survey articles".

Crandall (1996) specifically mentions how he started to deal with random vibrations: "I have had the good fortune to have encountered several…problems [exhibiting] new features which require basic research. The most significant was the problem of payload vibration of rocket-boosted vehicles. As a consultant to the Air Force during the early days of rocket development, I witnessed several fiery failures and was able to participate in the development of the discipline now called Random Vibration".

Crandall (1996) also emphasizes the importance of experiments: "Although Mechanics is now predominantly analytical discipline, it was developed to explain the results of experimental observation, and the experiment continues to be an inspiration for further development in the subject. I was fortunate to be associated with the MIT Acoustics and Vibration Laboratory for over 50 years, and often used unexpected experimental results as the starting point for subsequent investigation. The most dramatic was the accidental discovery of symmetric patterns of statistical response enhancement in structures excited by wide-band random excitation. Prior to making a routine measurement, we had sprinkled salt on a plate to test whether the exciter was working. The inverse Chladni patterns which appeared was a complete surprise, and provided the stimulation for a research program which lasted over ten years". It should be stressed that chladni patterns obtained by Crandall for random vibrations of rectangular plates (Figs. 3 and 5) show only the qualitative behavior. Chladni patterns could not detect the effect of cross-correlations, neglected by Crandall. This topic is the heart of this book, namely the dramatic effect the cross-correlations may play.

Here we come to Ernst Florens Friedrich Chladni (1756–1827) who was a German physicist and musician. He is famous for his most important research on vibrating plates for which he sometimes is labeled as the father of acoustics. According to Wikipedia, "Chladni's best-known achievements was inventing a technique to show the various modes of vibration on a rigid surface. When resonating, a plate or membrane is divided into regions that vibrate in opposite directions, bounded by lines where no vibration occurs (nodal lines). Chladni repeated the pioneering experiments of Robert Hooke (1635–1703) who, on July 8, 1680, had observed the nodal patterns associated with the vibrations of glass plates. Hooke ran a violin bow along the edge of a plate covered with flour and saw the nodal patterns emerge. Chladni's technique, first published in 1787 in his book *Entdeckungen über die Theorie des Klanges* ("Discoveries in the Theory of Sound"), consisted of drawing a bow over a piece of metal whose surface was lightly covered with sand. The plate was bowed until it reached resonance, when the vibration causes the sand to move and concentrate along the nodal lines where the surface is still, outlining the nodal lines. The patterns formed by these lines are what are now called *Chladni figures*".

Stöckmann (2007) describes in following words the meeting between Chladni and Napoleon: "On a February evening of the year 1809 in Paris, a carriage drove up before Tuilerie Palace, the official residence of Napoleon Bonaparte, emperor of the French people. The passengers were the Count de la Place, chancellor of the senate, La Cépède, great chancellor of the Legion of Honour, Bertholet, senator of the empire, and a Dr. Chladni from Wittenberg. Chladni had become known for his experiments on vibrating plates, and stopped for a longer stay in Paris while on a lecture tour through Europe. There he got to know the leading contemporary scientists in person. Apart from the three already mentioned, the mathematician Count de la Place, better known as Laplace, the chemist Bertholet, and the biologist La Cépède, those were among others Poisson, Savart, and Biot, as well as Alexander von Humboldt, who stayed in Paris at the same time. Napoleon was interested in a demonstration of Chladni's experiments and invited him to the Tuileries through the mediation of Laplace. A February evening of the year 1809 in Paris, a carriage drove up before Tuilerie Palace, the official residence of Napoleon Bonaparte, emperor of the French people. The passengers were the Count de la Place, chancellor of the senate, La Cépède, great chancellor of the Legion of Honour, Bertholet, senator of the empire, and a Dr. Chladni from Wittenberg. While performing artists were rather often invited to court, the invitation of a scientist was a singularity (Figs. 7–11).

Fig. 7 Chladni meets Napoleon Bonaparte (extracted from Kollbrunner and Meister, 1958)

Fig. 8 Lichtenberg used electrically charged plates with (magnetized) powder, being a forerunner of Chladni

Smiliansky and Stöckmann (2007) emphasize: "Chladni's discovery of the sound figures which now bear his name, belongs to the group of crucial discoveries which shaped the development of modern physics in its formative years. During his time, chladni's discovery caused quite a stir in both the professional and lay public. The impact of Chladni's work persists till this very day and the study of nodal structures continues to attract the attention of mathematicians and physicists alike". It must be noted that according to Bremer (1999), Chladni figures had a forerunner, namely Georg Christoph Lichtenberg (1742–1799): "When G. C. Lichtenberg (1779) published his famous figures …, he would probably not have guessed how much he would influence plate vibration research with his contribution. However, his visualization of electrically charged plates with (magnetized) powder inspired E. F. F. Chladni who asked himself whether dusted plates would also show figures once they were forced to vibrate" (see also the paper by Takahashi, 1979 which is reviewing two hundred years of Lichtenberg figures).

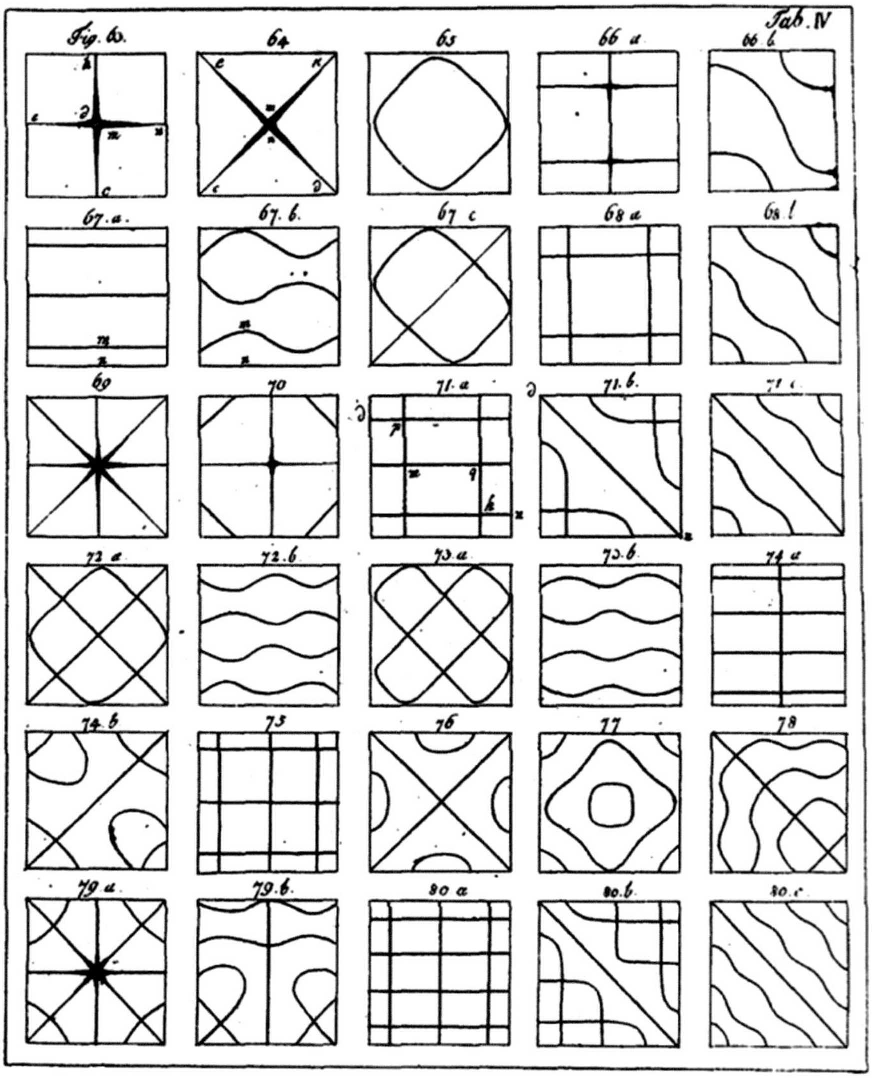

Fig. 9 Chladni patterns in rectangular plates, deterministic case

Chladni patterns are discussed extensively in the literature in numerous contexts. The interested reader can consult with papers by Wheatstone (1833), da C. Andrade and Smith (1931), Colwell (1932a, 1932b, 1934, 1936), Waller (1938, 1954, 1957, 1961), Stewart and Colwell (1939), Colwell, Atewarta, and Arnett (1940), Colwell, Stewart, and Friend (1939), Jensen (1955), Ravenhall and Som (1973), Ikushima H., and Ohji (1974), Langley and Taylor (1979), Rossing (1982, 2005), Grinsted (1952), Bohringer, Bhatt, and Goldberg (1995), Comer, Shepard, Hendriksen, and Ramsier (2004), Acikalin, Bietoch, Dorrestijn, Gerber, Hegner,

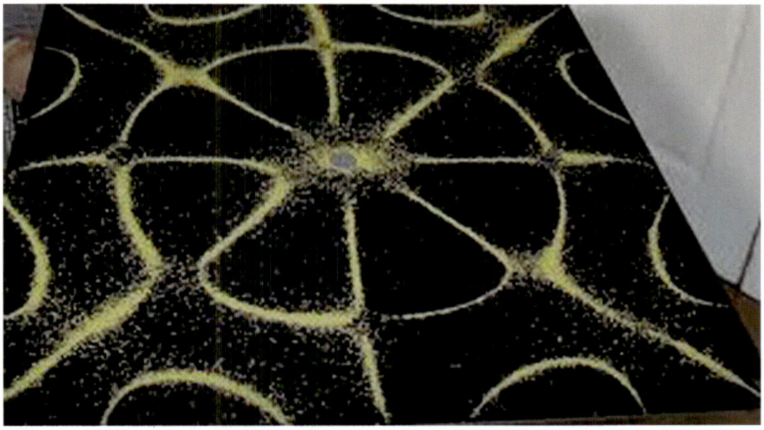

Fig. 10 Chladni pattern in a square plate, deterministic case

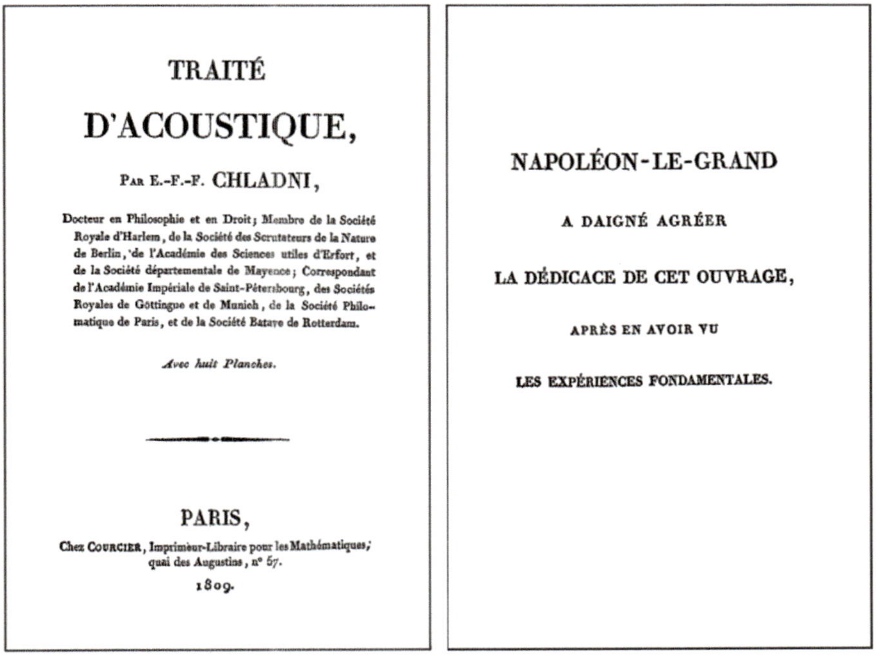

TRAITÉ

D'ACOUSTIQUE,

PAR E.-F.-F. CHLADNI,

Docteur en Philosophie et en Droit; Membre de la Société Royale d'Harlem, de la Société des Scrutateurs de la Nature de Berlin, de l'Académie des Sciences utiles d'Erfort, et de la Société départementale de Mayence; Correspondant de l'Académie Impériale de Saint-Pétersbourg, des Sociétés Royales de Göttingue et de Munich, de la Société Philomatique de Paris, et de la Société Batave de Rotterdam.

Avec huit Planches.

PARIS,

Chez COURCIER, Imprimeur-Libraire pour les Mathématiques; quai des Augustins, n° 57.

1809.

NAPOLÉON-LE-GRAND

A DAIGNÉ AGRÉER

LA DÉDICACE DE CET OUVRAGE,

APRÈS EN AVOIR VU

LES EXPÉRIENCES FONDAMENTALES.

Fig. 11 Book by Chladni, dedicated to Napoleon Bonaparte

Meyer, and Raman (2007), Gough (2007), Stöckmann (2007), Ullmann (2007), Van Gerner, Van der Hoef, Van der Meer, and Van der Weele (2010), Gander and Kwok (2012), Gbur (2013), Müller (2013), Rieux (2013), Taillan, Combe, and Morillo (2015), Tuan, Wen, Chiang, Liang, Hunag, and Chen (2015),

Trejo-Mandujano, Mijares-Bernal, and Ordoñez-Casanova (2015), Skrodzki, Reitebuch, and Polthier (2016), Arango and Reyes (2016), Pala and Azgın (2016), Zhou, Sariola, Latifi, and Liimatainen (2016), Mould (2016), Grabec (2017), Latifi, Wijayaand Zhou (2017), Hunter (2017), Irving and Patel (2018), and possibly others.

Bart Kosco concludes his essay with these bold words: "Science needs to take seriously its favorite answer: It depends". One surely has to agree with his statement that science should retire the "fiction of statistical independence".

As Lancaster (1959) notes, "it is well known that zero correlation between two variables does not imply their mutual independence. Examples are easily constructed in which X is a function of Y and yet two variables are uncorrelated; the bivariate distribution consisting of a uniform distribution of mass over the circle with unit radius is such an example. Often it is said that uncorrelated normal errors are independent. This statement is rather too general. It is true, however, if the errors are independently distributed or more generally if the errors have a joint normal distribution ... But it is not enough to specify that the marginal variables are normal...".

Fig. 12 Stephen Harry Crandall (1920–2013)—uniformly considered as the father of the discipline of Random Vibrations

Fig. 13 Cooperation with S. H. Crandall (left) of the present writer (right) resulted in the paper on random vibrations of cylindrical shells (Elishakoff, van Zanten, and Crandall) with attendant dramatic effect of cross-correlations

Fig. 14 V. V. Bolotin (1926–2008) (in the middle) developed criterion of neglect of cross-correlations in case of viscous damping; present writer (on the left) provided criterion for structural damping; Y. K. Lin (b. 1923) mentioned effect of cross-correlations in his text book (Lin, 1967)

How does this translate to structures be they discrete or continuous? Extremely many researchers neglect the cross-correlation terms in the random vibration analysis because *sometimes* they are very small. We show that in many cases, the effect of cross-correlations is *dramatically* changing the structural response of discrete systems, as well as beams, plates, and shells. Thus, not less important conclusion, than that made by Kosko (Brockman, 2015), is shown in this book to be in order: "Science and engineering should retire the acceptability of omission of cross-correlations between different mode shapes of structures" (Figs. 12–14).

Boca Raton, USA Isaac Elishakoff

References

Acikalin, T., Bietoch, A., Dorrestijn, M., Gerber, C., Hegner, M., Meyer, E., et al. (2007). Chladni figures revisted Based on nanomechanics. *Physical Review*, *98*, article 026102.

da C. Andrade, E. N., & Smith, D. H. (1931). The method of formation of sand figures on a vibrating plate. *Proceedings of the Physical Society*, *43*(4), 405.

Arango, J., & Reyes, C. (2016). Stochastic models for chladni figures. *Proceedings of the Edinburgh Mathematical Society*, *59*(2), 287–300.

Bohringer, K. F., Bhatt, V., & Goldberg, R. Y. (1995), Sensorless manipulation using transverse vibrations of a plate. In *Robotics and Automations Proceedings, 1995 IEEE International Conference on Robotics and Automations* (Vol. 2, pp. 1989–1996). New York: IEEE Press.

Bolotin, V. V., & Elishakoff, I. (1971). Random vibrations of elastic shells containing an acoustic medium. *Mechanics of Solids*, *6*, 99–107.

Bremer, H. (1999). On the dynamics of elastic multibody systems. *Applied Mechanics Reviews*, *52* (9), 275–296.

Brockman, J. (Ed.). (2015). This idea must die: Scientific theories that are blocking progress. New York: Harper Perennial.

Clarkson, J. K., Hammond, P. J., Holmes, & Kistner, A., (Eds.). (1977). *Stochastic problems in dynamics*. London: Pitman.

Colwell, R. C. (1936). Chladni figures on square plates. *Journal of the Franklin Institute*, *221*(5), 635–652.

Colwell, R. C. (1934). Chladni plates at high frequencies. *Nature*, *133*, 258.

Colwell, R. C. (1932a). Higher vibrations of chladni plates. *Nature*, *130*, 701.

Colwell, R. C. (1932b). Related nodal lines on chladni plates. *The Journal of the Acoustical Society of America*, *4*(1A), 3.

Colwell, R. C., Stewart, J. K., & Friend, A. W. (1939). Symmetrical figures on circular plates and membranes. *Philosophical Magazine*, *27*(7), 123.

Comer, J. R., Shepard, M. J., Hendriksen, P. N., & Ramsier, R. D. (2004). Chladni plates revisted. *American Journal of Physics*, *72*, 1345–1346.

Crandall, S. H., & Wittig, L. E. (1972). Chladni's patterns for random vibration of a plate. In G. Herrmann & N. Perrone, (Eds.), *Dynamic response of structures*, (pp. 35–71), New York: Pergamon Press.

Crandall, S. H. (1996). Selection of successful research subjects: Message from Prof. S.H. Crandall. *Journal of the Society of Mechanical Engineers (Japan)*, *99*(7), No. 932, 527–529.

Crandall, S. H. (1977). Structured response patterns due to wide—Band random excitation. In B. L. Clarkson, (Ed.), *Stochastic Problems in Dynamics*, Pitman, London, 366–389.

Crandall, S. H. (1974). Wide-band random vibration of structures, In *Proceedings of the 7th U.S. National Congress of Applied Mechanics*, (pp. 131–138), New York: ASME Press.

Elishakoff, I., Van Zanten, A. Th., & Crandall, S. H. (1979). Wide-band random axisymmetric vibration of cylindrical shells. *Journal of Applied Mechanics*, *46*(2), 417–423.

Gander, M., & Kwok, F. (2012). Chladni figures and the Tacoma Bridge: Motivating PFE eigenvalue problems via vibrating plates. *SIAM Review*, *54*(3), 573–596.

Gough, C. (2007). The violin: Chladni patterns, plates shells and sounds. *European Physics Journal-Special Topics*, *145*, 77–101.

Grabec, I. (2017). Vibration driven random walk in a Chladni experiment. *Physics Letters A.*, *381* (2), 59–64.

Grinsted, B. (1952). Nodal pattern analysis, *Proceedings of the Institution of Mechanical Engineers*, Series A, (Vol. 166, pp. 309–326).

Hollander, M., & Proschan, F. (1984). *The statistical exorcist: Dispelling statistics anxiety*. New York: Marcel Dekher.

Ikushima, H., & Ohji, K. (1974). Observation of Chladni figures in piezoelectric ceramic plates of rectangular form, *Japanese Journal of Applied Physics*, *13*(1), 28.

Irving, A. D., & Patel, E. L. (2018). Chladni componenets, *Mathematics Today*, *54*(4), 158–161.

Jensen, H. C. (1955). Production of Chladni figures on vibrating plates using continous excitation. *American Journal of Physics*, *23*(8), 503–505.

Kollbrunner, C. F., & Meister, M. (1958). *Ausbeulen: Theorie und Brechnung von Blechen* (p. 4). Berlin: Springer (in German).

Lancaster, H. O. (1959). Zero correlation and independence. *Australian Journal of Statistics*, *1*. 53–56.

Langley, A. J., & Taylor, P. H. (1979). Chladni patterns in random vibration. *International Journal of Engineering Science*, *17*, 1039–1047.

Latifi, K., Wijaya, H., & Zhou, Q. (2017). Multi-particle acoustic manipulation on a chladni plate, In *Manipulation, Automation and Robotics at Small Scales (MARSS), 2017 International Conference* (pp. 1–7). New York: IEEE Press.

Lichtenberg, G. C. (1779). On a new research method for the nature and the motion of electric material. In *Commentationes Societatis Regiae Scientiarum Gottingensis, Classis Mathematicae*. (Vol. 1, pp. 65–79). (in Latin) (see also lso in German, Leipzig: Akademie Verlag, 1956).

Lin, Y. K. (1967). *Probabilistic theory of structural dynamics*, (Vol. 7, 203–252), New York: McGraw-Hill.

Müller, T. (2013). Numerical Chladni figures, *European Journal of Physics*, Vol. *34*(4), 1067–1074.

Pala, S., & Azgın, K. (2016). A MEMS square Chladni plate resonator, *Journal of Micromechanics and Microengineering*, *26*(10), article 105016.

Ravenhall, F. W. & Som, A. K. (1973). Some recent observations on Chladni's figures, *Acustica*, *29*(1), 14–21.

Rieux, F. (2013). Discrete simulation of a Chladni experiment, *Mathematical Morphology and Its Application to Signal and Image Processing*, pp. 496–507.

Rossing, T. D. (1982). Chladni law for vibrating plates, *American Journal of Physics*, *50*(3), 271–274.

Rossing, T. D. (2005). Comment on Chladni plates revisted, *American Journal of Physics*, *73*, 282–283.

Skrodzki, M., Reitebuch, U., & Polthier, K. (2016). Chladni figures revisited: A peek into the third dimension. In *Proceedings of Bridges 2016: Mathematics, Music, Art, Architecture, Education, Culture* (pp. 481–484). Tessellations Publishing.

Smiliansky, U., & Stöckmann, H. J. (2007). Preface, *European Physical, Special Topics*, *145*, 1–2.

Stöckmann, H. J. (2007). Chladni meets Napoleon, *The European Physical Journal-Special Topics*, *145*(1), 15–232.

Taillan, C., Combe, N., & Morillo, J. (2015). Chladni figures at the nanoscale, *The European Physical Journal B*, *88*(12), 317.

Takahashi, Y. (1979). Two hundred years of lichtenberg figures, *Journal of Electrostatics*, *6*(1), 1–13.

Trejo-Mandujano, H. A., Mijares-Bernal, G., & Ordoñez-Casanova, E. G. (2015). Alternate model of Chladni figures for the cicular homogeneous thin plate case with open boundaries, *Journal of Physics Conference Series*, *582*(1), article 012022.

Tuan, P. H., Wen, C. P., Chiang, P. Y., Yu, Y. T., Liang, H. C., Huang, K. F. et al. (2015). Exploring the resonant vibration of thin plates: Reconstruction of Chladni patterns and determination of resonant wave numbers, *The Journal of the Acoustical Society of America*, *137*(4), 2123.

Ullmann, D. (2007). Life and work of E.F.F. Chladni, *European Physics Journal Special Topics*, *145*, 25–32.

Van Gerner, H. J., Van der Hoef, M. A., Van der Meer, R. M., & Van der Weele, J. R. (2010). Inversion of Chladni patterns by tunning vibrational acceleration. *Physical Review E*, *82*(1) article 012301.

Waller, M. D. (1954). Critical review of earlier Chladni figures. *Acta Acustica United with Acustica*, *4*(6), 677–680.

Waller, M. D. (1961). *Chladni figures*. London: G Bell.

Waller, M. D. (1957). Interpreting Chladni figures. *American Journal of Physics*, *25*(3), 157–158.

Waller, M. D. (1938). Vibrations of free circular plates. Part 3: A study of Chladni's original figures. *Proceedings of the Physical Society*, *50*(1), 83.

Wheatstone, C. (1833). On the figures obtained by strewing sand on vibrating surfaces, commonly called acoustic figures, *Philosophical Transaction of the Royal Society*, *123*, 593–637.

Young, L. E. (1975). On the dynamic response of damped membrane to a point stochastic load. *Acoustics and Vibration Laboratoty Report 81678-2*. Cambridge, MA: MIT.

Zhou, Q., Sariola, V., Latifi, K., & Liimatainen, V. (2016). Controlling the motion of multiple objects on a Chladni plate. *Nature Communications*, *7*, article 12764.

Contents

Chapter 1
Normal-Mode Approach in Random Vibrations

1.1 Introductory Comments

The normal-mode method (see e.g. Bolotin, 1969; Crandall & Mark, 1963; Elishakoff, 1983, 1999, 2017; Lin, 1967) yields the formula shown in Eq. 1.1 for the mean-square displacement at any point ρ of the structure

$$E\left[W^2(\rho,t)\right]=\sum_{j=1}^{\infty}a_{jj}\psi_j^2(\rho)+\sum_{\substack{j=1\\ j\neq k}}^{\infty}\sum_{k=1}^{\infty}a_{jk}\psi_j(\rho)\psi_k(\rho) \tag{1.1}$$

where ρ is a vector defining the position of points on the surface, $\psi_j(\rho)$ and $\psi_k(\rho)$ are the mode shapes of the structure in vacuo; a_{jk} are generally complex-valued numbers, depending on the probabilistic characteristics of the external loading and the deterministic characteristics of the structure itself; j and k are single indices for one-dimensional continuous structures (such as string or beam) and replaced by the paired indices for two-dimensional structures (such as plates or shells).

The first sum in Eq. 1.1 represents the contribution of identical modes in formulating the mean-square displacement (referred to as direct correlation terms). The second, double sum represents the contribution of the interactions between separate modes (referred to as cross-correlations).

The a_{jk} are given by Eq. 1.2

$$a_{jk}=\int_{-\infty}^{\infty}S_{jk}(\omega)H_j^*(\omega)H_k(\omega)d\omega \tag{1.2}$$

where S_{jk} is defined by Eq. 1.3, for a one-dimensional structure,

I. Elishakoff, *Dramatic Effect of Cross-Correlations in Random Vibrations of Discrete Systems, Beams, Plates, and Shells*,
https://doi.org/10.1007/978-3-030-40394-2_1

$$S_{jk} = \frac{1}{v_j^2 v_k^2} \int_{-\infty}^{\infty} \int_{-\infty}^{\infty} S_q(\rho_1, \rho_2, \omega) \psi_j(\rho_1) \psi_k(\rho_2) d\rho_1 \rho_2 \tag{1.3}$$

$H_j(\omega)$ is the complex frequency response function and v_j^2 is the norm of the function $\psi_j(\rho)$. In the particular case where the excitation cross-spectral density $S_q(\rho_1, \rho_2, \omega)$ is a separable function of spatial and frequency coordinates, i.e.

$$S_q(\rho_1, \rho_2, \omega) = \Phi_1(\rho_1, \rho_2) \Phi_2(\omega) \tag{1.4}$$

and moreover, the loading is a spatial white noise or "rain-on-the-roof-excitation"

$$\Phi_1(\rho_1, \rho_2) = \delta(\rho_2 - \rho_1) \tag{1.5}$$

the cross-terms ($j \neq k$) in Eq. 1.3 identically vanish due to the orthogonality of the normal modes; $\delta(x)$ in Eq. 1.5 is Dirac's delta function. This is presumably the only case when the cross-correlation terms a_{jk} disappear in Eq. 1.1 and the mean-square displacement contains only direct correlations.

Another possibility of the cross-terms not playing any role is as follows: suppose hypothetically that the acceptability condition for actual exploitation of the structure is determined not by the maximum mean-square response, but rather by the surface-averaged mean-square response (Eq. 1.6, where R is the surface area of the structure). Here too, due to the orthogonality of the normal modes, only direct correlations remain (Eq. 1.7).

$$\langle E[w^2(\rho, t)] \rangle = \int_R E[w^2(\rho, t)] d\rho. \tag{1.6}$$

$$\langle E[w^2(\rho, t)] \rangle = \sum_{j=1}^{\infty} a_{jj} \int_R \psi_j^2(\rho) d\rho. \tag{1.7}$$

Such an integral characterization is obviously not enough, and usually the cross-correlations terms are present. They are, however, often neglected in random vibration literature because of the numerical difficulties involved in their inclusion combined with the general premise that the model density of the natural frequencies should not be high. This premise, however, is not borne out by practice: in shell-type structures, the modal density is high either at the low end of the frequency spectrum (e.g. in a spherical shell) or in the vicinity of some specific frequencies (e.g. in a cylindrical shell). It has been shown qualitatively by Bolotin (1969) that the integrand in Eq. 1.2 is large, when two resonant peaks overlap within the excitation band. This happens when natural frequencies crowd together, or the damping is sufficiently large. The dramatic consequences of omission of the cross-correlations were demonstrated for the shallow spherical shell by Elishakoff (1977a, 1977b, 1977c, 1977d). Of particular interest in this respect is generalization of Crandall's problem (see Crandall 2001;

Elishakoff, van Zanten, & Crandall 1979) (that of beams and plates driven by a point force, random in time) to the cylindrical shell under ring loading. It was shown that inclusions of the cross-correlation terms change the probabilistic response both qualitatively and quantitatively. The span-wise response, which is symmetric in the absence of the cross-correlations, is in basis of its direct correlations along would be underdesigned by factor of three or more (Elishakoff et al., 1979), with the attendant dangerous consequences for its performance.

In the above examples, the natural frequencies of the structure happen to lie very close together, in some structures, they may actually coalesce, in which case, a large error would be introduced if their joint effects are omitted. Note that two-degree-of-freedom systems were studied without recourse to the normal-mode method (which implies automatic inclusion of the cross-correlations) by Crandall and Mark (1963) and many-degree-of-freedom systems were studied recently via the normal-mode method, with investigation of the attendant effect of cross-correlation, by Elishakoff (1983, 1999) in particular, for a two-degree-of-freedom analog of the Crandall problem with only direct correlations included, the resulting error was 50% for the first mass, and the three natural frequencies coalesce, the corresponding error was 30%. In another recent example (Lubliner & Elishakoff, 1986) of an *N*-degree-of-freedom system, it was shown that as the number of degrees of freedom tends to infinity, the error introduced by omission of the cross-correlation term reaches 70%, for the driven mass response.

Despite these findings, many researchers still neglect the cross-correlations in their analyses. In order to "convince" the random vibration analyst of the necessity to check first whether cross-correlation terms are negligible, before they are actually omitted, it appeared to the writer that a striking, but a simple example should be constructed to elucidate this point.

Bolotin and Chirkov (1978, p. 294) reiterate: "We assume, that the following conditions are met $\epsilon_j \sim \epsilon_k \sim \epsilon$, $\left|\omega_j - \omega_k\right| \ll \left(\omega_j + \omega_k\right)/2$. If α is large, then for small damping condition

$$K_{u_j u_k}(0) \ll K_{u_j}(0)(j \neq k) \tag{1.8}$$

is holding, if $\epsilon^2 \ll \left(\omega_j - \omega_k\right)^2$, i.e. natural frequencies should not be too close to each other.

If $\left|\omega_j - \omega_k\right| \sim \left(\omega_j + \omega_k\right)/2$, then condition (1.8) will be satisfied for small damping.

If $\alpha \sim \epsilon$, then condition (1.8) is holding if frequency ω_0 is close to one of the natural frequencies and (or) when natural frequencies are significantly far away from each other.

In other cases, cross-correlations cannot be neglected. Investigations show that for small damping, the contribution of the imaginary part of cross-spectral density of generalized forces can be neglected if the following condition holds $\epsilon \ll \left|\omega_j - \omega_k\right|$."

Preumont (1994, p. 111) writes: "In order to examine the relative importance of the cross-correlations as compared to the autocorrelations, let us introduce

- the mean frequency

$$\omega_n = \frac{\omega_1 + \omega_2}{2} \tag{1.9}$$

- the reduced frequency spacing

$$\alpha = \frac{\omega_2 - \omega_1}{2\zeta\omega_n} \tag{1.10}$$

α expresses the spacing between the natural frequencies in terms of the bandwidth of the system. With these notations, we find

$$\begin{aligned}\frac{\text{cross-correlation}}{\text{auto-correlation}} &= 2\zeta\omega_n^3 \frac{8\xi}{(\omega_1 + \omega_2)\left[(\omega_1 - \omega_2)^2 + 4\xi^2\omega_1\omega_2\right]} \\ &= \frac{1}{1 + \alpha^2\left(1 - \xi^2\right)} \simeq \frac{1}{1 + \alpha^2}\end{aligned}. \tag{1.11}$$

This ratio is small if α is large, that is if *the spacing between the natural frequencies as much larger than the bandwidth of the oscillators.*

Clough and Penzien (1993, p. 546), when discussing random vibration of multi-degree-of-freedom systems, note: "assuming a low-damped system, that is $\xi < 0.1$, the cross-terms in [this] equation will be relatively small and can be neglected." They do not treat neither the case of large damping nor the closely spaced natural frequencies.

If the contribution of the cross-correlations is neglected, the overall mean-square response is the sum of the modal mean-square responses. This amounts to considering the modal responses as *statistically independent*, this is acceptable if the modes are *well separated* ($\alpha \gg 1$). This assumption is at the origin of the method known as the *SRSS rule (Square Root of the Sum of the Squares)*, which is widely used in seismic analysis. According to that rule, the maximum response of each mode, z_i, is computed separately using response spectra; next, they are combined according to

$$z_{\max}^2 = \sum_{i=1}^{m} z_i^2. \tag{1.12}$$

It is well known that this rule may lead to serious errors for closely spaced modes (e.g. see Der Kiureghian, 1981)."

Soong and Grigoriu (1993, pp. 190-191) state: "The modes with nearly the same frequency can be weakly correlated when the damping ratio is small, and the modes with quite different frequencies can be strongly correlated for large modal dampings. This observation suggests that present design practice in which the use of modal combination rules such as the absolute value sum (ABS) and square-root-of-the-sum-of-square procedure (SRSS), which is based only on modal frequency ratios, may not be satisfactory. According to the ABS and SRSS rules, maximum system response is estimated by $\sum_{i=1}^{n}|d_j|$ and $\left(\sum_{i=1}^{n} d_i^2\right)^{1/2}$, where d_i denotes maximum system response in mode i. Alternative modal combination rules have been proposed so that one can account for the correlation between modal coordinates (Grigoriu, 1980)." Grigoriu (1980) writes: "Conventional seismic design is usually based on approximations of the largest structural response obtained as a combination of modal response maxima. The absolute sum (ABS) or the square root of the sum of the squares (RSS) of modal response maxima are typical combination rules. They are recommended in building codes when the modal frequencies differ by less than 10% (*closely spaced modes*) or by more than 10% (*widely spaced modes*), respectively. Modal response maxims are given by the response spectrum. Recent results indicate that this *response spectrum method* for seismic design is not *risk consistent*, because it produces structures with very different safety levels (Grigoriu, 1979)." Grigoriu (1980, p. 113) stressed: "...the structures designed by RSS and ABS rules can be rational (e.g., less than 10% overdesigned or 5% underdesigned with respect to d_{RC}) if these rules are applied when [coefficient of correlation] $\rho < 0.1$ and $\rho > 0.8$, respectively. However, the ABS and RSS rules can be quite inaccurate for correlation coefficients in the range (0.1, 0.8). In this intermediate range of correlation, the design rule (Grigoriu, 1979, 1980; Ruiz, 1970)

$$d_{RRSS} = \sqrt{d_i^2 + d_j^2 + 2\rho d_i d_j},\ \left(\rho = \tilde{\rho}_{ij}\right) \tag{1.13}$$

is very accurate... Its use also involves a knowledge of the correlation coefficient in addition to the "modal response maxima." Here is a somewhat lengthy quote from the book by Wirsching, Paez, and Ortiz (1995, pp. 246).

It is generally thought if two natural frequencies are "close," then there is a significant dependency in q_1 and q_2 because of congruence of the gain functions. And if the two frequencies are "far apart," q_1 and q_2 will tend to be independent. No general operational definition for closeness exists.

Various methods of "combining modes" to compute the RMS response have been proposed. For example, STARDYNE has a selection of 15 mode combinations available. Given that the analyst must choose which modes are close, define the following:

N = total modes considered.
Z = total number of modes considered to be close.
H = number of sets equal or close eigenvalues.
K_j = number of close frequencies in any set.
σ_p^2 = mean-square value of $q_p(t)$.
ϕ_{ip} = mode vector, coordinate i in pth mode.

Some of the various methods used for computing the RMS of X_i from modal responses are as follows:

1. *Square Root of the Sum of Squares (SRSS)*

$$\sigma_{X_i} = \left[\sum_{p=1}^{N} \left(\phi_{ip}^2 \sigma_p^2\right)\right]^{1/2}. \tag{1.14}$$

   This method assumes that all modal responses are independent (see example in Sect. 9.3.9).
2. *Absolute Method*

$$\sigma_{X_i} = \sum_{p=1}^{N} \left(\left|\phi_{ip}\sigma_p\right|\right). \tag{1.15}$$

   This method assumes all modal responses have correlation coefficients of $\rho = 1, -1$.
3. *Naval Research Laboratory (NRL) Method*

$$\sigma_{X_i} = \left|\phi_{11}\sigma_{q1}\right| + \left[\sum_{p=2}^{N} \phi_{ip}^2 \sigma_p^2\right]^{1/2}. \tag{1.16}$$

   The first mode is assumed to be correlated with the sum (by SRSS) of all of the others.
4. *Close Method*

$$\sigma_{X_i} = \left\{\sum_{j=1}^{H}\left[\sum_{p=1}^{K_j}\left(\left|\phi_{ip}\sigma_p\right|\right)\right]^2 + \sum_{p=Z+1}^{N-Z}\left(\phi_{ip}^2\sigma_p^2\right)\right\}^{1/2}. \tag{1.17}$$

   It is assumed that there is a group of modes that is close and another group that is separated.
5. *Modified Root Sum of Squares (MRSS)*

$$\sigma_{X_i} = \left\{\sum_{j=1}^{H}\left[\sum_{p=1}^{K_j}\left(\left|\phi_{ip}\sigma_p\right|\right)\right]^2 + \sum_{p=Z+1}^{N-Z}\left(\phi_{ip}^2\sigma_p^2\right)\right\}^{1/2}. \tag{1.18}$$

   A more general form of the close method. Here, there are H groups of modes that are close. All other modes are separated (Richard et al., 1988).

6. *Combined Quadratic Combination (CQC)*

$$\sigma_{X_i} = \left(\sum_{j=1}^{N} \sum_{k=1}^{N} \sigma_{ij} \rho_{jk} \sigma_{ki} \right)^{1/2} \tag{1.19}$$

where

$$\begin{aligned} \sigma_{ij} &= \phi_{ij} \sigma_j \\ \rho_{jk} &= \frac{8\sqrt{\zeta_j \zeta_k}\left(\zeta_j + r\zeta_k\right) r^{3/2}}{\left(1 - r^2\right)^2 + 4\zeta_j \zeta_k r\left(1 + r^2\right) + 4\left(\zeta_j^2 + \zeta_k^2\right) r^2}. \\ r &= \frac{\omega_k}{\omega_j} \qquad k > i \end{aligned} \tag{1.20}$$

For example, for two modes

$$\sigma_{X_i}^2 = \sigma_{i1\rho 11}^2 + \sigma_{i1} \rho_{12} \sigma_{i2} + \sigma_{i2} \rho_{21} \sigma_{i1} + \sigma_{i2}^2 \rho_{22} \tag{1.21}$$

or

$$\sigma_{X_i}^2 = \sigma_{i1}^2 \sigma_1^2 + \phi_{i2}^2 \sigma_2^2 + 2\rho_{12} \phi_{i1} \phi_{i2} \sigma_1 \sigma_2 \tag{1.22}$$

(see Der Kiuregian, 1980; Wilson, Der Kiureghian, & Bayo, 1981)

Note that SRSS and other methods are extensively discussed by Singh (1980), Grigoriu (1980), Borino and Muscolinio (1986), Yiachao (1992), Der Kiureghian and Nakamura (1993), Wilson, Suharwardy, and Habibullah (1995), Sinha and Igusa (1995), Menun and Der Kiureghian (1998), Hernandes and Lopez (1998), Falsone (1999), Falsone and Muscolino (1999), Falsone and Neri (2001), Falsone and Muscolino (2004), Cocciola, Colajanni, and Muscolino (2004), Palmeri (2006), Walker and Stafford (2010), Cocciola and Muscolino (2011), Heredia-Zavoni (2011), Rojas, Kalashnykova, Rojas, Martinez, and Sifuentes (2012), Muscolino and Versaci (2014), Menum, Reyes, and Chopra (2015), and other authors.

Despite the fact that the book of Bolotin (1965) was uniformly available, many authors continue to neglect cross-correlations in random vibrations, overlooking case of equal or closely spaced natural frequencies. The reason probably is the fact that Bolotin never exemplified the importance of cross-correlations in his papers or books.

One of the vibration texts (Meirovitch, 2010) associates effects of direct and cross-correlations with the following integrals:

$$\int_{-\infty}^{\infty} |G_r|^2 d\omega = \int_{-\infty}^{\infty} \frac{d\omega}{\left[1 - [\omega/\omega_r)^2\right]^2 + [2\zeta_r \omega/\omega_r]^2}, \quad r = 1, 2 \tag{1.23}$$

$$\int_{-\infty}^{\infty}(\bar{G}_1G_2+G_1\bar{G}_2)d\omega$$
$$=\int_{-\infty}^{\infty}\frac{\{[1-\omega/\omega_1)^2][1-(\omega/\omega_2)^2]+(2\zeta_1\omega/\omega_1)(2\zeta_2\omega/\omega_2)\}d\omega}{\{[1-(\omega/\omega_1)^2]^2+(2\zeta_1\omega/\omega_1)^2\{[1-(\omega/\omega_2)^2]^2+(2\zeta_2\omega/\omega_2)^2\}}. \tag{1.24}$$

Then he notes: "The first integral can be evaluated using results from Sect. 12.11, but the second integral is likely to cause a great deal of difficulties. Because no new knowledge is gained from the evaluation of the integrals, we do not pursue the subject any further."

This statement of "no new knowledge" being gained from the evaluation of integral (1.24) is not correct. For closely spaced natural frequencies this term plays a dramatic role, changing the response both qualitatively and quantitatively. It is interesting that analogous mistake was made by famous French mathematician and mechanician Joseph Louis Lagrange (1736–1813). In his famous classical book *Analytical Mechanics,* Lagrange (1811, 1997) dealt with vibration theory of mechanical systems. According to Panovoko and Gubanova (1965, p. 169), "Lagrange touched briefly upon the particular case where some of the roots of the frequency equation are equal. In such cases, Lagrange writes, terms occur in the solution which contain the time outside of the sine function ("perennial" or "secular" terms). For example, when two roots are equal, we have a solution of the form

$$q_k=(A_k+B_kt)\sin(\rho t+\varphi). \tag{1.25}$$

Thus we see that as the time t increases, the first factor increases without limit, and the oscillations take on something of a resonant character, in that the amplitude approaches infinity." (Fig. 1.1).

In this context, the quote of Thomson and Tait (1879) appears to be relevant: "It is remarkable that both Lagrange and Laplace fall into the error of supposing that equality among roots necessarily implies that for stability the roots must be all unequal. This we find in the *Mécanique Analytique* of the second edition of 1811 published three years before Lagrange's death, and repeated without change in the posthumous edition of 1853. It would be curious if such an error had remained for twenty-three years in Lagrange's mind. It could scarcely have existed even during the writing and printing of the Article for his last edition if he had been in the habit of considering particular applications of his splendid analytical work: if he had he would have seen that a proposition which asserted that the equilibrium of a particle in the bottom of a frictionless bowel is unstable if the bowl be a figure of revolution with its axis vertical, cannot be true."

Lord Rayleigh (1894, 1945) repeats the statement made by Thomson and Tait (1879): "It is remarkable that both Lagrange and Laplace fell into the error of supposing that equality among roots necessarily implies terms containing t as a factor"

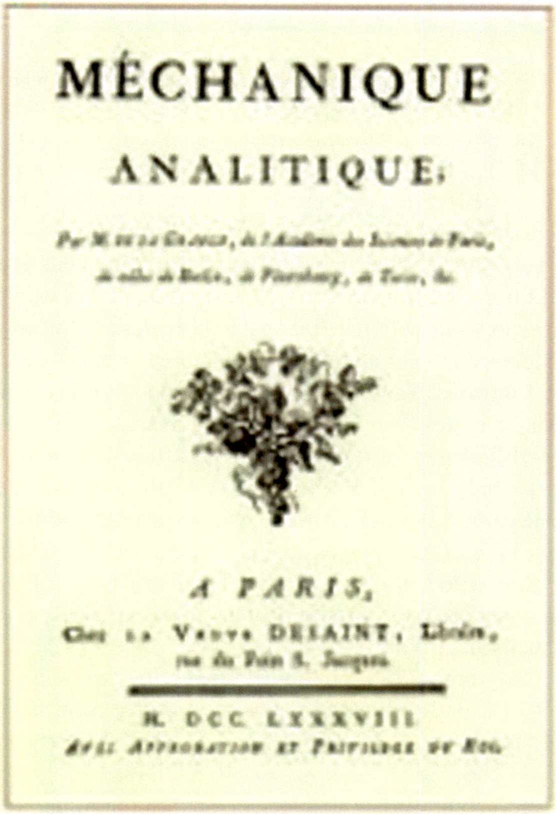
MÉCHANIQUE

ANALITIQUE.

A PARIS,

M. DCC. LXXXVIII.

Fig. 1.1 The title page of the book by Joseph Lagrange (1811) *Analytical Mechanics* where he stated that the case of equal roots meant instability, and hence beyond the need of consideration

Moreover, Panovko and Gubanova (1965, p. 170) stress: “the case of equal roots is easily realized physically, although the secular terms in the solution are inconsistent with common sense and the law of conversation of energy. However, Lagrange avoided discussing the paradox in the following words ‘But since the development of these cases would serve no useful purpose for the present investigation, we will not elaborate on it’., we shall not stop to consider them.” Readers may note that the statement “no new knowledge” being gained and “no useful purpose” sounds very similar! The present writer recalls, that one of the investigators (was it Lord Rayleigh?), later one remarked that had Lagrange conducted a thought-experiment on a deep spherical bowl allowing for a concentrated mass the oscillations in two directions, he would have discovered that the natural frequencies were identical, but there would be no instability!

Hurty and Rubinstein (1964, pp. 404-405) in their textbook claim: “The magnification factors $|H_r(\Omega)|$ have regions of pronounced peaks in the neighborhood of the corresponding natural frequencies ω_r. The products $|H_r(\Omega)||H_s(\Omega)|$ for $r \neq s$ are seen to be small in comparison with the same products for $r = s$. In addition, terms with $r \neq s$ may be negative as well as possible depending upon the sign of the product $\Phi_r(x)\Phi_s(x)\Gamma_r\Gamma_s$, while terms with $r = s$ are always positive. The contribution of cross product terms ($r \neq s$) to the mean-square response will be small. Hence, we approximate $y^2(t)$ …by disregarding the cross product terms." Authors overlook the case when ω_r and ω_s ($r \neq s$) are in close vicinity.

Somewhat analogous reasoning is employed by Argyris and Mlejnek (1991, p. 309): "The two functions V_i and V_j possess a peak at $\Omega \approx \omega_i$ and $\Omega \approx \omega_j$, i.e. at the locations of eigenfrequencies. Providing that the eigenfrequencies are well separated and $i \neq j$, one function displays a peak value while the other only has a very small value. In contrast, if $i = j$, the peak amplitudes are multiplied and in addition, the function $\cos(\varphi_i - \varphi_j)$ assumes a value of 1. Therefore, it often suffices to work with the diagonal values alone." Authors do not address the case when eigenfrequencies are not well separated.

Likewise, Yang (1985) in his textbook writes, when considering random vibration of plates: "we assume that the average modal overlap ratio r = (modal bandwidth)/(average modal spacing) is sufficiently small to permit approximations, that is we only include in the summation those terms of I_{jpdq} for which $j = k$, $p = q$." One cannot assume, however, that the average modal overlap ratio in plates is small, since plates may possess coinciding natural frequencies!

Lin (1967, p. 215) states: "If the structure is *lightly damped* and if the natural frequencies in the various normal modes are *well separated,* then the cross-terms are usually negligible."

Maymon (1998, pp. 59–61) treats the two-degree-of-freedom system shown in Fig. 1.2 with following data $m_1 = 1\,\mathrm{kg\,s^2/cm}$, $m_2 = 0.1\,\mathrm{kg\,s^2/cm}$, $\omega_p = \omega_s = 125\,\mathrm{rad/s}$, $\zeta_p = \zeta_s = 0.01$, and $S_0 = 550\,(\mathrm{kg})^2/(\mathrm{rad/s})$. Natural frequencies are $\omega_1 = 106.78862\,\mathrm{rad/s}$, $\omega_2 = 146.31709\,\mathrm{rad/s}$.

The mean-square displacements turn out to be

$$\begin{aligned} E\left(x_1^2\right) &= 0.07391 + 0.003937 + 0.0001677 \\ &= 0.0780147\,\mathrm{cm}^2 \end{aligned} \quad (1.26)$$

$$\begin{aligned} E\left(x_2^2\right) &= 0.10127 + 0.02873 - 0.0001678 \\ &= 0.129547\,\mathrm{cm}^2 \end{aligned} \quad (1.27)$$

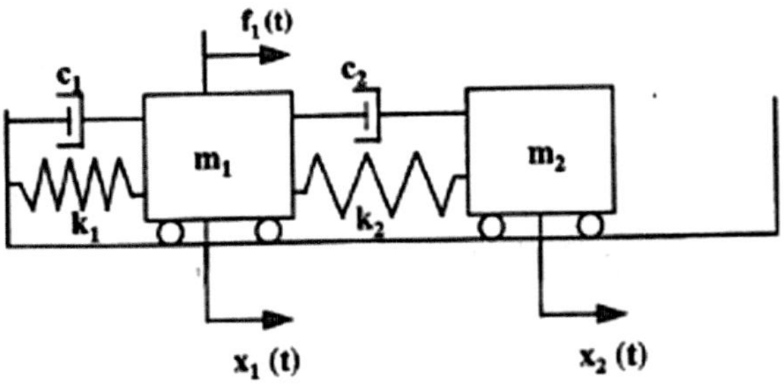

Fig. 1.2 Maymon's example: two-degree-of-freedom system subjected to random excitation $f_1(t)$

In Eq. (1.22), first and second terms represent the contributions of the direct correlations, in the first and second modes, respectively, whereas the third term constitutes the effect of cross-correlations. Maymon (1998, p. 62) provides the following discussion: "In the preceding solutions for the mean-square value of the responses, the three terms described earlier are presented separately. It can be seen that the contribution of the first resonance is larger than that of the second one. For modal dampings that are similar (though not equal), this is always the case, because of the term ω_i^3 in the denominator, the contribution of the (third) interaction (cross-correlation—I.E.) term is much smaller. This happens when the resonance frequencies are well separated from each other, due to the term $(\omega_i^2 - \omega_j^2)^2$ in Eq. (A.6)." In Maymon's (1998) example, the effect of cross-correlations turned out to be negligible, namely just under 1.5% for estimation of $E(x_1^2)$ and 0.13% for $E(x_2^2)$, due to the fact that the natural frequencies lie far away from each other, with ratio ω_2/ω_1 being about 1.37.

1.2 Analysis of a Two-Degree-of-Freedom System with Coinciding Natural Frequencies

Consider a mass m, which can vibrate in a given plane, as shown in Fig. 1.3. The spring constants are k_1 and k_2, respectively, and the damping coefficients c_1 and c_2. The mass is subjected to a pair of random forces, which form a jointly stationary

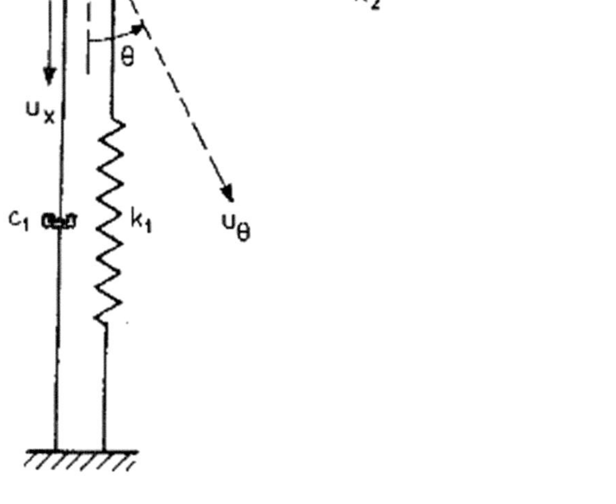

Fig. 1.3 Two-degree-of-freedom system with coalescing natural frequencies

random vector process with a given spectra matrix (Eq. 1.28),

$$S_{11} = S_0,\ S_{12} = S_{21} = \rho S_0,\ S_{22} = \alpha^2 S_0 \tag{1.28}$$

so that the forces $f_1(t)$ and $f_2(t)$ are ideal white noise processes with zero means and spectral densities S_0 and $\alpha^2 S_0$, and the correlation coefficient is ρ, such that $\alpha^2 \geq \rho^2$. The equations of motion read:

$$m\ddot{u}_x + c_1\dot{u}_x + k_1 u_x = f_1(t) \tag{1.29}$$

$$m\ddot{u}_y + c_2\dot{u}_y + k_2 u_y = f_2(t) \tag{1.30}$$

Using random vibration formalism (Crandall & Mark, 1963; Elishakoff, 1983), we have for the displacement variance–covariance responses

$$\mathrm{Var}(u_x) = E\left(u_x^2\right) = S_0 \int_{-\infty}^{\infty} |H_1(\omega)|^2 d\omega \tag{1.31}$$

$$\mathrm{Var}\left(u_y\right) = E\left(u_y^2\right) = \alpha^2 S_0 \int_{-\infty}^{\infty} |H_2(\omega)|^2 d\omega \tag{1.32}$$

$$\mathrm{Cov}\left(u_x u_y\right) = E\left(u_x u_y\right) = \rho S_0\ \mathrm{Re} \int_{-\infty}^{\infty} H_1^*(\omega) H_2(\omega) d\omega \tag{1.33}$$

where $H_1(\omega)$ and $H_2(\omega)$ are the frequency response functions

$$H_j(\omega) = 1/m^2\left[\omega_j^2 - \omega^2 + \left(c_j/m\right)i\omega\right)\ (j = 1, 2) \tag{1.34}$$

ω_1 and ω_2 are in vacuo natural frequencies

$$\omega_j = \left(k_j/m\right)^{1/2}\ (j = 1, 2). \tag{1.35}$$

Consider the displacement in the direction at angle θ to the horizontal axis:

$$u_\theta = u_x \cos\theta + u_y \sin\theta. \tag{1.36}$$

Since $u_x(t)$ and $u_y(t)$ have zero means, so $u_\theta(t)$. The variance of $u_\theta(t)$ reads

$$\begin{aligned} \mathrm{Var}(u_\theta) &= E\left(u_\theta^2\right) - [E(u_\theta)]^2 \\ &= E\left(u_x^2\right)\cos^2\theta + E\left(u_y^2\right)\sin^2\theta + E\left(u_x u_y\right)\sin 2\theta \end{aligned} \tag{1.37}$$

We are interested in the angle θ^* which corresponds to the maximum displacement variance. This is found from the requirement

$$\frac{d}{d\theta}\mathrm{Var}(u_\theta) = 0 \tag{1.38}$$

which yields

$$\tan 2\theta^* = \frac{E\left(u_x u_y\right)}{E\left(u_x^2\right) - E\left(u_y^2\right)} \tag{1.39}$$

Remarkably, Eq. 1.17 is analogous to the expression for the stress component $\sigma_{x'} = \sigma_x(\theta)$ on a face which passes through a point in question and whose normal lies in the xy plane at an arbitrary angle θ with the x axis with $E\left(u_x^2\right)$ replacing σ_x, $E\left(u_y^2\right)$ replacing σ_y, and $E\left(u_x u_y\right)$ replacing the shear stress τ_{xy} (Fig. 1.4).

To carry the above analogy even further, we formulate the mass displacement $\upsilon_\theta(t)$ in the direction perpendicular to $u_\theta(t)$:

$$\upsilon_\theta(t) = u_{\theta+90^\circ}(t) = -u_x \sin\theta + u_y \cos\theta. \tag{1.40}$$

In particular, $\upsilon_{0^\circ}(t) = u_y(t)$. The appropriate variances become

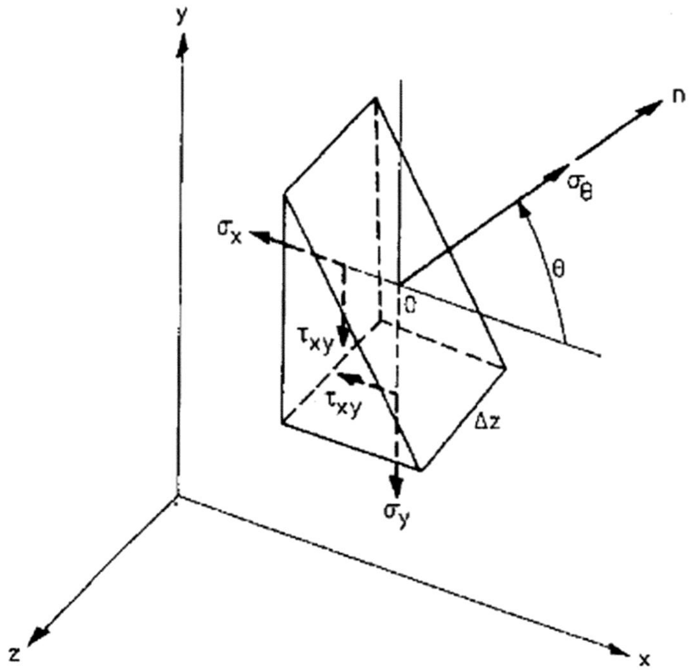

Fig. 1.4 Familiar stress transformation derivation

$$\mathrm{Var}(\upsilon_\theta) = E\left(u_x^2\right)\sin^2\theta + E\left(u_y^2\right)\cos^2\theta - E\left(u_x u_y\right)\sin 2\theta \tag{1.41}$$

and the covariance $E(u_\theta \upsilon_\theta)$ is

$$\mathrm{Cov}(u_\theta, \upsilon_\theta) = -\frac{1}{2}\left[E\left(u_x^2\right) - E\left(u_y^2\right)\right]\sin 2\theta + E\left(u_x u_y\right)\cos 2\theta. \tag{1.42}$$

As is seen, $\mathrm{Var}(\upsilon_\theta)$ corresponds to $\sigma_{y'} = \sigma_y(\theta)$ (normal stress) and $\mathrm{Cov}(u_\theta, \upsilon_\theta)$ to $\tau_{x'y'}$ (shear stress).

The identity of the transformation formulas implies that the variances and covariances of the mass response form a second-order tensor. In particular, the displacement variance has its maximum (which we will refer to as the principal displacement variance) when the covariance $\mathrm{Cov}(u_\theta, \upsilon_\theta)$ vanishes.

The principal (extremal) displacement variances are immediately found, using the analogous expression for the principal normal stress

$$\mathrm{Var}\left(u_{1,2}\right) = E\left(u_{1,2}^2\right) = \frac{1}{2}\left[E\left(u_x^2\right) + E\left(u_y^2\right)\right] \pm \sqrt{\frac{1}{4}\left[E\left(u_y^2\right)\right]^2 + E^2\left(u_x u_y\right)}. \tag{1.43}$$

These values are also readily obtained by constructing the appropriate Mohr circle for the displacement responses (Fig. 1.5).

For the quantities in Eq. 1.23, we have (see Eq. e.g. 9.67 and 9.148, respectively, Elishakoff, 1983, 1999)

$$E\left(u_x^2\right) = \frac{\pi S_0}{2\zeta_1\omega_1^3 m^2}, \quad E\left(u_y^2\right) = \frac{\pi\alpha^2 S_0}{2\zeta_2\omega_2^3 m^2} \tag{1.44}$$

$$E\left(u_x u_y\right) = \frac{4\pi(\zeta_1\omega_1 + \zeta_2\omega_2)\rho S_0/m^2}{\left(\omega_1^2 - \omega_2^2\right)^2 + 4\left[\zeta_1\zeta_2\omega_1\omega_2\left(\omega_1^2 + \omega_2^2\right) + \left(\zeta_1^2 + \zeta_2^2\right)\omega_1^2\omega_2^2\right]} \tag{1.45}$$

where

$$\zeta_1 = c_j/2m\omega_m \quad (j = 1, 2). \tag{1.46}$$

As is seen from formulas (1.39), (1.44), and (1.45), the angle at which maximum displacement variance is reached depends both on the properties of the structure and the probabilistic characteristics of the loading. The conclusion is that this angle does not generally coincide with the resultant of the forces, whose direction is itself a random function.

To gain more insight, we form the response correlation coefficient, defined as

$$\kappa = \frac{E\left(u_x u_y\right)}{\sqrt{E\left(u_x^2\right)E\left(u_y^2\right)}} \tag{1.47}$$

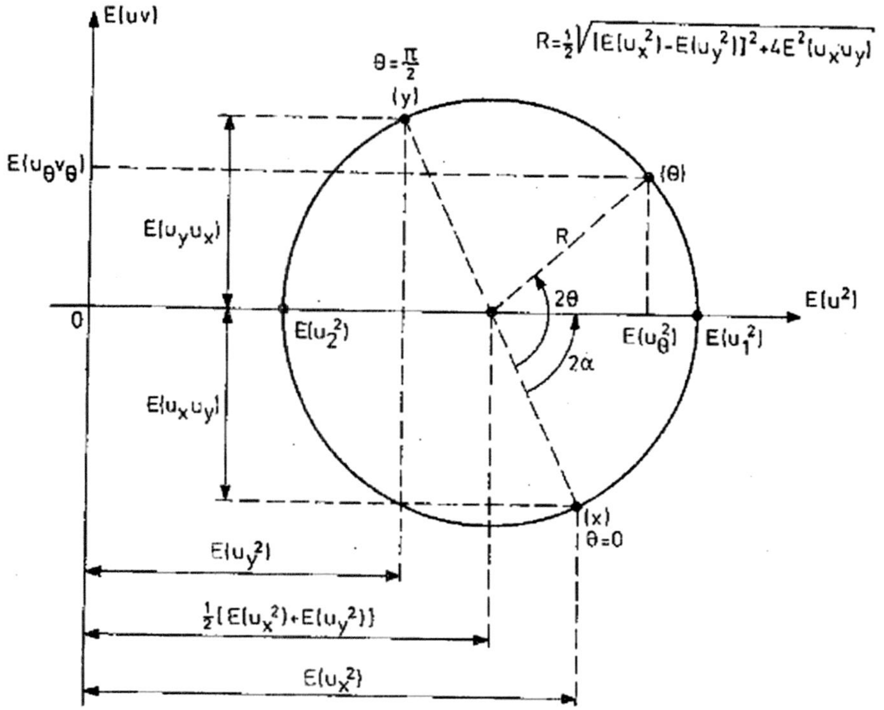

Fig. 1.5 Mohr's circle for mean-square values

Bearing in mind Eqs. 1.44 and 1.45, we get κ:

$$\kappa = \frac{8\rho}{\alpha} \frac{\zeta_1^2(1+c_{21})(c_{21}k_{21})^{1/2}}{\left(1-k_{21}^2\right)^2 + 4\zeta_1^2\left[c_{21}\left(1+k_{21}^2\right) + \left(1+c_{21}^2/k_{21}\right)k_{21}^2\right]} \tag{1.48}$$

where the damping ratio c_{21} and stiffness ratio κ_{21} are defined as

$$c_{21} = c_2/c_1, \quad k_{21} = k_2/k_1 \tag{1.49}$$

If the excitation correlation coefficient ρ vanishes identically, so does the response correlation coefficient κ. For deeper insight, consider the case of perfectly positive correlation, $\rho = 1$, and equality of the spectral density of excitation

$$S_{F_1}(\omega) = S_{F_2}(\omega) \tag{1.50}$$

that is

$$\alpha = 1 \tag{1.51}$$

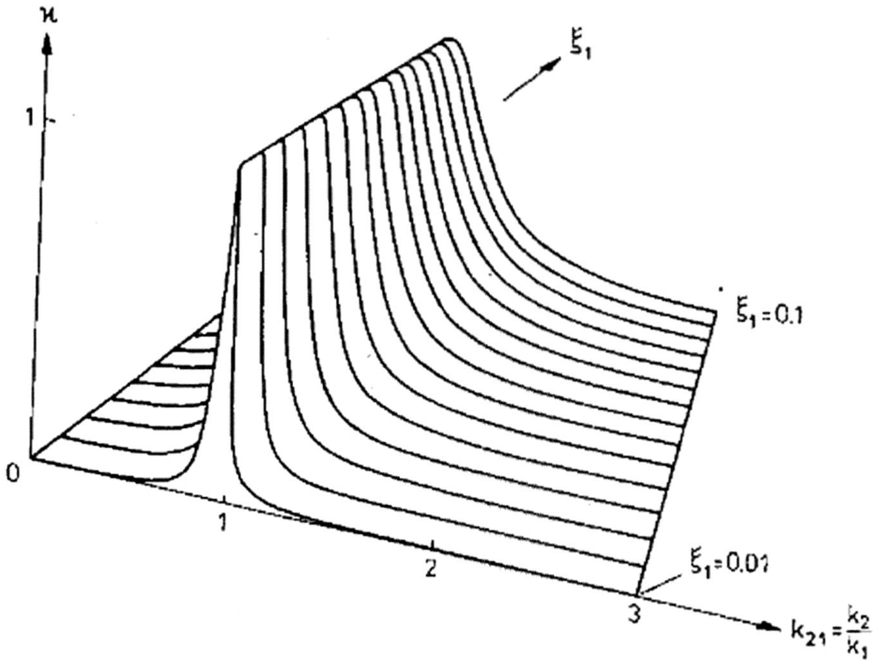

Fig. 1.6 The dependence of κ versus k_{21}

We fix the ratio of the damping coefficients at $c_{21} = 1$ and vary that of the stiffness coefficients over $0 < k_{21} \leq 1$. The dependence of κ versus k_{21} is shown in Fig. 1.6.

Consider now some particular examples.

(i) *Uncorrelated forces*

Let $\rho = 0$, then

$$E\left(u_x u_y\right) = 0, \quad \tan 2\theta^* = 0 \tag{1.52}$$

that is

$$\theta^* = 0, \quad \text{or } \theta^* = 90°. \tag{1.53}$$

The maximum mean-square displacement is then

$$E\left(u_1^2\right) = \max\left\{E\left(u_x^2\right), E\left(u_y^2\right)\right\}, \quad E\left(u_2^2\right) = \min\left\{E\left(u_x^2\right), E\left(u_y^2\right)\right\}. \tag{1.54}$$

(ii) *Equal displacement variances in directions of spring axes*

Let now $E\left(u_x^2\right) = E\left(u_y^2\right)$ or

$$\frac{1}{c_1 k_1} = \frac{\alpha^2}{c_2 k_2} \tag{1.55}$$

which implies that $\theta^* = 45°$. Then

$$E\left(u_1^2\right) = E\left(u_x^2\right) + E\left(u_x u_y\right) \tag{1.56}$$

$$E\left(u_2^2\right) = E\left(u_x^2\right) - E\left(u_x u_y\right) \tag{1.57}$$

The appropriate Mohr circle is shown in Fig. 1.7a. The situation here is equivalent to consideration of an element with stress components equal numerically (Fig. 1.7b),

$$\sigma_x = E\left(u_x^2\right), \quad \sigma_y = E\left(u_y^2\right), \quad \rho_{xy} = E\left(u_x u_y\right). \tag{1.58}$$

An element oriented along the principal axes is shown in Fig. 1.7c. From Eq. 1.36 and its graphical representation in Fig. 1.7c, it is immediately seen that the response of the two-degree-of-freedom system may exceed that of a single-degree one, $E\left(u_x^2\right)$, under white noise in the direction of motion. In these circumstances, omission of the term associated with the cross-correlation, $E\left(u_x u_y\right)$ could badly underestimate the actual response.

To check this numerically, we consider a particular case meeting the requirement (Eq. 1.55). Let

$$c_1 = c_2, \quad k_1 = k_2, \quad \alpha = 1. \tag{1.59}$$

Then

$$\omega_1 = \omega_2 = k/m \equiv \omega_0, \quad \zeta_1 = \zeta_2 = c/2m\omega_0 \tag{1.60}$$

and

$$E\left(u_x u_y\right) = \rho \frac{\pi S_0}{ck} = \rho E\left(u_x^2\right). \tag{1.61}$$

Hence

$$d_{\text{exact}} = E\left(u_1^2\right) = (1+\rho) E\left(u_x^2\right). \tag{1.62}$$

Omission of the cross-correlation would yield

$$d_{\text{approx}} = E\left(u_x^2\right) \tag{1.63}$$

with the associated error

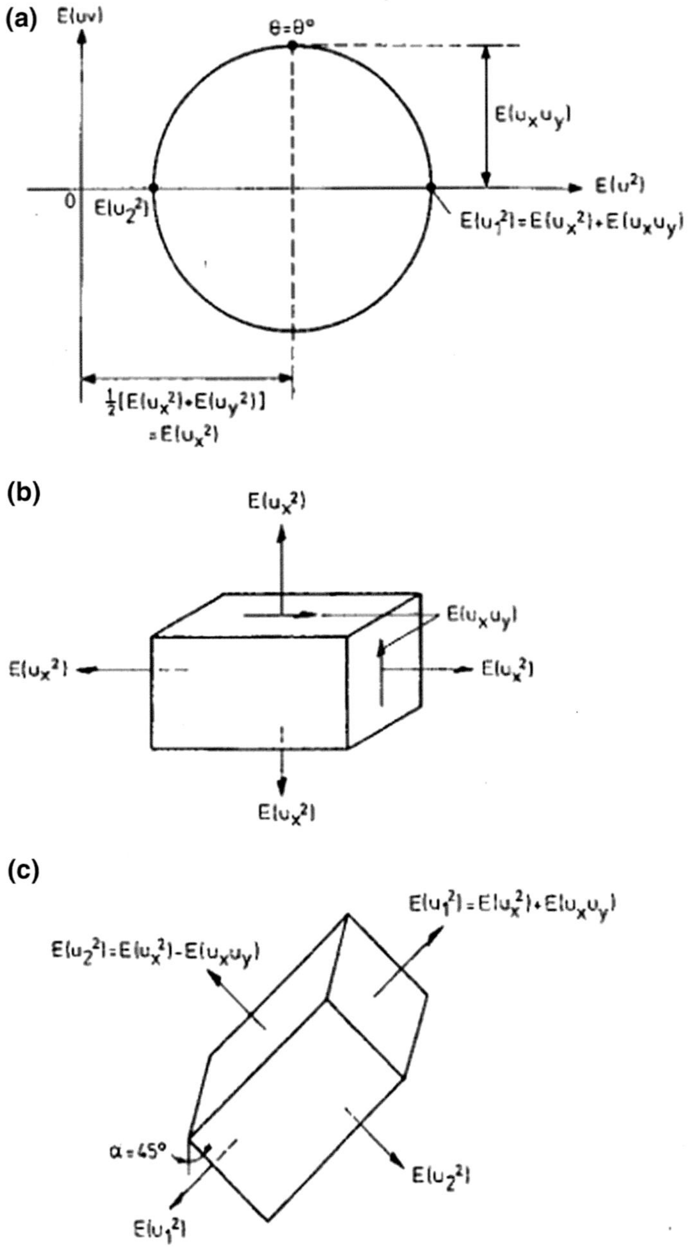

Fig. 1.7 Mohr' circle and associated mean-square values

$$\eta = \frac{|d_{\text{exact}} - d_{\text{approx}}|}{d_{\text{exact}}} \times 100\% = \frac{100|\rho|}{1+\rho}\% \tag{1.64}$$

In the case of perfectly positive correlation, $f_1(t) = f_2(t)$ with probability one, we have

$$\rho = 1, \quad \eta = 50\% \tag{1.65}$$

It is interesting to note that exact maximum response is here double that of a single-degree-of-freedom system.

$$E\left(u_1^2\right) = 2E\left(u_x^2\right). \tag{1.66}$$

In the case of perfectly negative correlation, $f_1(t) = -f_2(t)$ with probability one. Hence,

$$\rho = -1, \quad \eta \to \infty \tag{1.67}$$

Here, the percentage error tends to infinity. These large crosses are due to the fact that for close (or coincident) natural frequencies, the modal cross-correlations cannot be neglected. The situation associated with perfect correlation is shown in the Mohr circle in Fig. 1.8d. It is analogous to treating an element subject to stresses numerically equal (Fig. 1.8b)

$$\sigma_x = \sigma_y = \tau_{xy} = E\left(u_x^2\right) \tag{1.68}$$

The element oriented along the principal axes is shown in Fig. 1.8c. Figure 1.8d illustrates the mechanical interpretation of omission of the cross-correlation: the latter implies treatment of an element under stresses as per that figure, rather than as per Fig. 1.8b.

1.3 Simple Model Illustrating Crandall's Random Vibration Results

Some interesting results have been reported in literature on the random vibration of various continuous systems under point excitation representing a stationary wide-band random process in time (Crandall, 1979; Elishakoff et al., 1979). In some one-dimensional systems, the distribution of the mean-square response velocity turned out to be surprisingly uniform, except for small zones or narrow bands of enhanced response. For example, if the force acts on a string or beam of length l at $x = a$, these zones are $x = a$ and $x = l - a$. As the bandwidth of the excitation increases unboundedly, the relative enhancement tends asymptotically to 3/2. This pattern

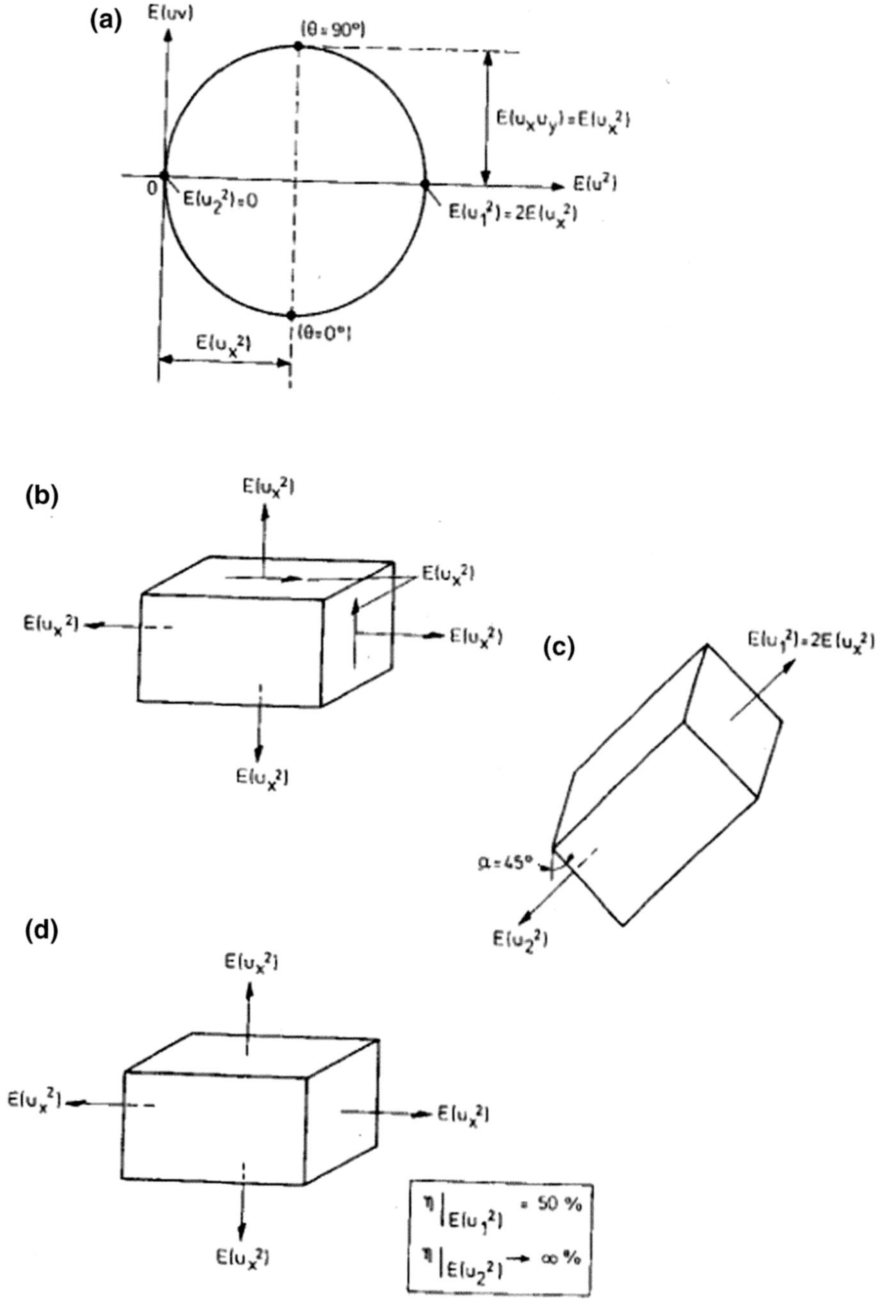

Fig. 1.8 Mohr's circles for various cases

was explained by normal-mode analysis (Crandall, 1979) which showed that the cross-correlations between different vibrational modes are negligible compared with the direct ones between identical vibrational modes due to the small modal overlap (light damping and widely separated natural frequencies). It was demonstrated that this centrally symmetrical distribution is not characteristic of all structures with symmetrically distributed mass and stiffness under symmetric boundary conditions (Elishakoff et al., 1979) . The response of homogeneous cylindrical shells subjected to ring loading, uniform along the circumference, and random in time, was found to differ radically from that of point-excited homogeneous strings and beams; owing to the large overlap at the low end of the spectrum, it is no longer centrally symmetrical. The structures under several random forces are considered and the zones or band of enhanced or diminished response identified their number and strength depending on the correlation of the input force processes by Crandall and Kulvets (1979).

In the current chapter, we elaborate on the simple model, that of a three-degree-of-freedom system undergoing a stationary excitation. The properties of the system depend on the governing non-negative parameter ϵ, so that for ϵ tending to zero, two natural frequencies coalesce. For very small ϵ, the structure resembles a shell with the symmetry of the mean-square response distribution distorted; for large ϵ, the cross-correlations between different vibrational modes become negligible and the structure has a symmetric random response like a string or a beam. In addition, the importance of the cross-correlations for the structure under several forces is demonstrated and the simplest results of the type reported are given by Crandall and Kulvets (1977). The model considered here is an extension of the two-degree-of-freedom system presented by Elishakoff (1983, 1999).

1.4 Theoretical Analysis

Consider a three-degree-of-freedom system, as shown in Fig. 1.9. m denotes the mass, k the stiffness, and c the damping. ϵ is a non-negative governing parameter, which may vary from zero to infinity. Forces $f_1(t)$, $f_2(t)$, and $f_3(t)$ are stationary, stationarily connected random processes with a given spectral matrix. We are interested in the distribution of the mean-square velocity response as a function of ϵ.

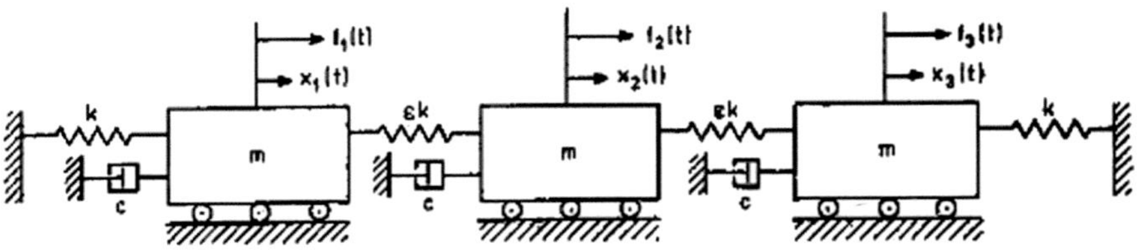

Fig. 1.9 A three-degree-of-freedom system

The equation of motion reads

$$[M]\{\ddot{x}\} + [C]\{\dot{x}\} + [K]\{x\} = \{f\} \tag{1.69}$$

where $[M]$, $[C]$, and $[K]$ are the mass, damping, and stiffness matrices, respectively, $\{f_i\}$ is a column excitation vector, and $\{x\}$ is a column displacement vector. Further

$$\begin{aligned} &[M] = m[I], \quad [C] = c[I] \\ &k_{11} = k_{31} = (1+\epsilon)k, \quad k_{22} = 2\epsilon k, \\ &k_{12} = k_{21} = k_{23} = k_{32} = -\epsilon k, \quad k_{13} = k_{31} = 0 \end{aligned} \tag{1.70}$$

where $[I]$ is an identity matrix.

The problem is formulated as follows: The excitation represents a weakly stationary vector function with vanishing mean $E[\{f(t)\}] = 0$, and given correlation matrix

$$E\big[\{f(t)\}\{f(t+\tau)\}^T\big] = [C(\tau)]_{3\times 3}; \tag{1.71}$$

find the variance–covariance matrix $[R]$ of the velocity vector $x(t)$

$$[R] = E[\{\dot{x} - E\{\dot{x}\}][x - E\{\dot{x}\}]^T \tag{1.72}$$

under a stationary regime.

In solving this problem, we utilize the classical normal-mode method, for illustration of some recent random vibration results, and to investigate the error introduced by neglecting cross-correlations between different vibrational modes. The natural frequencies are

$$\begin{aligned} &\omega_1^2 = [1 + 3\epsilon - \sqrt{(1-\epsilon)^2 + 8\epsilon^2}](k/2m) \\ &\omega_2^2 = (1+\epsilon)(k/m) \\ &\omega_3^2 = [1 + 3\epsilon + \sqrt{(1-\epsilon)^2 + 8\epsilon^2} \end{aligned} \tag{1.73}$$

with associated normal modes

$$\begin{aligned} &\{\phi_1\}^T = \lfloor 1 \; a \; 1 \rfloor, \quad \{\phi_2\}^T = \lfloor 1 \; 0 \; -1 \rfloor \\ &\{\phi_3\}^T = \lfloor 1 \; -b \; 1 \rfloor \end{aligned} \tag{1.74}$$

where

$$a = (1 - \epsilon + \sqrt{(1-\epsilon)^2 + 8\epsilon^2})/(2\epsilon)b = \left(\epsilon - 1 + \sqrt{(1-\epsilon)^2 + 8\epsilon^2}\right)/(2\epsilon) \tag{1.75}$$

so that

$$ab = 2, \quad a - b = (1 - \epsilon)/\epsilon \tag{1.76}$$

The modal matrix is

$$[\Phi] = \begin{bmatrix} 1 & 1 & 1 \\ a & 0 & -b \\ 1 & -1 & 1 \end{bmatrix} \tag{1.77}$$

The calculation yields diagonal matrices

$$\begin{aligned} &\left[B^{(1)}\right] = [\Phi]^T[M][\Phi], \quad \left[B^{(2)}\right] = (c/m)\left[B^{(1)}\right] \\ &\left[B^{(3)}\right] = [\Phi]^T[K][\Phi] \end{aligned}. \tag{1.78}$$

$$\begin{aligned} &B_{11}^{(1)} = \left(2 + a^2\right)m, \quad B_{22}^{(1)} = 2m, \quad B_{33}^{(1)} = \left(2 + b^2\right)m \\ &B_{11}^{(3)} = \left(2 + a^2\right)m\omega_1^2, \quad B_{22}^{(3)} = 2m\omega_2^2, \quad B_{33}^{(3)} = \left(2 + b^2\right)m\omega_3^2 \end{aligned}.$$

Substituting in Eq. (1.49)

$$\{x\} = [\Phi]u \tag{1.79}$$

and premultiplying the result by $[\phi]^T$, we obtain in view of Eq. (1.56)

$$u_i + \frac{c}{m}u_i + \omega_i^2 u_i = F_i/B_{11}^{(1)} \tag{1.80}$$

where

$$F = [\Phi]^T\{f\} = \begin{Bmatrix} f_1 + af_2 + f_3 \\ f_1 - f_3 \\ f_1 - bf_2 + f_3 \end{Bmatrix}. \tag{1.81}$$

Since $E\{f\} = 0$, we have $E\{F\} = 0$ and therefore $E\{\dot{u}\} = 0$ for the stationary regime. Then $E\{\dot{x}\} = 0$. Denoting the variance–covariance matrix $E\left[\{\dot{u}\}\{\dot{u}\}^T\right]$ of the velocities of normal coordinates by $[r]$,

$$[r] = \begin{bmatrix} E\left(\dot{u}_1^2\right) & E(\dot{u}_1\dot{u}_2) & E(\dot{u}_1\dot{u}_3) \\ E(\dot{u}_2\dot{u}_1) & E\left(\dot{u}_2^2\right) & E(\dot{u}_2\dot{u}_3) \\ E(\dot{u}_1\dot{u}_3) & E(\dot{u}_2\dot{u}_3) & E\left(\dot{u}_3^2\right) \end{bmatrix} = \begin{bmatrix} r_{11} & r_{12} & r_{13} \\ r_{12} & r_{22} & r_{23} \\ r_{13} & r_{23} & r_{33} \end{bmatrix} \tag{1.82}$$

the variance–covariance matrix of the velocities of original coordinates $[R] = E\left[\{\dot{x}\}\{\dot{x}\}^T\right]$ becomes, bearing in mind Eq. (1.77)

$$[R] = [\Phi][r][\Phi]^T \tag{1.83}$$

with elements

$$\begin{aligned}
R_{11} &= \sum_{j=1}^{3}\sum_{k=1}^{3} r_{jk} \\
R_{12} &= a(r_{11} + r_{21} + r_{31}) - b(r_{13} + r_{23} + r_{33}) \\
R_{13} &= r_{11} + r_{21} + r_{31} - r_{12} - r_{22} - r_{32} + r_{13} + r_{23} + r_{33} \\
R_{21} &= a(r_{11} + r_{12} + r_{13}) - b(r_{31} + r_{32} + r_{33}) \\
R_{22} &= a^2 r_{11} - ab(r_{13} + r_{31}) + b^2 r_{33} \\
R_{23} &= a(r_{11} - r_{12} + r_{13}) - b(r_{31} - r_{32} + r_{33}) \\
R_{31} &= r_{11} + r_{12} + r_{13} - r_{21} - r_{22} - r_{23} + r_{31} + r_{32} + r_{33} \\
R_{32} &= a(r_{11} - r_{21} + r_{31}) - b(r_{13} - r_{23} + r_{33}) \\
R_{33} &= r_{11} - r_{12} + r_{13} - r_{21} + r_{22} - r_{23} + r_{31} - r_{32} + r_{33}
\end{aligned} . \tag{1.84}$$

1.5 One Particular Case: $f_2 = f_3 = 0$ with Probability One

In addition, assume that $f_1(t)$ is a white noise with intensity S_0:

$$R_{f_1}(\tau) = 2\pi S_0 \delta(\tau). \tag{1.85}$$

Then classical analysis yields

$$r_{jk} = \frac{\pi S_0 m}{c m_j^2} \quad \text{for} \quad j = k \tag{1.86}$$

$$r_{jk} = \frac{4\pi(\zeta_i\omega_k + \zeta_k\omega_j)\omega_j\omega_k}{\left(\omega_j^2 - \omega_k^2\right)^2 + 4\left[\zeta_j\zeta_k\omega_j\omega_k\left(\omega_j^2 + \omega_k^2\right) + \left(\zeta_j^2 + \zeta_k^2\right)\omega_j^2\omega_k^2\right]} \frac{S_0}{m_j m_k}, \quad \text{for} \quad j \neq k \tag{1.87}$$

with

$$\zeta_j\omega_j = c/2m. \tag{1.88}$$

The variation of the natural frequencies with ϵ is shown in Fig. 1.10. As is seen in this figure, for large ϵ, the natural frequencies are widely separated and the contribution of cross-correlations $r_{jk} (j \neq k)$ may be neglected.

Then the formulas (1.62) become

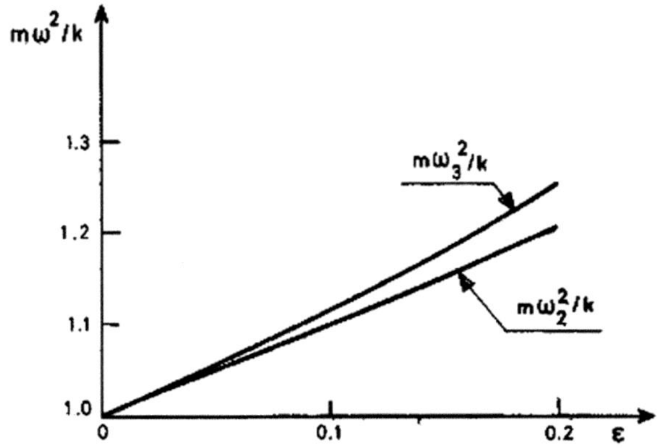

Fig. 1.10 Variation of natural frequencies ω_2^2 and ω_3^2 versus ϵ

$$R_{11} \simeq R_{33} \simeq r_{11} + r_{22} + r_{33}$$
$$= 0.25\sigma^2\Big[\big(1+0.5a^2\big)^{-2} + 1 + \big(1+0.5b^2\big)^{-2}\Big] \tag{1.89}$$

$$R_{22} \simeq a^2 r_{11} + b^2 r_{33}$$
$$= 0.25b^2\Big[a^2\big(1+0.5a^2\big)^{-2} + b^2\big(1+0.5b^2\big)^{-2}\Big], \quad \sigma^2 = \pi S_0/mc \tag{1.90}$$

In the special case

$$\epsilon = 1, \quad a = b = \sqrt{2} \tag{1.91}$$

we have

$$\begin{aligned} R_{11} &\simeq R_{33} \simeq 0.375\sigma^2 \\ R_{22} &\simeq 0.25\sigma^2 \end{aligned} \tag{1.92}$$

That is, for large ϵ, the mean-square velocity response distribution is symmetric about the axis of symmetry, as in the series of one-dimensional structures described by Crandall (1979). Moreover, $R_{11} \simeq (3/2)R_{22}$. This, however, does not imply that the value 3/2 is the relative enhancement (as defined by Crandall for some one-dimensional system, 1979) since the latter is defined not at the ratio R_{11}/R_{22} but differently. For elucidating this point, the random vibration of a massless beam is considered with an increasingly large number of equally spaced lumped masses, for which the mode shapes and the natural frequencies are well known in literature (Filippov, 1970).

Consider now the case of small ϵ . The two natural frequencies ω_2 and ω_3 tend to coalesce, and the contribution of the cross-correlation term r_{23} cannot be neglected, whereas r_{12} and r_{13} remain negligible. For $\epsilon \to +0$, we have

$$\begin{aligned} R_{11} &\simeq r_{11} + r_{22} + r_{33} + 2r_{23} \\ &= 0.25\sigma^2\left[\left(1+0.5a^2\right)^{-2} + 1 + \left(1+0.5b^2\right)^{-2} + 2\left(1+0.5b^2\right)^{-1}\right]. \end{aligned} \tag{1.93}$$

However, $a \to +\infty$, $b \to 0$ and

$$R_{11} \simeq \sigma^2. \tag{1.94}$$

Omission of the cross-correlation term r_{23} would result in

$$R_{11} \simeq 0.5\sigma^2 \tag{1.95}$$

constituting, 50% difference in evaluating R_{11}. For R_{33}, we have

$$R_{33} \simeq r_{11} + r_{22} + r_{33} - 2r_{23} \tag{1.96}$$

Comparison with Eq. (1.73) shows that the cross-correlation term now appears with the minus sign and the symmetry $R_{11} \simeq R_{33}$. observed for large ϵ is now distorted. Indeed,

$$R_{33} \simeq 0.25\sigma^2\left[\left(1+0.5a^2\right)^{-2} + 1 + \left(1+0.5b^2\right)^{-2} - 2\left(1+0.5b^2\right)^{-1}\right] = 0. \tag{1.97}$$

Omission of the cross-correlation term would yield $R_{11} \simeq 0.5\sigma^2$. The results obtained in this section resemble the behavior of a cylindrical shell (Elishakoff et al., 1979) where for thickness-to-length ratio 0.0003, the contribution of cross-correlations is thrice as much as that of direct correlations. In the system under consideration, this contribution might be even larger.

Note as well that for $\epsilon \to 0$, $R_{11} \to \sigma^2$, which equals the mean-square velocity response of a single-degree-of-freedom system. Indeed, for $\epsilon \to 0$, the three-degree-of-freedom system resolves into three uncoupled, single-degree-of-freedom ones, with the force $f_1(t)$ acting only on one of them; therefore R_{33} approaches zero with ϵ.

In order to investigate the contribution of the cross-correlation term, we form the ratio

$$\begin{aligned} &\frac{2r_{23}}{r_{22} + r_{33} + 2r_{23}} \times 100\% \\ &\qquad = 100\%\left\{1 + \frac{1}{2}\left(\frac{m_2}{m_3} + \frac{m_3}{m_2}\right)\left[1 + \frac{(\omega_2+\omega_3)^2}{2\left(\omega_2^2+\omega_3^2\right)_3^2}\right]\right\}^{-1} \end{aligned} \tag{1.98}$$

where Ω_j is defined as the local modal overlap ratio

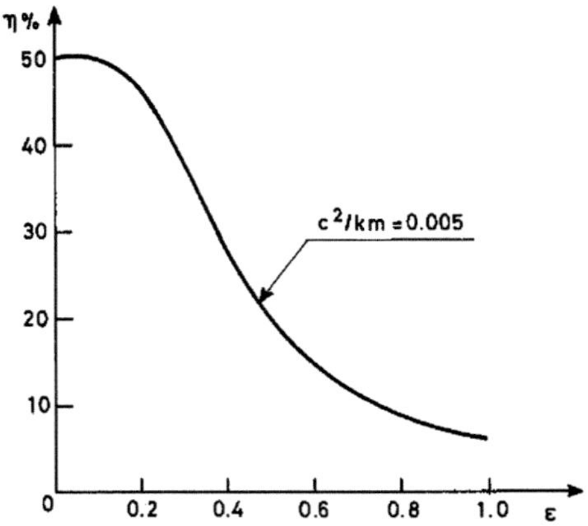

Fig. 1.11 The percentage contribution of the cross-correlation term

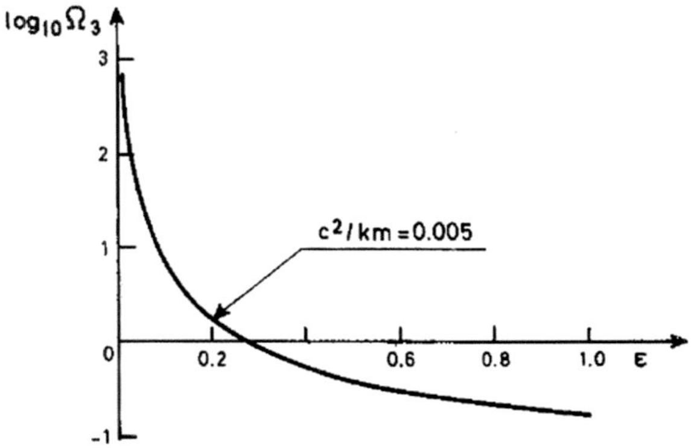

Fig. 1.12 Local modal overlap ratio versus ϵ

$$\Omega_j = \frac{\text{modal bandwidth}}{\text{modal separation}} = \frac{c/m}{\omega_j - \omega_{j-1}} \tag{1.99}$$

$$\lim_{\epsilon \to 0} \frac{200 r_{23}}{r_{22} + r_{33} + r_{23}} \% = \frac{100 \Omega_3^2}{1 + 2\Omega_3^2} \% \tag{1.100}$$

and for Ω_j far exceeding unity, the contribution of the term to the mean-square velocity of the loaded mass approaches 50%. For not too small ϵ, expression (1.99) should be used. The percentage contribution is shown as a function of ϵ in Fig. 1.11. For $\epsilon \geq 0.3$, the cross-correlations contribute not more than five percent to the total mean-square velocity response. The value of $\log_{10}\Omega_3$ is shown in Fig. 1.12.

1.6 $f_1(t) = f_3(t) = 0$ with Probability One

Assume now also that the force represents an ideal white noise

$$R_{f_2}(\tau) = 2\pi S_0 \delta(\tau) \tag{1.101}$$

then

$$r_{11} = \frac{a^2}{\left(2+a^2\right)^2}\sigma^2, \quad r_{22} = 0, \quad r_{33} = \frac{b_2}{\left(2+b^2\right)^2}\sigma^2 \tag{1.102}$$

and for not too small ϵ

$$R_{11} = R_{33} \simeq 0.25\sigma^2 \left[\frac{a^2}{\left(1+0.5a^2\right)^2} + \frac{b^2}{\left(1+0.5b^2\right)}\right] \tag{1.103}$$

while

$$R_{22} \simeq 0.25\sigma^2 \left[\frac{a^4}{\left(1+0.5a^2\right)^2} + \frac{b^4}{\left(1+0.5b^2\right)}\right]; \tag{1.104}$$

for $\epsilon = 1$

$$R_{11} = R_{33} \simeq 0.25\sigma^2, \quad R_{22} \simeq 2R_{11} = 0.5\sigma^2 \tag{1.105}$$

In the lecture, the pair of correlated forces acting on the system will be considered as well. Particular cases of perfectly positive correlation ($\alpha = 1$), perfectly negative correlation ($\alpha = -1$), and vanishing correlation will be discussed. The above model is apparently the simplest way of explaining the experimental observations and analytical calculations (Crandall and Kulvets, 1977). In an experiment, a square aluminum plate was excited by loudspeaker voice coils driven by wide-band random noise generator. Powdered salt (or sugar) sprinkled on the vibrating plate was seen to migrate away from zones of strong mean-square response and accumulate in those weak response zones. For diagonally symmetric driving forces, the powder clearly shifted away from the diagonal, while with $\alpha = -1$, the powder accumulated along it. (Figs. 1.13, 1.14, 1.15, 1.16 and 1.17)

Tongue (1996) devotes an entire section to the topic of repeated natural frequencies. He considers the two-degree-of-freedom system, shown in figure. According to Tongue (1996, p. 232) , "the x, θ set tracks the vertical motion of the beam's center of mass (x) and the beam's rotation (θ)." Equations of motion read

$$m\ddot{x} + (k_1 + k_2)x + l(k_2 - k_1)\theta = 0$$
$$\bar{I}\ddot{\theta} + l(k_2 - k_1)x + l^2(k_1 + k_2)\theta = 0$$

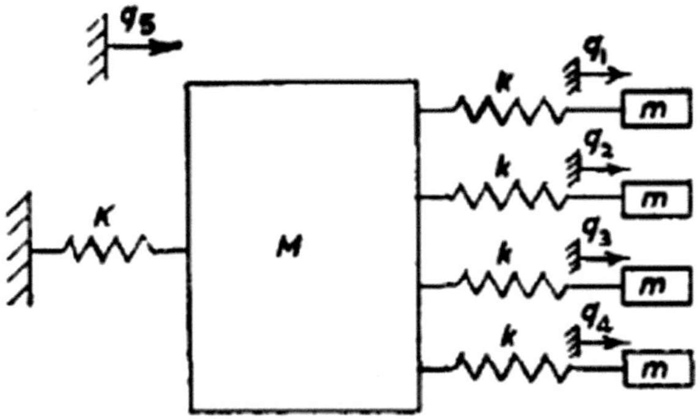

Fig. 1.13 Gladwell's (1961, p. 180) example of a system with equal natural frequencies consists of 4 equal masses m connected with mass M by springs which all have stiffness k; three of the natural frequencies of the system are the same and are $\sqrt{k/m}$

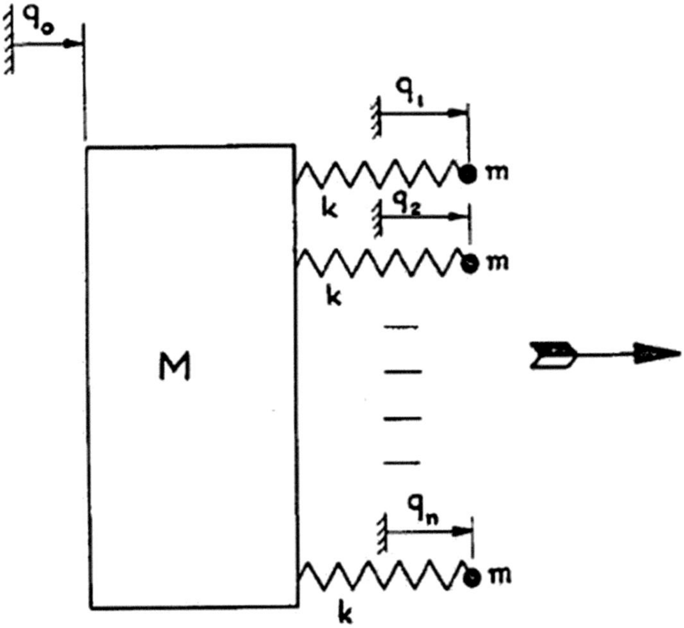

Fig. 1.14 Johnson and Bishop's (1956, p. 379) mechanical system having $(n - 1)$ equal natural frequencies

lending to natural frequencies squared

$$\omega_1^2 = \frac{2k}{m}, \quad \omega_2^2 = \frac{2kl^2}{\bar{I}}$$

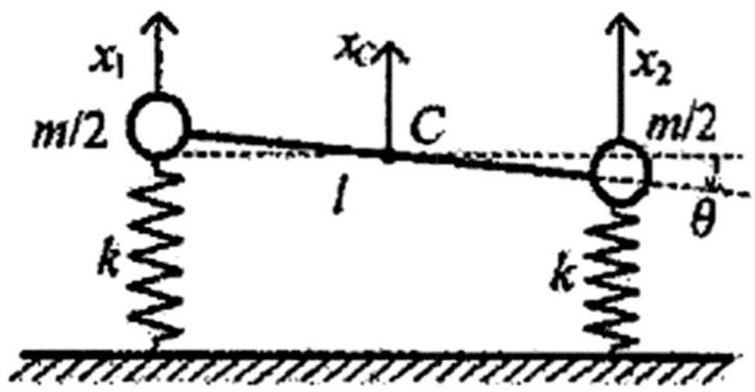

Fig. 1.15 The example used by Thomson (1981) of a simple system in which coincident natural frequencies may appear, based on a rigid connecting bar, whose weight is negligible

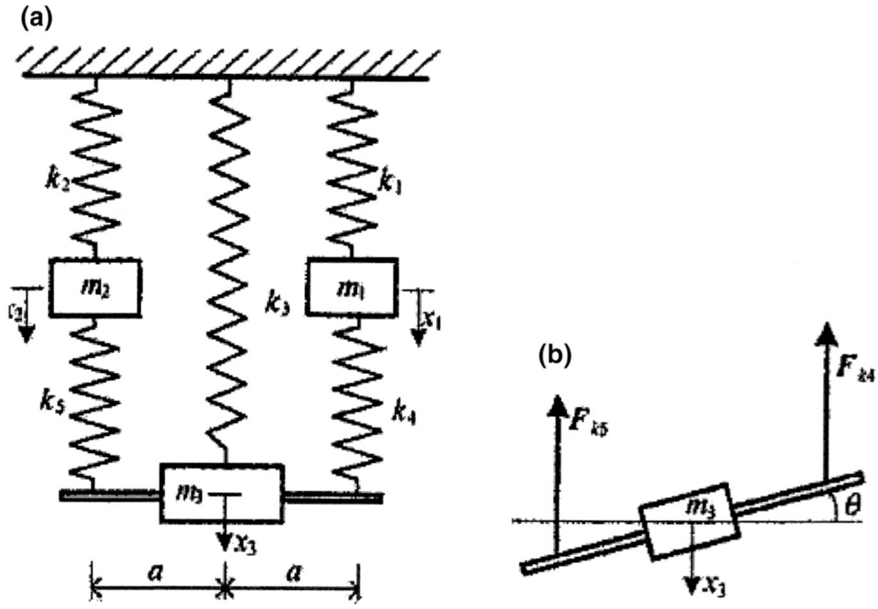

Fig. 1.16 **a** Example of a mass–spring system with three DOF, **b** free body diagram for mass m_3 in the three-DOF example (from Chen, Wang, & Zhou, 2009)

This clearly is the case when we have a repeated natural frequency (Fig. 1.18).

Anuta (1996a, 1996b) investigated discrete, constrained, linear elastic systems in free vibration, by means of their modal equations, in order to establish the necessary conditions for equal frequencies. Author notes: "Equal frequencies of vibration in discrete systems only occur under certain special conditions. For the semi-definite system problem, a series of trial-and-error computer runs [were] applied to the general requirements for equal frequencies, and a subsequent mathematical analysis

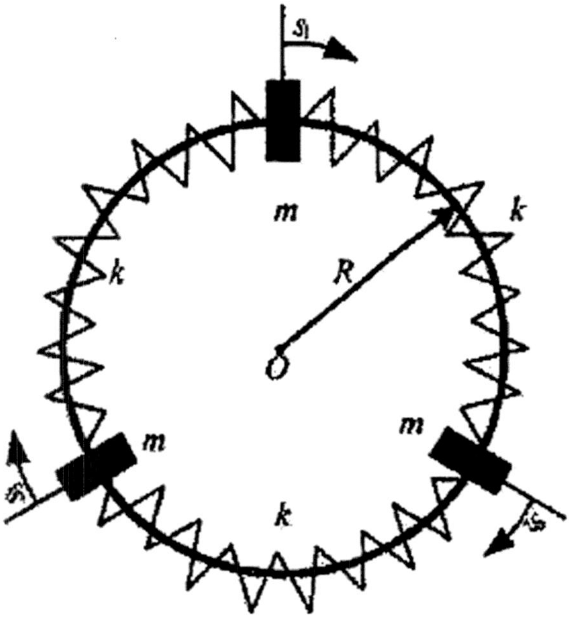

Fig. 1.17 The mass–spring ring with three DOF (from Chen et al., 2009)

of the modal equations for the particular configurations then determines the specific requirements for their attainment of equal frequencies (Anuta, 1994)". Anuta (1996a, 1996b) noted that the results of a series of exploratory computer solutions involving various configurations of definite systems in free vibration find two problems of primary interest: (a) close-coupled systems and (b) all-coupled systems. He arrived at a conclusion that one has to consider also (c) hybrid systems consisting of two all-coupled groups connected together in a close-coupled arrangement. Anuta (1996a, 1996b, p. 358) notes: "Any symmetric all-coupled system of three or more degrees of freedom always develops a unique fundamental frequency and $n - 1$ equal high frequencies." He considers a specific system exhibiting fundamental natural frequency equal to 1.2247 rad/s, and two other frequencies that coincide with each other, namely 1.7320 rad/s. This is in agreement with much earlier results by Elishakoff (1983, p. 375) with first natural frequency squatted equals *1.0* k/m, where two other squared frequencies equal *4* k/m. More general case of n-degree-of-freedom system where n-1 natural frequencies coincide is considered in the next chapter, and was first obtained by Lubliner and Elishakoff (1986).

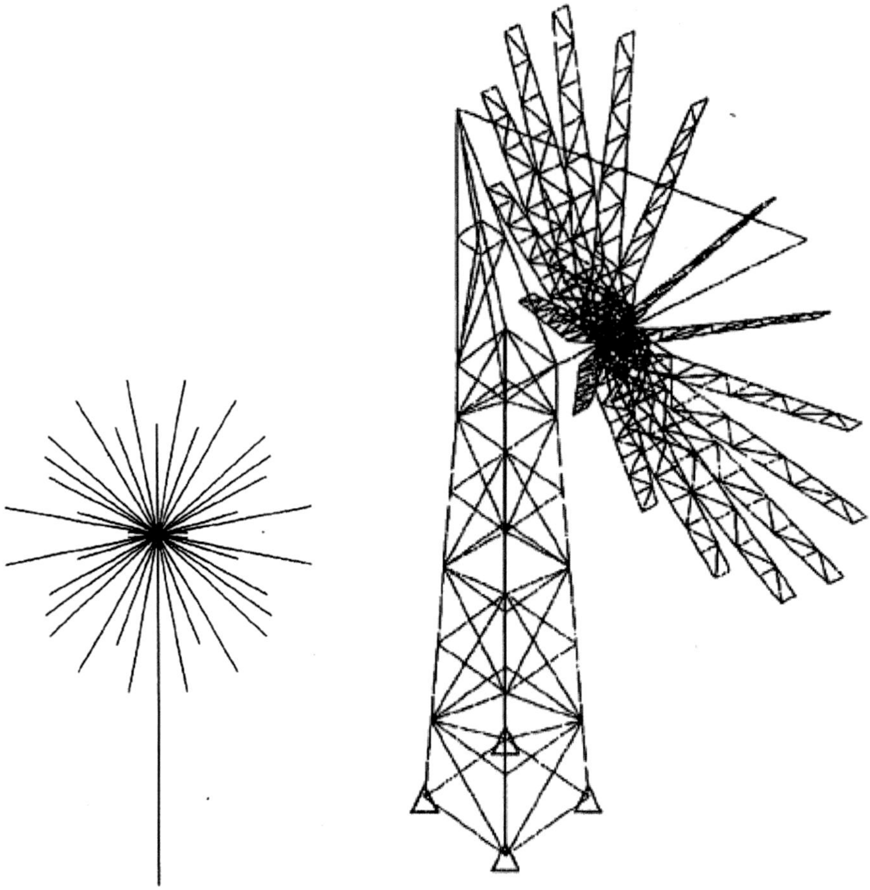

Fig. 1.18 Symmetry may lead to coinciding natural frequencies. Left figure shows man-made fountain model; right figure shows antenna model (courtesy of AIAA Journal, Ren, Zheng, Cheng, Vol. 35(2), p. 356, 1997)

References

Anuta, A. E. (1996a). The model equations and semi-definite systems. *Computers and Structures, 51,* 325–329.

Anuta, A. E. (1996b). The model equations and equal frequencies in definite system. *Computers and Structures*, *60*(3), 357–367.

Argyris, J., & Mlenjenek, H.-P. (1991). *Dynamics of structures*. Amsterdam: North-Holland.

Bolotin, V. V. (1965). The density of eigenvalues in vibration problems of elastic plates and shells. *Proceedings of Vibration Problems, Warsaw, 6*(5), 342–351.

Bolotin, V. V. (1969). *Statistical methods in structural mechanics* (English translation by S. Aroni, of the second edition) (Chap. 5, pp. 102–153), San Francisco: Holden-Day Inc.

Bolotin, V. V., & Chirkov, V. P. (1978). Random vibrations of systems with finite degrees of freedom. In V. V. Bolotin (Ed.), *Vibrations in technology* (Vol. 1, pp. 286–299), Moscow: "Mashinostroenie", Publishing House (in Russian).

Borino, G., & Muscolino, G. (1986). Mode-superposition methods in dynamic analysis of classically and non-classically damped linear systems. *Earthquake Engineering and Structural Dynamics, 14*(5), 705–717.

Cacciola, P., Colajanni, P., & Muscolino, G. (2004). Combination of modal responses consistent with seismic input representation. *Journal of Structural Engineering, 130*(1), 47–55.

Cacciola, P., & Muscolino, G. (2011). Stochastic seismic analysis of large linear structural systems under fully non-stationary spectrum compatible ground motion. *Computational Methods in Stochastic Dynamics.*

Chen, K. F., Wang, Y. G., & Zhou, Z. (2009). Systems with few degrees of freedom and repeated natural frequencies. *International Journal of Mechanical Engineering Education, 37*(1), 67–77.

Clough, R. W., & Penzien, J. (1993). *Dynamics of structures* (2nd ed., Section 23-3: Normal Mode Forcing Function due to Discrete Loadings, pp. 543–547). New York: McGraw-Hill.

Crandall, S. H. (2001). The Future of structural dynamics. In D. J. Ewins & D. J. Inman (Eds.), *Structural dynamics @ 2000: current status and future directions* (pp. 441–443), Baldock, UK: Research Saludes Press Ltd.

Crandall, S. H., & Kulvets, A. P. (1977). Source correlation effects on structural response. In P. R. Krishnaiah (Ed.), *Applications in statistics* (pp. 163–182). Amsterdam, North-Holland.

Crandall, S. H., & Kulvets, A. P. (1979). Random vibration of mechanisms on plates. In *Proceedings of the Fifth World Congress on Theory of Machines and Mechanisms* (pp. 1568–1571). New York: ASME Press.

Crandall, S. H., & Mark, W. D. (1963). *Random vibration in mechanical systems.* New York: Academic Press.

Der Kiureghian, A. (1980). Structural response to stationary excitation. *Journal of Engineering Mechanics, 106,* 1195–1213.

Der Kiureghian, A. (1981). A Response spectra method for random vibration analysis of MDF system. *Earthquake Engineering and Structural Dynamics, 9,* 419–435.

Der Kiureghian, A., & Nakamura, Y. (1993). Complte quadratic combination (CQC) modal combination rule for high-frequency modes. *Earthquake Engineering and Structural Dynamics, 22,* 943–956.

Elishakoff, I. (1977a). Random vibrations of orthotropic plates clamped or simply supported all round. *Acta Mechanica, 28*(1–4), 165–176.

Elishakoff, I. (1977b). On the role of cross—correlations in the random vibrations of shells. *Journal of Sound and Vibration, 50*(2), 239–252.

Elishakoff, I. (1977c). Random vibrations of orthotropic plates with all edges clamped or simply supported. *Acta Mechanica, 28,* 165–176.

Elishakoff, I. (1977d). Flutter and random vibration in plates. In B.L. Clarkson (Ed.), *Stochastic problems in dynamics* (pp. 390–410). London: Pitman.

Elishakoff, I. (1983). *Probabilistic theory of structures.* New York: Wiley (Dover, 1999; World Scientific, 2017).

Elishakoff, I. (1999). *Probabilistic theory of structures* (pp. 317–380), Wiley: New York (second edition, Dover Publications: Mineola, 1999; third edition, World Scientific: Singapore, 2017).

Elishakoff, I., Van Zanten, A. Th., & Crandall, S. H. (1979). Wide-band random axisymmetric vibration of cylindrical shells. *Journal of Applied Mechanics, 46*(2), 417–423.

Falsone, G. (1999). Stationary and non-stationary modal cross-correlation coefficients for seismic analysis of MDOF structures. In *Proceedings of Eurodyn '99, Prague* (pp. 305–310). Rotterdam: Balkema.

Falsone, G., & Muscolino, G. (1999). Cross-correlation coefficients and modal combination rules for non-classically damped systems. *Earthquake Engineering and Structural Dynamics, 28,* 1669–1684.

Falsone, G., & Muscolino, G. (2004). New real-value modal combination rules for non-classically damped structures. *Earthquake Engineering and Structural Dynamics, 33*(12), 1187–1209.
Falsone, G., & Neri, F. (2001). Modal cross-correlation coefficients compatible with eurocode & prescriptions on seismic structural analysis. *European Earthquake Engineering, 2*(1), 51–60.
Filippov, A. P. (1970). *Vibration of deformarble systems* (pp. 40–44). Moscow: "Mashinostroenie" Publishing House.
Gladwell, G. M. L. (1961). Vibration systems with equal natural frequencies. *Journal of Mechanical Engineering Science, 3*(2), 174–185.
Grigoriu, M. (1979). Safety of structural systems with correlated resistances. *Applied Mathematical Modeling, 3,* 130–136.
Grigoriu, M. (1980). New criteria for selecting modal combination rules. In M. Petyt (Ed.), *Recent advances in structural dynamics* (Vol. 1, pp. 109–121). Institute of Sound and Vibration Research, University of Southampton.
Panovko, Ya. G., & Gubanova, I. I. (1965). *Stability and oscillations of elastic systems: paradoxes, fallacies, and new concepts*, New York: Consultants Bureau.
Heredia-Zavoni, E. (2011). The complete SRSS modal combination rule. *Earthquake Engineering and Structural Dynamics, 40*(11), 1181–1196.
Hernandes, J. J., & Lopez, O. A. (1998). Discussion of "A Replacement for the 30%, 40%, and SRSS Rules for Multicomponent Seismic Analysis" by Charles Menun and Armen Der Kiureghian. *Earthquake Spectra, 14*(4), 713–715.
Hurty, W. C., & Rubinstein, M. F. (1964). *Dynamics of structures*. Englewood Cliffs, N.Y.: Prentice-Hall.
Johnson, D. C., & Bishop, R. E. D. (1956). The modes of vibration of a certain system having a number of equal frequencies. *Journal of Applied Mechanics, 23,* 379–384.
Lagrange, J. L. (1997). *Mécanique analytique,* 1811 (in French) (English translation *Analytical Mechanics,* translated by A. Boissonnade and V.N.Vagliente, Dordrecht: Kluwer Academic Publishers).
Lin, Y. K. (1967). *Probabilistic theory of structural dynamics* (Chap. 7, pp. 203–252), New York: McGraw-Hill.
Lubliner, E., & Elishakoff, I. (1986). Random vibration of system with finitely many degrees of freedom and several coalescent natural frequencies. *International Journal of Engineering Science, 24*(4), 461–470.
Maymon, G. (1998). *Some engineering applications in random vibrations and random structures*. Washington, D.C.: AIAA Press.
Meirovitch, L. (2010). *Fundamentals of vibration* (p. 743). Long Grove, IL: Waveland.
Menun, C., & Der Kiureghian, A. (1998). A replacement for the 30%, 40%, and SRSS rules for multicomponent seismic analysis. *Earthquake Spectra, 14*(1), 153–163.
Menun, C., Reyes, J. C., & Chopra, A. K. (2015). Errors caused by peak factor assumptions in response-spectrum-based analyses. *Earthquake Engineering and Structural Dynamics, 44*(11), 1729–1746.
Muscolino, G., & Versaci, C. (2014). Interval complete quadratic combination rule for the response spectrum analysis of buildings with accidental eccentricity. *Earthquake Engineering and Structural Dynamics, 43*(15), 2217–2235.
Palmeri, A. (2006). Correlation coefficients for structures with viscoelastic dampers. *Engineering Structures, 28*(8), 1197–1208.
Preumont, A. (1994). *Random Vibration and spectral analysis*. Dordrecht: Kluwer Academic Publishers.
Rojas, A. L., Kalashnykova, N. I., Rojas, R. L., Martinez, F. C., & Sifuentes, A. C. U. (2012). Vibrations in pipe systems, taking into account shear deformations. *International Journal of Innovative Computing, Information and Control, 8*(12), 8435–8453.
Ruiz, P. (1970). On the maximum response of structures subjected to earthquake excitations. In *Proceedings of the Symposium on Earthquake Engineering* (pp. 272–277). India: University of Rookee.

Singh, M. P. (1980). Seismic design input for secondary structures. *Journal of Structural Division, 106*(2), 505–517.

Sinha, R., & Igusa, T. (1995). CQC and SRSS methods for non-classically damped structures. *Earthquake Engineering and Structural Dynamics, 24*(4), 615–619.

Soong, T. T., & Grigoriu, M. (1993). *Random vibration of mechanical and structural systems*. Englewood Cliffs, NJ: Prentice Hall.

Strutt J. W. (Baron Rayleigh), (1945) *The Theory of sound*, Vol 1, Macmillan, 1894 Dover, New york. p. 109

Thomson, W. and Tait, PG. (1879) *Treatise on Natural philosophy*, Vol. 1, part 1, Cambridge university press, Cambridge, UK, pp. 381–382. (Available at Google Books site)

Thomson, W. T. (1981). *Theory of vibrations and application*. Englewood Cliffs, N.J.: Prentice-Hall.

Tongue, B. H. (1996). *Principles of vibration*, Section 4.14: Repeated Frequencies and Zero Frequencies, (pp. 237–241). New York: Oxford University Press.

Walker, C., & Stafford, P. J. (2010). The use of modal-combination rules with cable-stayed bridges. *Proceedings of the ICE-Bridge Engineers, 163*(BE4), 225–240.

Wilson, E. L., Der Kiureghain, A., & Bayo, E. P. (1981). A replacement for the SRSS method in seismic analysis. *Earthquake Engineering and Structural Dynamics, 9,* 187–192.

Wilson, E. L., Suharwardy, I., & Habibullah, A. (1995). A clarification of the orthogonal effects in a three-dimensional seismic analysis. *Earthquake Spectra, 11*(4), 659–666.

Wirsching, P. H., Paez, T. L., & Ortiz, H. (1995). *Random Vibration: theory and practice*. New York: Wiley.

Yang, C. Y. (1985). *Random vibrations of structures*. New York: Wiley-Interscience.

Yiahao, L. A. (1992). Fast CQC algorithm of PSD matrices for random seismic responses. *Computers and Structures, 44*(3), 683–687.

Chapter 2
Random Vibration of System with Finitely Many Degrees of Freedom and Several Coalescent Natural Frequencies

2.1 Introductory Comments

As is well known, in the framework of the normal-mode method, every probabilistic characteristic of a mechanical system is the sum of two terms representing the respective interactions of *like* and *unlike* modes of vibration and referred to as *direct correlations* and *cross-correlations*. The latter are usually neglected because of the numerical difficulties involved in their inclusion—combined with the general premise that the modal density of the natural frequencies should not be high. This premise, however, is not borne out in practice: in shell-type structures, the modal density is high either at the low end of the frequency spectrum (e.g. in a spherical shell) or in the vicinity of some specific frequencies (e.g. in a cylindrical shell). In fact, it is shown by Elishakoff (1977a, 1977b, 1977c, 1977d, 1979) that inclusion of the cross-correlation terms changes the probabilistic response both qualitatively and quantitatively. The span-wise response, which is obtained as symmetric in the absence of the cross-correlations, is in reality non-symmetric in a cylindrical shell (Elishakoff, 1979); moreover, a structure evaluated on the basis of its direct correlations alone would be underdesigned by a factor of three or more (Elishakoff, 1977a, 1977b, 1977c, 1977d, 1979), with the attendant dangerous consequences for its performance.

In the above examples, the natural frequencies of the shell happen to lie very close together; in some structures, they may actually coalesce, in which case a large error would be introduced if their joint effects are omitted. Note that two-degree-of-freedom systems were studied without recourse to the normal-mode method (which implies automatic inclusion of the cross-correlations) by Crandall and Mark (1963); many-degree-of-freedom systems were studied recently via the normal-mode method, with investigation of the attendant effect of cross-correlation, by Elishakoff (1983, 1985): In particular, for a two-degree-of-freedom system with only direct correlations included, the resulting error was 50% for the first mass, and tended to infinity for the second mass, in a three-degree-of-freedom system where two of the

I. Elishakoff, *Dramatic Effect of Cross-Correlations in Random Vibrations of Discrete Systems, Beams, Plates, and Shells*,
https://doi.org/10.1007/978-3-030-40394-2_2

three natural frequencies coalesce, the corresponding error constituted 30%. A case in point is a rectangular plate with commensurable side lengths in certain ratios. For example, in a square plate, there are double natural frequencies of type $\omega_{\alpha\beta} = \omega_{\beta\alpha}$, where α and β are the number of half-waves in the directions of the plate axes, and for specific plates, there may be triple, quadruple, etc., natural frequencies. Such structures are considered by Dimentberg (1961), Elishakoff (1977a, 1977b, 1977c, 1977d), Crandall and Zhu (1983a, 1983b, 1983c).

This study deals with random vibration of a discrete system with finitely many degrees of freedom and coalescent natural frequencies, whose behavior resembles that of a plate. In fact, it represents a generalization of the three-degree-of-freedom case by Elishakoff (1983).

2.2 Analysis of a Multi-Degree-of-Freedom System with Many Coinciding Frequencies

Consider the n-degree-of-freedom system shown in Fig. 2.1. The system consists of n identical masses m, and identical spring stiffnesses k. Each mass is connected both with the ground and with the other masses. Damping proportional to the mass is provided by a dashspot attached to each mass, the damping coefficient being c. The first mass is subjected to ideal white noise loading with intensity S_0. We are interested in the mean-square responses of the system. The equations of motion read

$$\begin{aligned}
&m\ddot{x}_1 + c\dot{x}_1 + kx_1 + k(x_1 - x_2) + k(x_1 - x_3) + k(x_1 - x_4)\\
&\qquad + \cdots + k(x_1 - x_{n-1}) + k(x_1 - x_n) = f_1,\\
&m\ddot{x}_2 + c\dot{x}_2 + kx_2 + k(x_2 - x_3) + k(x_2 - x_4)\\
&\qquad + \cdots + k(x_2 - x_{n-1}) + k(x_2 - x_n) = 0,\\
&m\ddot{x}_3 + c\dot{x}_3 + kx_3 + k(x_3 - x_1) + k(x_3 - x_2) + k(x_3 - x_4)\\
&\qquad + \cdots + k(x_3 - x_{n-1}) + k(x_3 - x_n) = 0,\\
&m\ddot{x}_4 + c\dot{x}_4 + kx_4 + k(x_4 - x_1) + k(x_4 - x_2) + k(x_4 - x_3)\\
&\qquad + \cdots + k(x_4 - x_{n-1}) + k(x_4 - x_n) = 0\\
&\vdots\\
&m\ddot{x}_{n-1} + c\dot{x}_{n-1} + kx_{n-1} + k(x_{n-1} - x_1) + k(x_{n-1} - x_2) + k(x_{n-1} - x_3)\\
&\qquad + k(x_{n-1} - x_4) + \cdots + k(x_{n-1} - x_n) = 0,\\
&m\ddot{x}_n + c\dot{x}_n + kx_n + k(x_n - x_1)\\
&\qquad + k(x_n - x_2) + k(x_n - x_3) + k(x_n - x_4) + \cdots + k(x_n - x_{n-1}) = 0
\end{aligned} \tag{2.1}$$

which can also be put in the form

$$[M]\{\ddot{x}\} + [C]\{\dot{x}\} + [K]\{x\} = \{F\} \tag{2.2}$$

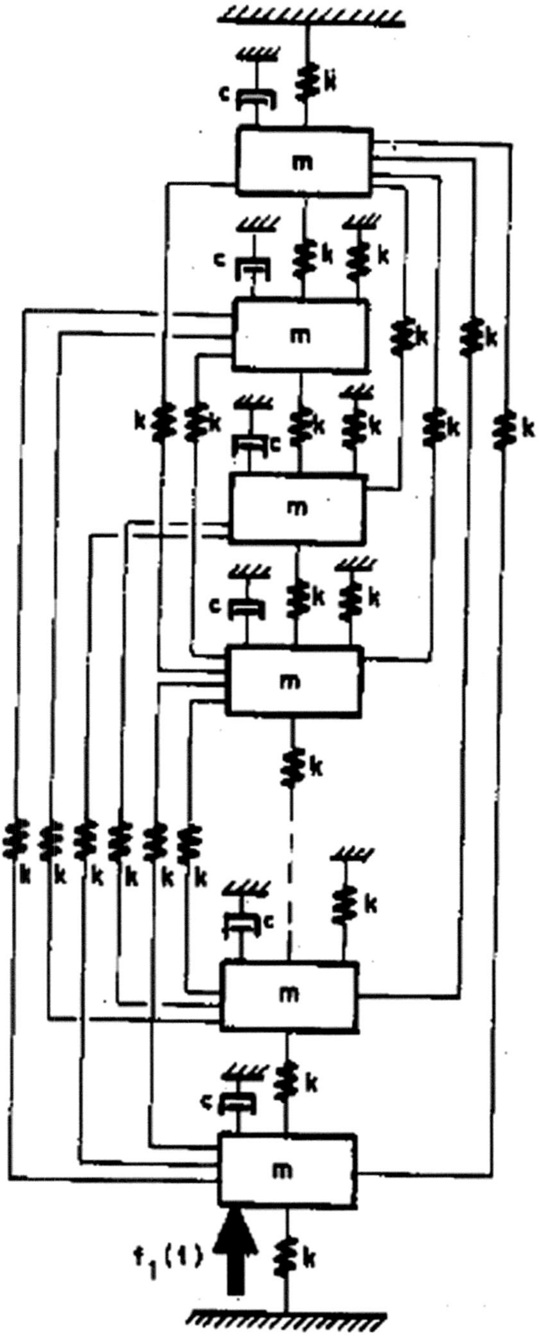

Fig. 2.1 *n*-degree-of-freedom system with mass-proportional damping

where the mass, damping, and stiffness matrices are

$$[M] = [m], \quad [C] = [c], \tag{2.3}$$

$$[K] = \begin{bmatrix} nk & -k & -k & -k & \cdots & -k & -k \\ -k & nk & -k & -k & \cdots & -k & -k \\ -k & -k & nk & -k & \cdots & -k & -k \\ -k & -k & -k & nk & \cdots & -k & -k \\ \vdots & \vdots & \vdots & \vdots & & \vdots & \vdots \\ -k & -k & -k & -k & \cdots & nk & -k \\ -k & -k & -k & -k & \cdots & -k & nk \end{bmatrix}. \tag{2.4}$$

The natural frequencies are found from the equation $\det\left[K - \omega^2 M\right] = 0$, or, in other words, from

$$\begin{bmatrix} nk - m\omega^2 & -k & -k & -k & \cdots & -k & -k \\ -k & nk - m\omega^2 & -k & -k & \cdots & -k & -k \\ -k & -k & nk - m\omega^2 & -k & \cdots & -k & -k \\ -k & -k & -k & nk - m\omega^2 & \cdots & -k & -k \\ \vdots & \vdots & \vdots & \vdots & & \vdots & \vdots \\ -k & -k & -k & -k & \cdots & nk - m\omega^2 & -k \\ -k & -k & -k & -k & \cdots & -k & nk - m\omega^2 \end{bmatrix} = 0. \tag{2.5}$$

Subtracting the last column successively from the other columns, we obtain

$$\begin{bmatrix} (n+1)k - m\omega^2 & 0 & 0 & 0 & \cdots & 0 & -k \\ 0 & (n+1)k - m\omega^2 & 0 & 0 & \cdots & 0 & -k \\ 0 & 0 & (n+1)k - m\omega^2 & 0 & \cdots & 0 & -k \\ 0 & 0 & 0 & (n+1)k - m\omega^2 & \cdots & 0 & -k \\ \vdots & \vdots & \vdots & \vdots & & \vdots & \vdots \\ 0 & 0 & 0 & 0 & \cdots & (n+1)k - m\omega^2 & -k \\ -(n+1)k + m\omega^2 & (n+1)k + m\omega^2 & -(n+1)k + m\omega^2 & -(n+1)k + m\omega^2 & \cdots & -(n+1)k + m\omega^2 & nk - m\omega^2 \end{bmatrix} = 0. \tag{2.6}$$

The last row is now added to the other rows to yield

$$
\left[\begin{array}{ccc}
(n+1)k - m\omega^2 & 0 & 0 \\
0 & (n+1)k - m\omega^2 & 0 \\
0 & 0 & (n+1)k - m\omega^2 \\
0 & 0 & 0 \\
\vdots & \vdots & \vdots \\
0 & 0 & 0 \\
0 & 0 & 0
\end{array}\right.
$$

$$
\left.\begin{array}{cccc}
0 & \cdots & 0 & -k \\
0 & \cdots & 0 & -k \\
0 & \cdots & 0 & -k \\
(n+1)k - m\omega^2 & \cdots & 0 & -k \\
\vdots & & \vdots & \vdots \\
0 & \cdots & (n+1)k - m\omega^2 & -k \\
0 & \cdots & 0 & k - m\omega^2
\end{array}\right]. \tag{2.7}
$$

Expanding the above determinant with respect to the last row, we finally obtain

$$\left[(n+1)k - m\omega^2\right]^{n-1}\left(k - m\omega^2\right) = 0.$$

The natural frequencies are thus represented by two sets of eigenvalues: One comprising the separate first natural frequency

$$\omega_1^2 = \frac{k}{m} \tag{2.8}$$

which can be interpreted as that of a single-degree-of-freedom system, and the other the remaining (coalescent) $n - 1$ natural frequencies

$$\omega_2^2 = \cdots = \omega_n^2 = \frac{(n+1)k}{m}. \tag{2.9}$$

The modal matrix is

$$[V] = \frac{1}{\sqrt{m}} \begin{bmatrix} \frac{1}{\sqrt{n}} & \frac{1}{\sqrt{2}} & \frac{1}{\sqrt{6}} & \frac{1}{\sqrt{12}} & \cdots & \frac{1}{\sqrt{(n-1)(n-2)}} & \frac{1}{\sqrt{n(n-1)}} \\ \frac{1}{\sqrt{n}} & \frac{-1}{\sqrt{2}} & \frac{1}{\sqrt{6}} & \frac{1}{\sqrt{12}} & \cdots & \frac{1}{\sqrt{(n-1)(n-2)}} & \frac{1}{\sqrt{n(n-1)}} \\ \frac{1}{\sqrt{n}} & 0 & \frac{-2}{\sqrt{6}} & \frac{1}{\sqrt{12}} & \cdots & \frac{1}{\sqrt{(n-1)(n-2)}} & \frac{1}{\sqrt{(n-1)(n-2)}} \\ \frac{1}{\sqrt{n}} & 0 & 0 & \frac{-3}{\sqrt{12}} & \cdots & \frac{1}{\sqrt{(n-1)(n-2)}} & \frac{1}{\sqrt{n(n-1)}} \\ \vdots & \vdots & \vdots & \vdots & & \vdots & \vdots \\ \frac{1}{\sqrt{n}} & 0 & 0 & 0 & \cdots & \frac{-(n-2)}{\sqrt{(n-1)(n-2)}} & \frac{1}{\sqrt{n(n-1)}} \\ \frac{1}{\sqrt{n}} & 0 & 0 & 0 & \cdots & 0 & \frac{-(n-1)}{\sqrt{n(n-1)}} \end{bmatrix}. \tag{2.10}$$

It satisfies the orthogonality condition

$$[V]^T[M][V] = [I] \tag{2.11}$$

where $[I]$ is an identity matrix, and T denotes the transpose operation.

We now proceed with the normal-mode analysis. Accordingly, we effect the transformation

$$\{x\} = [V]\{q\} \tag{2.12}$$

and substitute this in Eq. (2.1), premultiplying on the left by $[V]^T$ and using the properties

$$[V]^T[C][V] = \frac{c}{m}[I], \tag{2.13}$$

$$[V]^T[K][V] = \left[\omega^2\right] \tag{2.14}$$

where $\left[\omega^2\right]$ is a diagonal matrix of the natural frequencies squared ω_j^2. We thus obtain the decoupled equations of motion

$$\{\ddot{q}\} + \frac{c}{m}\{\dot{q}\} + \left[\omega^2\right]\{q\} = \{\Phi\} \tag{2.15}$$

where

$$\{\Phi\} = [V]^T\{F\} \tag{2.16}$$

analogous to that for a single-degree-of-freedom system.

Denoted by $[S_F(\omega)]$ the special density matrix of excitation in the original coordinate system, its counterpart $[S_\Phi(\omega)]$ in normal coordinate reads

$$[S_\Phi(\omega)] = [V]^T[S_F(\omega)][V]. \tag{2.17}$$

The spectral density matrix $\left[S_q(\omega)\right]$ of the generalized displacements is then found by Elishakoff (1983)

$$\left[S_q(\omega)\right] = [H(\omega)][S_\Phi(\omega)][H(\omega)]^* \tag{2.18}$$

the asterisk denoting the complex conjugate. In Eq. (2.18),

$$H_j(\omega) = \frac{1}{(i\omega)^2 + 2\varsigma_j\omega_j(i\omega) + \omega_j^2} \tag{2.19}$$

and

$$\varsigma_j = \frac{c}{2m\omega_j}. \tag{2.20}$$

The Wiener–Khintchine relationship yields the correlation function of the generalized displacements

$$\left[R_q(0)\right] = \int_{-\infty}^{\infty} \left[S_q(\omega)\right] e^{i\omega t} d\omega \tag{2.21}$$

where $\left[S_q(\omega)\right]$ is as per Eq. (2.15). The mean-square response matrix is obtained by setting $\tau = 0$ in Eq. (2.21):

$$\left[R_q(0)\right] = \int_{-\infty}^{\infty} \left[S_q(\omega)\right] d\omega. \tag{2.22}$$

Its counterpart in the original coordinates is obtained as

$$\begin{aligned} [R_x(0)] &= E([V]\{q\}\langle[V]\{q\}^*\rangle^T) = E\left([V]\{q\}\{q\}^{T*}[V]^T\right) \\ &= [V]E\left(\{q\}\{q\}^{T*}\right)[V]^T = [V]\left[R_q(0)\right][V]^T \end{aligned} \tag{2.23}$$

or, component-wise

$$R_{x_j x_k}(0) = \sum_{\alpha=1}^{n}\sum_{\beta=1}^{n} v_{j_\alpha} v_{k_\beta} R_{q_\alpha q_\beta}(0) \tag{2.24}$$

where

$$R_{q_\alpha q_\beta}(0) = \int_{-\infty}^{\infty} H_\alpha(\omega) H_\beta^*(\omega) S_{\Phi_\alpha \Phi_\beta}(\omega) d\omega. \tag{2.25}$$

Equation (2.24) can be rewritten as

$$R_{x_j x_k}(0) = \sum_{\alpha=1}^{n} v_{j_\alpha} v_{k_\alpha} R_{q_\alpha q_\alpha}(0) + \sum_{\substack{\alpha=1 \\ \alpha \neq \beta}}^{n} \sum_{\beta=1}^{n} v_{j\alpha} v_{k_\beta} R_{q_\alpha q_\beta}(0) \tag{2.26}$$

and due to the Hermitian property

$$R_{q_\alpha q_\beta}(0) = R^{*}_{q_\beta q_\alpha}(0) \tag{2.27}$$

we obtain

$$v_{j_\alpha} v_{k_\beta} R_{q_\alpha q_\beta}(0) + v_{j_\beta} v_{k_\alpha}(0) = 2 v_{j_\alpha} v_{k_\beta} Re\left[R_{q_\alpha q_\beta}(0)\right] \tag{2.28}$$

$Re[\cdots]$ denoting the real part of $[\cdots]$.

Finally, for the mean-square displacement, we have

$$E\left[x_j^2(t)\right] = R_{x_j x_j}(0) = \sum_{\alpha=1}^{n} v_{j\alpha}^2 R_{q_\alpha q_\alpha}(0) + 2 \sum_{\substack{\alpha=1 \\ \alpha < \beta}}^{n} \sum_{\beta=1}^{n} v_{j\alpha} v_{j\beta} R_{q_\alpha q_\beta}(0) \tag{2.29}$$

the first summation taking care of the direct correlations, and the second of the cross-correlations.

In complete analogy, for the mean-square velocities, we arrive at

$$E\left[\dot{x}_j^2(t)\right] = R_{\dot{x}_j \dot{x}_j}(0) = \sum_{\alpha=1}^{n} v_{j\alpha}^2 R_{\dot{q}_\alpha \dot{q}_\alpha}(0) + 2 \sum_{\substack{\alpha=1 \\ \alpha < \beta}}^{n} \sum_{\beta=1}^{n} v_{j\alpha} v_{j\beta} R_{\dot{q}_\alpha \dot{q}_\beta}(0) \tag{2.30}$$

where

$$R_{\dot{q}_\alpha \dot{q}_\beta}(0) = \int_{-\infty}^{\infty} \omega^2 H_\alpha(\omega) H_\beta(\omega) S_{\Phi_\alpha \Phi_\beta}(\omega) d\omega. \tag{2.31}$$

In the case under consideration,

$$S_{f_1 f_1}(\omega) = S_0, \quad S_{f_j f_k}(\omega) = 0, \quad \text{for} \quad j \neq k \neq 1. \tag{2.32}$$

For $R_{\dot{q}_\alpha \dot{q}_\beta}(0)$, we adapt the value from Elishakoff (1983), Eq. (9.220) by replacing A_j there by v_{i_j}, and arrive at

$$R_{\dot{q}_\alpha \dot{q}_\beta}(0) = \begin{cases} \frac{\pi s_0 m}{c} v_{1_\alpha}^2, & \text{for}\, \alpha = \beta \\ \frac{4\pi(\varsigma_\alpha \omega_\beta + \varsigma_\beta \omega_\alpha \omega_\beta S_0 v_{1_\alpha} v_{1_\beta})}{\left(\omega_\alpha^2 - \omega_\beta^2\right)^2 + 4\left[\varsigma_\alpha \varsigma_\beta \omega_\alpha \omega_\beta \left(\omega_\alpha^2 + \omega_\beta^2\right) + \left(\varsigma_\alpha^2 + \varsigma_\beta^2\right)\omega_\alpha^2 \omega_\beta^2\right]}, & \text{for}\, \alpha \neq \beta \end{cases} \tag{2.33}$$

Here, for v_{1_α}, we have

$$v_{1\alpha} = \begin{cases} \frac{1}{\sqrt{nm}}, & \text{for } \alpha = 1 \\ \frac{1}{\sqrt{\alpha(\alpha-1)m}}, & \text{for } \alpha = 2, 3, \ldots, n. \end{cases} \tag{2.34}$$

Therefore

$$R_{\dot{q}_\alpha \dot{q}_\alpha}(0) = \begin{cases} \frac{\pi S_0}{nc}, & \text{for } \alpha = 1 \\ \frac{\pi S_0}{\alpha(\alpha-1)c}, & \text{for } \alpha = 2, 3, \ldots, n. \end{cases} \tag{2.35}$$

Let us calculate $R_{\dot{q}_\alpha \dot{q}_1}(0)$. Recalling that

$$\begin{aligned} &\omega_1^2 = k/m, \quad \omega_\alpha^2 = (n+1)k/m, \quad 2\varsigma_1\omega_1 = 2\varsigma_1\omega_1 = c/m, \\ &\varsigma_1 = c/2mk, \quad \varsigma_\alpha = c/2\sqrt{(n+1)mk}, \\ &v_{11} = \sqrt{nm}, \quad v_{1_\alpha} = \sqrt{\alpha(\alpha-1)m} \end{aligned} \tag{2.36}$$

we obtain, after the necessary algebraic manipulations,

$$R_{\dot{q}_1 \dot{q}_\alpha}(0) = \frac{\pi S_0}{c} \frac{4(n+2)}{2mk/c^2 + 4(n+2)} \frac{1}{\sqrt{\alpha(\alpha-1)n}}. \tag{2.37}$$

Other elements of the matrix $\left[R_q(0)\right]$ are calculated in perfect analogy. The final result reads

$$R_{\dot{q}_\alpha \dot{q}_\beta}(0) = \begin{cases} \frac{\pi S_0}{c} \frac{1}{1+mk/2(n+2)c^2} \frac{1}{\sqrt{\beta(\beta-1)n}}, & \text{for } \begin{matrix} \alpha = 1 \\ \beta \geq 2 \end{matrix} \\ \frac{\pi S_0}{c} \frac{1}{\sqrt{\alpha\beta(\alpha-1)(\beta-1)}}, & \text{for } \begin{matrix} \alpha \geq 2 \\ \beta \geq 3. \end{matrix} \end{cases} \tag{2.38}$$

Having calculated $R_{\dot{q}_\alpha \dot{q}_\beta}(0)$, we obtain the mean-square velocity matrix $[R_{\dot{x}}(0)]$ by applying Eq. (2.30).

We next consider a few examples.

2.3 System with Two Degrees of Freedom

As is seen from Eqs. (2.8) and (2.9), the system does not possess the coalescent natural frequencies but its study instructive in elucidating the role of the cross-correlations. The modal matrix has a form

$$[V] = \frac{1}{\sqrt{m}} \begin{bmatrix} \frac{1}{\sqrt{2}} & \frac{1}{\sqrt{2}} \\ \frac{1}{\sqrt{2}} & \frac{1}{\sqrt{2}} \end{bmatrix}. \tag{2.39}$$

Elements of $R_{\dot{q}_\alpha \dot{q}_\beta}(0)$ are given by

$$
\begin{aligned}
R_{\dot{q}_1\dot{q}_1}(0) &= \frac{\pi S_0}{2c}, \quad R_{\dot{q}_2\dot{q}_2}(0) = \frac{\pi S_0}{2c}, \\
R_{\dot{q}_1\dot{q}_2}(0) &= \frac{1}{1 + mk/8c^2}\frac{\pi S_0}{2c}
\end{aligned} \tag{2.40}
$$

so that the mean-square velocities are

$$R_{\dot{x}_1\dot{x}_1}(0) = v_{11}^2 R_{\dot{q}_1\dot{q}_1}(0) + v_{12}^1 R_{\dot{q}_2\dot{q}_2}(0) + 2v_{11}v_{12}R_{\dot{q}_1\dot{q}_2}(0), \tag{2.41}$$

$$R_{\dot{x}_2\dot{x}_2}(0) = v_{21}^2 R_{\dot{q}_1\dot{q}_1}(0) + v_{22}^2 R_{\dot{q}_2\dot{q}_2}(0) + 2v_{21}v_{22}R_{\dot{q}_1\dot{q}_2}(0). \tag{2.42}$$

Substitution of Eqs. (2.39) and (2.40) in Eqs. (2.41) and (2.42) yields

$$R_{\dot{x}_1\dot{x}_1}(0) = \left[\frac{1}{2} + \frac{1}{2\left(1 + mk/8c^2\right)}\right]\frac{\pi S_0}{c}, \tag{2.43}$$

$$R_{\dot{x}_2\dot{x}_2}(0) = \left[\frac{1}{2} - \frac{1}{2\left(1 + mk/8c^2\right)}\right]\frac{\pi S_0}{c}. \tag{2.44}$$

As is seen from these expressions, for very light damping, namely $c^2/mk \ll 1$, the effect of the cross-correlations may well be disregarded. The following approximate relationship then holds

$$R_{\dot{x}_1\dot{x}_1}(0) \approx R_{\dot{x}_2\dot{x}_1}(0) \tag{2.45}$$

that is, the responses of both masses are the same although the force is applied directly at the first mass. This remarkable phenomenon was observed by Crandall in various uniform continuous systems by Elishakoff (1983) (see for a detailed discussion).

However, as damping in the structure increases, the effect of the cross-correlations ceases to be negligible. For example, at the damping coefficient level of $\varsigma_1 = 0.0417$ (corresponding to $c^2/\text{km} = 0.00691$), the error

$$\eta = \left|1 - \frac{\bar{R}_{\dot{x}_1\dot{x}_1}(0)}{R_{\dot{x}_1\dot{x}_1}(0)}\right| \times 100\% \tag{2.46}$$

due to omitting the cross-correlations constitutes 5%, and reaches 37.5% for $\varsigma_1 = 0.224$ ($c^2/\text{km} = 0.2$). The cross-correlations are included, the response turns out to be asymmetric. For the latter value of ς_1, we have

$$R_{\dot{x}_1\dot{x}_1}(0) = 0.807\pi S_0/c, \quad R_{\dot{x}_2\dot{x}_2}(0) = 0.193\pi S_0/c \tag{2.47}$$

whereas

$$\bar{R}_{\dot{x}_1\dot{x}_1}(0) = \bar{R}_{\dot{x}_2\dot{x}_2}(0) = 0.5\pi S_0/c. \tag{2.48}$$

Thus the error in deriving the mean-square response of the second mass amounts to 160%. We thus conclude that omission of the cross-correlations makes for a distorted response distribution both qualitatively and quantitatively.

2.4 Systems with Three and More Degrees of Freedom

The modal matrix $[V]$ reads

$$[V] = \frac{1}{\sqrt{m}}\begin{bmatrix} \frac{1}{\sqrt{3}} & \frac{1}{\sqrt{2}} & \frac{1}{\sqrt{6}} \\ \frac{1}{\sqrt{3}} & -\frac{1}{\sqrt{2}} & \frac{1}{\sqrt{6}} \\ \frac{1}{\sqrt{3}} & 0 & -\frac{2}{\sqrt{6}} \end{bmatrix}. \tag{2.49}$$

The element of $R_{\dot{q}_\alpha\dot{q}_\beta}(0)$ becomes

$$\begin{gathered} R_{\dot{q}_1\dot{q}_1}(0) = \tfrac{\pi S_0}{3c}, \quad R_{\dot{q}_2\dot{q}_2}(0) = \tfrac{\pi S_0}{2c}, \quad R_{\dot{q}_3\dot{q}_3}(0) = \tfrac{\pi S_0}{6c}, \\ R_{\dot{q}_1\dot{q}_2}(0) = \tfrac{\pi S_0}{\sqrt{6}c}\tfrac{1}{1+mk/10c^2}, \\ R_{\dot{q}_1\dot{q}_3}(0) = \tfrac{\pi S_0}{3\sqrt{2}c}\tfrac{1}{1+mk/10c^2}, \quad R_{\dot{q}_2\dot{q}_3}(0) = \tfrac{\pi S_0}{2\sqrt{3}c}. \end{gathered} \tag{2.50}$$

For the mean-square velocity of the first mass (subjected to the ideal white noise), we have

$$\begin{aligned} R_{\dot{x}_1\dot{x}_1}(0) = {} & v_{11}^2 R_{\dot{q}_1\dot{q}_1}(0) + v_{12}^2 R_{\dot{q}_2\dot{q}_2}(0) + v_{13}^2 R_{\dot{q}_3\dot{q}_3}(0) + 2v_{11}v_{12}R_{\dot{q}_1\dot{q}_2}(0) \\ & + 2v_{11}v_{13}R_{\dot{q}_1\dot{q}_3}(0) + 2v_{12}v_{13}R_{\dot{q}_2\dot{q}_3}(0). \end{aligned} \tag{2.51}$$

Closer examination of the values $R_{\dot{q}_1\dot{q}_2}(0)$ and $R_{\dot{q}_1\dot{q}_3}(0)$ shows that they can be neglected for $c^2/\text{km} \ll 1$, but such is not the case with $R_{\dot{q}_2\dot{q}_3}(0)$, which represents the interaction of the modes associated with the coalescent natural frequencies. Hence

$$R_{\dot{x}_1\dot{x}_1}(0) \simeq \frac{\pi S_0}{\text{mc}}\left(\frac{1}{9} + \frac{1}{4} + \frac{1}{36} + \frac{1}{6}\right) = \frac{5}{9}\frac{\pi S_0}{\text{mc}}. \tag{2.52}$$

Omission of the cross-correlation term $R_{\dot{q}_2\dot{q}_2}(0)$ yields an approximate value

$$\bar{R}_{\dot{x}_1\dot{x}_1} = \frac{\pi S_0}{mc}\left(\frac{1}{9} + \frac{1}{4} + \frac{1}{36}\right) = \frac{7}{18}\frac{\pi S_0}{mc} \tag{2.53}$$

associated with the percentage-wise error

$$\eta = 30\% \quad \text{for} \quad c^2/\text{km} \ll 1. \tag{2.54}$$

This result coincides with that obtained by Elishakoff (1983, p.376). The mean-square velocity for the second mass is

$$\begin{aligned} R_{\dot{x}_2\dot{x}_2}(0) = &\, v_{21}^2 R_{\dot{q}_1\dot{q}_1}(0) + v_{22}^2 R_{\dot{q}_2\dot{q}_2}(0) + v_{23}^2 R_{\dot{q}_3\dot{q}_3}(0) + 2v_{21}v_{22}R_{\dot{q}_1\dot{q}_2}(0) \\ &+ 2v_{21}v_{32}R_{\dot{q}_1\dot{q}_3}(0) + 2v_{22}v_{23}R_{\dot{q}_2\dot{q}_3}(0) \end{aligned} \tag{2.55}$$

or

$$\begin{aligned} R_{\dot{x}_2\dot{x}_2}(0) = &\, \frac{1}{m}\left[\frac{1}{3}R_{\dot{q}_1\dot{q}_1}(0) + \frac{1}{2}R_{\dot{q}_2\dot{q}_2}(0) + \frac{1}{6}R_{\dot{q}_3\dot{q}_3}(0)\right. \\ &\left. - \frac{2}{\sqrt{6}}R_{\dot{q}_1\dot{q}_2}(0) + \frac{2}{\sqrt{18}}R_{\dot{q}_1\dot{q}_3}(0) - \frac{2}{\sqrt{12}}R_{\dot{q}_2\dot{q}_3}\right]. \end{aligned} \tag{2.56}$$

Neglecting again $R_{\dot{q}_1\dot{q}_2}(0)$ and $R_{\dot{q}_1\dot{q}_3}(0)$, but retaining $R_{\dot{q}_2\dot{q}_3}(0)$, we have

$$R_{\dot{x}_2\dot{x}_2}(0) \simeq \frac{\pi S_0}{mc}\left(\frac{1}{9} + \frac{1}{4} + \frac{1}{36} - \frac{1}{6}\right) = \frac{2}{9}\frac{\pi S_0}{mc} \tag{2.57}$$

whereas the approximation is again

$$\bar{R}_{\dot{x}_2\dot{x}_2}(0) = \frac{7}{18}\frac{\pi S_0}{mc} \tag{2.58}$$

with an associated error of

$$\eta = \left|1 - \frac{R_{\dot{x}_2\dot{x}_2}(0)}{R_{x_2x_2}(0)}\right| \times 100\% = 75\%. \tag{2.59}$$

It is remarkable that the approximate predictions of the mean-square velocities of the first and second masses coincide

$$\bar{R}_{\dot{x}_1\dot{x}_1}(0) = \bar{R}_{\dot{x}_2\dot{x}_2}(0) \tag{2.60}$$

as though their responses were equal, despite the fact that the external force is applied to the first mass alone. By contrast, the more exact values of these quantities, with $R_{\dot{q}2\dot{q}3}(0)$ included, do *not* coincide. Moreover, comparison of Eqs. (2.42) and (2.43) reveals that the approximation predicts lower values for the response and makes for interdesign.

For cases with larger numbers of degree of freedom, a computer program was written, and the corresponding errors introduced by omitting the cross-correlations are shown in Fig. 2.2, the circles representing the value $c^2/\text{km} = 0.2$ and the squares $-c^2/\text{km} = 2 \times 10^{-6}$. (The points are linked for purposes of illustration only, in reality, the pattern is only meaningful for integer values of n). As is seen, the error approaches 70% and extreme care should be exercised in considering their omission.

Fig. 2.2 Error in mean-square velocity $R_{\dot{x}1\dot{x}1}(0)$ due to omitted modal cross-correlations

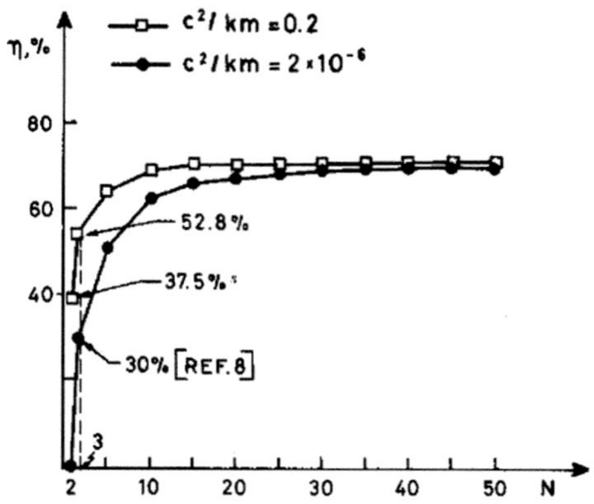

Further details can be found in the paper by Lubliner and Elishakoff (1986). In conclusion, the omission of the cross-correlation between different modes yields misleading results both quantitatively and qualitatively.

References

Crandall, S. H., & Mark, W. D. (1963). *Random vibration in mechanical systems*. New York: Academic Press.

Crandall S. H., & Zhu W. Q. (1983a). Wide band random excitation of square plates. In K. Hennig (Ed.), *Random vibrations and reliability* (pp. 231–244). Berlin: Akademie-Verlag.

Crandall, S. H., & Zhu, W. Q. (1983b). Random vibration: A survey of recent developments. *Journal of Applied Mechanics, 50*(46), 953–962.

Crandall S. H., & Zhu W. Q. (1983c). Wide band random vibration of square plates. In K. Hennig (Ed.), *Random vibrations and reliability* (pp. 231–243). Berlin: Akademie-Verlag.

Dimentberg M. F. (1961). Forced vibrations of plates under load representing spatial-temporary random process. *Inzhenernyi Zhurnal, 1*(2), 97–105 (in Russian).

Elishakoff, I. (1977a). Random vibrations of orthotropic plates clamped or simply supported all round. *Acta Mechanica, 28*(1–4), 165–176.

Elishakoff, I. (1977b). On the role of cross—correlations in the random vibrations of shells. *Journal of Sound and Vibration, 50*(2), 239–252.

Elishakoff, I. (1977c). Random vibrations of orthotropic plates with all edges clamped or simply supported. *Acta Mechanica, 28,* 165–176.

Elishakoff I. (1977d). Flutter and random vibration in plates. In B.L. Clarkson (Ed.), *Stochastic problems in dynamics* (pp. 390–410). London: Pitman.

Elishakoff I. (1983). *Probabilistic theory of structures*. New York: Wiley (Dover, 1999; World Scientific, 2017).

Elishakoff I. (1985). Random vibration of multi-degree-of-freedom systems with associated effect of cross-correlations. In F. Kozin, & W. Wedig (Ed.), *Structural identification and parameter estimation*. Vienna: Springer.

Elishakoff I., & Lyon R. H. (eds.) (1986). *Random vibration—status and recent development*. Amsterdam: Elsevier.

Elishakoff I., Van Zanten A. Th., & Crandall S.H. (1979). Wide-band random axisymmetric vibration of cylindrical shells. *Journal of Applied Mechanics*, *46*(2), 417–423.

Chapter 3
Random Vibration of a Vehicle Model

3.1 Introductory Comments

The studied random vibration of systems with repeated frequencies is very important as symmetrical structures widely exist, such as structures with two or more planes of reflective or cyclic symmetry or in the case of bodies of revolution, such as cylindrical or spherical shells.

When solving random vibration problems by the use of normal-mode method, the probabilistic characteristics of a mechanical system are represented as the sum of direct correlations and cross-correlations. As indicated above the cross-correlations are usually neglected in the literature and in computer codes because of the numerical difficulties and associated increase in the needed time of program execution; additional reason for such neglect is a wrong assertion that their effect must be small. However, when the natural frequencies happen to be repeated or lie close to each other, a large error would be introduced if the cross-correlations were omitted. Dahlberg (1982) studied the random vibration when the narrow-band excitation falls close to an anti-resonance of the structure. Elishakoff (1977a, 1977b, 1977c, 1977d), Elishakoff et al. (1979), Elishakoff and Ducreux (2014) studied the random vibration of various structures where the frequencies happen to lie close together. Crandall (1993) noticed that localization of response occurred at the symmetric counterparts of the structure and discussed the role of the cross-spectral densities. The authors showed that the cross-correlation terms may dramatically change the probabilistic response both qualitatively and quantitatively. Random vibration of mass–spring system with three degrees of freedom with coalescent natural frequencies was studied by Elishakoff (1983), which is extended to the n-degree-of freedom system by Lubliner and Elishakoff (1986). A two-degree-of-freedom mode where the mass can vibrate in a plane was suggested for illustration of the significance of cross-correlations in random vibration analysis by Elishakoff (1986).

I. Elishakoff, *Dramatic Effect of Cross-Correlations in Random Vibrations of Discrete Systems, Beams, Plates, and Shells*,
https://doi.org/10.1007/978-3-030-40394-2_3

In the above works, the discrete systems' examples are usually special cases of mass or stiffness distributions, whereas in other cases continuous systems, specifically the shells are studied. This paper deals with the vehicle models including a realistic model of a full vehicle, in addition, some other pertinent discrete systems with repeated frequencies are reconsidered. The contribution of the present investigation is twofold: extension of our earlier investigations by Lubliner and Elishakoff (1986); and Elishakoff (1986) from discrete systems with special mass and stiffness distribution to practical engineering problem, and investigation of the effects of the nonunique eigenvectors associated with repeated frequencies on probabilistic characteristics of random vibration responses. It should be noted that examples considered by Lubliner and Elishakoff (1986) were of purely academic nature, whereas in the present work we consider a practical engineering problem that exhibits the peculiar behavior that was uncovered by Lubliner and Elishakoff (1986).

3.2 Analysis of the Friswell Model

The equation of motion of a system having multi-degree-of-freedom system is given by

$$\mathbf{M}\ddot{\mathbf{X}} + \mathbf{C}\dot{\mathbf{X}} + \mathbf{K}\mathbf{X} = \mathbf{F}(t) \tag{3.1}$$

where $\mathbf{M}$, $\mathbf{C}$, and $\mathbf{K}$ are the $n \times n$ mass, damping, and stiffness matrices, respectively; $\mathbf{X}$ and $\mathbf{F}$ are the vectors of the generalized displacements and generalized forces, respectively; $\dot{\mathbf{X}}$ is a velocity vector, $\ddot{\mathbf{X}}$ is an acceleration vector.

For free vibration we express the solution of Eq. (3.1) in the form

$$x_j = A_j e^{i\omega t} \tag{3.2}$$

where ω is the frequency and A_j is the vibration amplitude. Substitution of Eq. (3.2) in Eq. (3.1) yields

$$|\mathbf{K} - \omega\mathbf{M}| = \mathbf{0}. \tag{3.3}$$

Depending on system parameters, the natural frequency ω may turn out to be repeated. To elucidate the encountered problems, we will start with a simple discrete system suggested by Friswell (1996). Consider the mass–spring system shown in Fig. 3.1 with the following parameter values: $m_1 = 1\,\text{kg}$, $m_2 = 4\,\text{kg}$, $m_3 = 1\,\text{kg}$, $k_1 = 0\,\text{N/m}$, $k_2 = 8p\,\text{N/m}$, $k_3 = 0\,\text{N/m}$, $k_4 = 2p\,\text{N/m}$, $k_5 = 2p\,\text{N/m}$, $k_6 = 1p\,\text{N/m}$, $c_1 = 0.5\,\text{Ns/m}$, $c_2 = 2\,\text{Ns/m}$, $c_3 = 0.5\,\text{Ns/m}$. Note that the values adopted by Friswell (1996) are obtained when parameter p is unity. Modification in Friswell's values is introduced in order to have a greater flexibility, and in order to study the effect of the variation of the control parameter p.

The mass, damping, and stiffness matrices can be written as

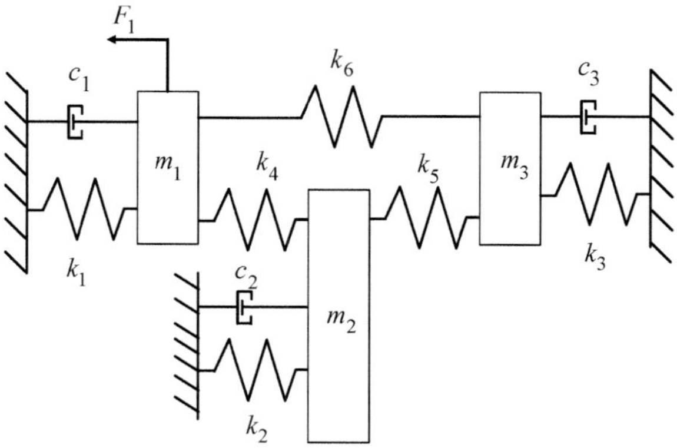

Fig. 3.1 Friswell's model

$$\mathbf{M} = \begin{bmatrix} 1 & 0 & 0 \\ 0 & 4 & 0 \\ 0 & 0 & 1 \end{bmatrix}, \mathbf{K} = \begin{bmatrix} 3 & -2 & -1 \\ -2 & 12 & -2 \\ -1 & -2 & 3 \end{bmatrix} p, \mathbf{C} = \begin{bmatrix} 0.5 & 0 & 0 \\ 0 & 2 & 0 \\ 0 & 0 & 0.5 \end{bmatrix}. \tag{3.4}$$

The natural frequencies are

$$\omega_1^2 = 1p\,\mathrm{rad}^2/\mathrm{s}^2, \omega_2^2 = \omega_3^2 = 4p\,\mathrm{rad}^2/\mathrm{s}^2. \tag{3.5}$$

Note that the Friswell system possesses repeated natural frequencies ω_2 and ω_3. The ω_3 modal matrix has a form

$$\mathbf{V} = \begin{bmatrix} \frac{\sqrt{3}}{3} & -\frac{\sqrt{6}}{6} & \frac{\sqrt{2}}{2} \\ \frac{\sqrt{3}}{6} & -\frac{\sqrt{6}}{12} & -\frac{\sqrt{2}}{4} \\ \frac{\sqrt{3}}{3} & \frac{\sqrt{6}}{3} & 0 \end{bmatrix} \tag{3.6}$$

and is independent of p.

Substituting in system equation $\mathbf{X} = \mathbf{VQ}$ and pre-multiplying the result by $\mathbf{v}^T$, we arrive at

$$\begin{aligned} \ddot{Q}_1 + \frac{1}{2}\dot{Q}_1 + pQ_1 &= \frac{\sqrt{3}}{3}F_1 \\ \ddot{Q}_2 + \frac{1}{2}\dot{Q}_2 + 4pQ_2 &= \frac{\sqrt{6}}{6}F_1. \\ \ddot{Q}_3 + \frac{1}{2}\dot{Q}_3 + 4pQ_3 &= \frac{\sqrt{2}}{2}F_1 \end{aligned} \tag{3.7}$$

Let F_1 be a stationary random process, specifically the white noise excitation of intensity S_0. According to the normal-mode method by Elishakoff (1983), the correlation matrix of generalized velocities $\dot{Q}_1$, $\dot{Q}_2$, and $\dot{Q}_3$ can be obtained as follows:

$$R_{\dot{q}_\alpha \dot{q}_\beta}(0) = \begin{cases} \frac{\pi S_0}{2\zeta_\alpha \omega_\alpha} v_{1\alpha}^2, & \text{for} \quad \alpha = \beta \\ \frac{4\pi(\zeta_\alpha \omega_\beta + \zeta_\beta \omega_\alpha)\omega_\alpha \omega_\beta S_0 v_{1\alpha} v_{1\beta}}{(\omega_\alpha^2 - \omega_\beta^2)^2 + 4\left[\zeta_\alpha \zeta_\beta \omega_\alpha \omega_\beta (\omega_\alpha^2 + \omega_\beta^2) + (\zeta_\alpha^2 + \zeta_\beta^2)\omega_\alpha^2 \omega_\beta^2\right]}, & \text{for} \quad \alpha \neq \beta \end{cases} \tag{3.8}$$

where $\zeta_\alpha = 1/4\omega_\alpha$, $v_{1\alpha}$ is the α th element of the first row of $\mathbf{V}$. Substituting Eqs. (3.5–3.6) into Eq. (3.8), the elements of the matrix $\mathbf{R}_{\dot{q}\dot{q}}(0) = \left[R_{\dot{q}\dot{q}}(0)\right]$ is derived as shown in Table 3.1.

Now we return to the sought quantities

$$\mathbf{R}_{\dot{x}\dot{x}}(0) = E\left(\dot{\mathbf{X}}\dot{\mathbf{X}}^T\right) = E\left(\mathbf{V}\dot{\mathbf{Q}}\dot{\mathbf{Q}}^T\mathbf{V}^T\right) = \mathbf{V}E\left(\dot{\mathbf{Q}}\dot{\mathbf{Q}}^T\right)\mathbf{V}^T = \mathbf{V}\mathbf{R}_{\dot{q}\dot{q}}(0)\mathbf{V}^T. \tag{3.9}$$

The expression for $R_{\dot{x}_1\dot{x}_1}(0)$ becomes

$$\begin{aligned} R_{\dot{x}_1\dot{x}_1}(0) = {} & v_{11}^2 R_{\dot{q}_1\dot{q}_1}(0) + v_{12}^2 R_{\dot{q}_2\dot{q}_2}(0) + v_{13}^2 R_{\dot{q}_3\dot{q}_3}(0) + v_{11}v_{12} R_{\dot{q}_1\dot{q}_2}(0) + v_{11}v_{12} R_{\dot{q}_2\dot{q}_1}(0) \\ & + v_{11}v_{13} R_{\dot{q}_1\dot{q}_3}(0) + v_{11}v_{13} R_{\dot{q}_3\dot{q}_1}(0) + v_{12}v_{13} R_{\dot{q}_2\dot{q}_3}(0) + v_{12}v_{13} R_{\dot{q}_3\dot{q}_2}(0) \end{aligned}. \tag{3.10}$$

Similarly, the expressions for $R_{\dot{x}_2\dot{x}_2}(0)$ and $R_{\dot{x}_3\dot{x}_3}(0)$ read

$$\begin{aligned} R_{\dot{x}_2\dot{x}_2}(0) = {} & v_{21}^2 R_{\dot{q}_1\dot{q}_1}(0) + v_{22}^2 R_{\dot{q}_2\dot{q}_2}(0) + v_{23}^2 R_{\dot{q}_3\dot{q}_3}(0) + v_{21}v_{22} R_{\dot{q}_1\dot{q}_2}(0) + v_{21}v_{22} R_{\dot{q}_2\dot{q}_1}(0) \\ & + v_{21}v_{23} R_{\dot{q}_1\dot{q}_3}(0) + v_{21}v_{23} R_{\dot{q}_3\dot{q}_1}(0) + v_{22}v_{23} R_{\dot{q}_2\dot{q}_3}(0) + v_{22}v_{23} R_{\dot{q}_3\dot{q}_2}(0) \end{aligned}, \tag{3.11}$$

$$\begin{aligned} R_{\dot{x}_3\dot{x}_3}(0) = {} & v_{31}^2 R_{\dot{q}_1\dot{q}_1}(0) + v_{32}^2 R_{\dot{q}_2\dot{q}_2}(0) + v_{33}^2 R_{\dot{q}_3\dot{q}_3}(0) + v_{31}v_{32} R_{\dot{q}_1\dot{q}_2}(0) + v_{31}v_{32} R_{\dot{q}_2\dot{q}_1}(0) \\ & + v_{31}v_{33} R_{\dot{q}_1\dot{q}_3}(0) + v_{31}v_{33} R_{\dot{q}_3\dot{q}_1}(0) + v_{32}v_{33} R_{\dot{q}_2\dot{q}_3}(0) + v_{32}v_{33} R_{\dot{q}_3\dot{q}_2}(0) \end{aligned}. \tag{3.12}$$

The first three terms' summations in formulates (3.10)–(3.12) represent the contribution of identical modes in formulating the mean-square response (referred as *direct correlations*). The terms can be interpreted as the contribution of interaction of the normal mode with itself. The remaining terms' summation represents the contribution of the interactions between separate modes (referred to as *cross-correlations*), which can be interpreted as the contribution of interaction of the normal mode with other modes.

Substituting Eqs. (3.6) and $\mathbf{R}_{\dot{q}\dot{q}}(0)$ into Eqs. (3.10–3.12) yields

Table 3.1 Results of the correlation matrix of generalized velocities

$R_{\dot{q}_1\dot{q}_1}(0)$	$R_{\dot{q}_2\dot{q}_2}(0)$	$R_{\dot{q}_3\dot{q}_3}(0)$	$R_{\dot{q}_1\dot{q}_2}(0)$	$R_{\dot{q}_2\dot{q}_1}(0)$
$\frac{2}{3}$	$\frac{1}{3}$	1	$-\frac{17\sqrt{2}}{1350p^2+51}$	$-\frac{17\sqrt{2}}{1350p^2+51}$
$R_{\dot{q}_1\dot{q}_3}(0)$	$R_{\dot{q}_3\dot{q}_1}(0)$	$R_{\dot{q}_2\dot{q}_3}(0)$	$R_{\dot{q}_3\dot{q}_2}(0)$	
$\frac{17\sqrt{2}}{1350p^2+51}$	$\frac{17\sqrt{2}}{1350p^2+51}$	$-\frac{\sqrt{3}}{3}$	$-\frac{\sqrt{3}}{3}$	

$$R_{\dot{x}_1\dot{x}_1}(0) = \frac{500p^2 + 34}{450p^2 + 17}\pi S_0,\ R_{\dot{x}_2\dot{x}_2}(0) = \frac{50p^2}{450p^2 + 17}\pi S_0,\ R_{\dot{x}_3\dot{x}_3}(0) = \frac{200p^2}{450p^2 + 17}\pi S_0. \tag{3.13}$$

If the cross-correlation terms are omitted, the mean square of velocities will be obtained as

$$\tilde{R}_{\dot{x}_1\dot{x}_1}(0) = \frac{7}{9}\pi S_0,\ \tilde{R}_{\dot{x}_2\dot{x}_2}(0) = \frac{7}{36}\pi S_0,\ \cdot\tilde{R}_{\dot{x}_3\dot{x}_3}(0) = \frac{4}{9}\pi S_0. \tag{3.14}$$

It is remarkable that the approximate predictions are independent of the parameter p.

The associated errors of mean-square velocity due to omitted modal cross-correlations are obtained as

$$\tilde{\eta}_1 = \left|\frac{\tilde{R}_{\dot{x}_1\dot{x}_1}(0) - R_{\dot{x}_1\dot{x}_1}(0)}{R_{\dot{x}_1\dot{x}_1}(0)}\right| \times 100\% = \frac{135000p^2 + 18700}{4500p^2 + 306}\%, \tag{3.15}$$

$$\tilde{\eta}_2 = \left|\frac{\tilde{R}_{\dot{x}_2\dot{x}_2}(0) - R_{\dot{x}_2\dot{x}_2}(0)}{R_{\dot{x}_2\dot{x}_2}(0)}\right| \times 100\% = \frac{135000p^2 + 11900}{1800p^2}, \tag{3.16}$$

$$\tilde{\eta}_3 = \left|\frac{\tilde{R}_{\dot{x}_3\dot{x}_3}(0) - R_{\dot{x}_3\dot{x}_3}(0)}{R_{\dot{x}_3\dot{x}_3}(0)}\right| \times 100\% = \frac{1700}{450p^2}\%. \tag{3.17}$$

For the first mass, when p tends to infinity, the error tends to be equal to 30%, while when p tends to zero, the maximum error appears and is equal to 61.11%. To the second mass, the minimum error is equal to 75% when p is infinite and the maximum error tends to infinity when p is zero. For the third mass, the error reaches zero when p is unbounded, and the error tends to infinity when p tends to zero. These large errors are due to the fact that the modal cross-correlations associated with repeated natural frequencies *cannot* be neglected.

3.3 Kim and Yoon's Four-Wheel Vehicle Model

Consider now the eight degrees of freedom vibration system of a vehicle are shown in Fig. 3.2. The suspensions and tires between the sprung mass and unsprung masses are modeled as linear proportional viscous dampers and spring elements. The parameters are shown in Table 3.2, in which the origin of the coordinate system is chosen to be the center of mass of the sprung mass This model is similar to those used by Zuo and Nayfeh (2003) and Kim and Yoon (1995).

Suppose that the vertical displacements of the connection between suspensions and vehicle body are Z_{fL0}, Z_{rL0}, Z_{fR0}, Z_{rR0} the vertical displacement the connection between driver's seat and vehicle body is Z_{c0}, thus

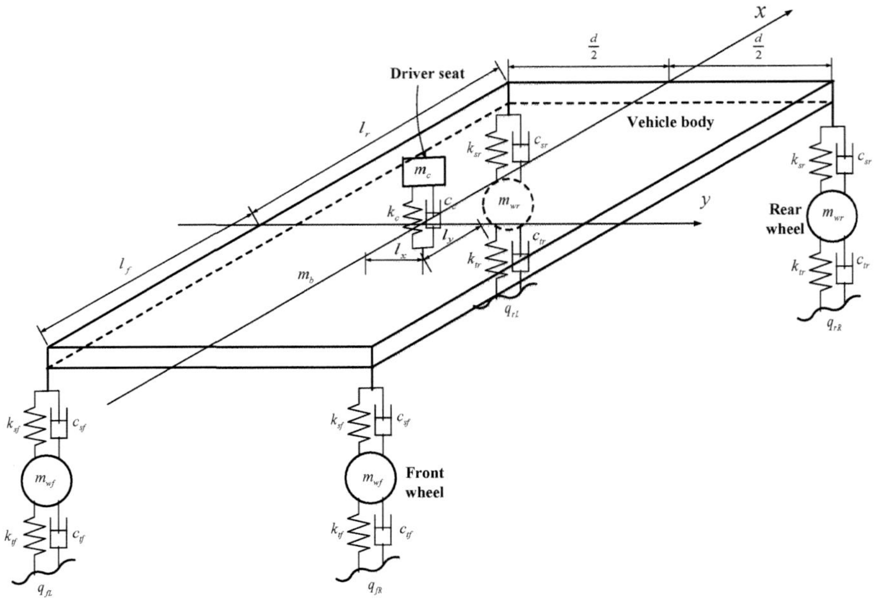

Fig. 3.2 A typical four-wheel vehicle model (after Long et al., 2014)

$$
\begin{aligned}
Z_{fL0} &= Z_b + \frac{d}{2}\theta_y - l_f\theta_x \\
Z_{rL0} &= Z_b + \frac{d}{2}\theta_y + l_r\theta_x \\
Z_{fR0} &= Z_b - \frac{d}{2}\theta_y - l_f\theta_x \\
Z_{rR0} &= Z_b - \frac{d}{2}\theta_y + l_r\theta_x \\
Z_{c0} &= Z_b + l_y\theta_y - l_x\theta_x
\end{aligned} \tag{3.18}
$$

Thus the forces of the suspensions and seat to the vehicle body are

$$
\begin{aligned}
&F_{fL} = k_{sf}\left(Z_{fL} - Z_{fL0}\right) + C_{sf}\left(\dot{Z}_{fL} - \dot{Z}_{fL0}\right) \\
&F_{rL} = k_{sr}(Z_{rL} - Z_{rL0}) + C_{sr}\left(\dot{Z}_{rL} - \dot{Z}_{rL0}\right) \\
&F_{fR} = k_{sf}\left(Z_{fR} - Z_{fR0}\right) + C_{sf}\left(\dot{Z}_{fR} - \dot{Z}_{fR0}\right). \\
&F_{rR} = k_{sr}(Z_{rR} - Z_{rR0}) + C_{sr}\left(\dot{Z}_{rR} - \dot{Z}_{rR0}\right) \\
&F_c = k_c(Z_{c0} - Z_c) + C_c\left(\dot{Z}_{c0} - \dot{Z}_c\right)
\end{aligned} \tag{3.19}
$$

After applying a force-balance analysis to the model in Fig. 3.2, the equations of motion are obtained as

Table 3.2 Ideal symmetrical vehicle model and corresponding parameters

Description	Symbol	Value
Unsprung mass (front/rear)	m_{wf} and m_{wr}	40/40 kg
Sprung mass	m_b	1375.9 kg
Sprung moment of inertia about x-/y-axis	J_x and J_y	484.4/2344.4 kg m^2
Mass of driver's seat	m_c	60 kg
Vertical stiffness of one tire (front/rear)	k_{tf} and k_{tr}	18208700/18208700 N/m
Suspension stiffness per wheel (front/rear)	k_{sf} and k_{sr}	20984.8/20984.8 N/m
Vertical stiffness of driver's seat	k_c	10507.1 N/m
Damping coefficient of one tire (front/rear)	c_{tf} and c_{tr}	182087/182087 Ns/m
Suspension damping per wheel (front/rear)	c_{sf} and c_{sr}	209.8/209.8 Ns/m
Damping coefficient of driver's seat	c_c	105.1 Ns/m
Distance between x-axis and driver's seat	l_x	0 m
Distance between y-axis and driver's seat	l_y	0 m
Distance between x-axis and front/rear tires	l_f and l_r	1.511/1.511 m
Distance between left and right tires	d	1.44 m
The ground displacement inputs	$q_{fL}, q_{rL}, q_{fR}, q_{rR}$	
The vertical displacements of four tires	$Z_{fL}, Z_{rL}, Z_{fR}, Z_{rR}$	
Attitude of vehicle body: roll, pitch, and heave	θ_x, θ_y, Z_b	
Vertical displacement of driver's seat	Z_c	

$$\begin{array}{c}
m_{wf}\ddot{Z}_{fL} - k_{tf}\left(q_{fL} - Z_{fL}\right) - F_{fL} = 0 \\
m_{wr}\ddot{Z}_{rL} - k_{tr}(q_{rL} - Z_{rL}) - F_{rL} = 0 \\
m_{wf}\ddot{Z}_{fR} - k_{tf}\left(q_{fR} - Z_{fR}\right) - F_{fR} = 0 \\
m_{wr}\ddot{Z}_{rR} - k_{tf}(q_{rR} - Z_{rR}) - F_{rR} = 0 \\
m_b\ddot{Z}_b - F_{fL} - F_{rL} - F_{fR} - F_{rR} + F_c = 0 \\
J_x\ddot{\theta}_x - \frac{d}{2}\left(F_{fL} + F_{rL}\right) + \frac{d}{2}\left(F_{fR} + F_{rR}\right) - l_y F_c = 0 \\
J_y\ddot{\theta}_y - l_r(F_{rL} + F_{rR}) + l_f\left(F_{fL} + F_{fR}\right) - l_x F_c = 0 \\
m_c\ddot{Z}_c - F_{c=0}
\end{array} \quad . \tag{3.20}$$

Substituting Eqs. (3.18) and (3.19) into Eq. (3.20), the model of the vehicle can be expressed in the form analogous to Eq. (3.1):

$$\mathbf{M}\ddot{\mathbf{X}} + \mathbf{C}\dot{\mathbf{X}} + \mathbf{K}\mathbf{X} = \mathbf{K}_t\mathbf{Q}. \tag{3.21}$$

The mass matrix reads

$$
\mathbf{M} = \begin{bmatrix}
m_{wf} & 0 & 0 & 0 & 0 & 0 & 0 & 0 \\
0 & m_{wr} & 0 & 0 & 0 & 0 & 0 & 0 \\
0 & 0 & m_{wf} & 0 & 0 & 0 & 0 & 0 \\
0 & 0 & 0 & m_{wr} & 0 & 0 & 0 & 0 \\
0 & 0 & 0 & 0 & m_b & 0 & 0 & 0 \\
0 & 0 & 0 & 0 & 0 & J_x & 0 & 0 \\
0 & 0 & 0 & 0 & 0 & 0 & J_y & 0 \\
0 & 0 & 0 & 0 & 0 & 0 & 0 & m_c
\end{bmatrix}. \tag{3.22}
$$

The stiffness matrix reads

$$
\mathbf{K} = \begin{bmatrix}
k_{tf}+k_{sf} & 0 & 0 & 0 & -k_{sf} & -\frac{d}{2}k_{sf} & k_{sf}l_f & 0 \\
0 & k_{tr}+k_{sr} & 0 & 0 & -k_{sr} & -\frac{d}{2}k_{sr} & -k_{sr}l_r & 0 \\
0 & 0 & k_{tf}+k_{sf} & 0 & -k_{sf} & \frac{d}{2}k_{sf} & k_{sf}l_f & 0 \\
0 & 0 & 0 & k_{tr}+k_{sr} & -k_{sr} & \frac{d}{2}k_{sr} & -k_{sr}l_r & 0 \\
-k_{sf} & -k_{sr} & -k_{sf} & -k_{sr} & 2(k_{sf}+k_{sr})+k_c & k_c l_y & 2(l_r k_{sr}-l_f k_{sf})-k_c l_x & -k_c \\
-\frac{d}{2}k_{sf} & -\frac{d}{2}k_{sr} & \frac{d}{2}k_{sf} & \frac{d}{2}k_{sr} & k_c l_y & \frac{d^2}{2}(k_{sf}+k_{sr})+k_c l_y^2 & -k_c l_x l_y & -k_c l_y \\
k_{sf}l_f & -k_{sr}l_r & k_{sf}l_f & -k_{sr}l_r & 2(l_r k_{sr}-l_f k_{sf})-k_c l_x & -k_c l_x l_y & 2\left(k_{sf}l_f^2+k_{sr}l_r^2\right)+k_c l_x^2 & k_c l_x \\
0 & 0 & 0 & 0 & -k_c & -k_c l_y & k_c l_x & k_c
\end{bmatrix}. \tag{3.23}
$$

The damping matrix is given by

$$
\mathbf{C} = \begin{bmatrix}
c_{tf}+c_{sf} & 0 & 0 & 0 & -c_{sf} & -\frac{d}{2}c_{sf} & c_{sf}l_f & 0 \\
0 & c_{tr}+c_{sr} & 0 & 0 & -c_{sr} & -\frac{d}{2}c_{sr} & -c_{sr}l_r & 0 \\
0 & 0 & c_{tf}+c_{sf} & 0 & -c_{sf} & \frac{d}{2}c_{sf} & c_{sf}l_f & 0 \\
0 & 0 & 0 & c_{tr}+c_{sr} & -c_{sr} & \frac{d}{2}c_{sr} & -c_{sr}l_r & 0 \\
-c_{sf} & -c_{sr} & -c_{sf} & -c_{sr} & 2(c_{sf}+c_{sr})+c_c & c_c l_y & 2(l_r c_{sr}-l_f c_{sf})-c_c l_x & -c_c \\
-\frac{d}{2}c_{sf} & -\frac{d}{2}c_{sr} & \frac{d}{2}c_{sf} & \frac{d}{2}c_{sr} & c_c l_y & \frac{d^2}{2}(c_{sf}+c_{sr})+c_c l_y^2 & -c_c l_x l_y & -c_c l_y \\
c_{sf}l_f & -c_{sr}l_r & c_{sf}l_f & -c_{sr}l_r & 2(l_r c_{sr}-l_f c_{sf})-c_c l_x & -c_c l_x l_y & 2\left(c_{sf}l_f^2+c_{sr}l_r^2\right)+c_c l_x^2 & c_c l_x \\
0 & 0 & 0 & 0 & -c_c & -c_c l_y & c_c l_x & c_c
\end{bmatrix}. \tag{3.24}
$$

Table 3.3 Natural frequencies of the ideal symmetrical vehicle model

Natural frequencies	$\omega_1^2(\text{rad}^2/\text{s}^2)$	$\omega_2^2(\text{rad}^2/\text{s}^2)$	$\omega_3^2(\text{rad}^2/\text{s}^2)$	$\omega_4^2(\text{rad}^2/\text{s}^2)$
Values	18.4586	55.9003	152.6109	397.1794
Natural frequencies	$\omega_5^2(\text{rad}^2/\text{s}^2)$	$\omega_6^2(\text{rad}^2/\text{s}^2)$	$\omega_7^2(\text{rad}^2/\text{s}^2)$	$\omega_8^2(\text{rad}^2/\text{s}^2)$
Values	4.5574×10^5	4.5574×10^5	4.5574×10^5	4.5574×10^5

The tires' stiffness matrix reads

$$\mathbf{K}_t = \begin{bmatrix} k_{tf} & 0 & 0 & 0 \\ 0 & k_{tr} & 0 & 0 \\ 0 & 0 & k_{tf} & 0 \\ 0 & 0 & 0 & k_{tr} \\ 0 & 0 & 0 & 0 \\ 0 & 0 & 0 & 0 \\ 0 & 0 & 0 & 0 \end{bmatrix}. \tag{3.25}$$

The excitations' vector reads

$$\mathbf{Q}^T = \left[q_{fL} \; q_{rL} \; q_{fR} \; q_{rR} \right]. \tag{3.26}$$

The response vector reads

$$\mathbf{X}^T = \left[Z_{fL} \; Z_{rL} \; Z_{fR} \; Z_{rR} \; Z_b \; \theta_x \; \theta_y \; Z_c \right]. \tag{3.27}$$

We first consider a symmetric vehicle system whose parameters are shown in Table 3.2. For simplicity, let q_{fL} be a stationary random process represented as a white noise with intensity S_0, and other inputs vanish. Solving Eq. (3.3), the natural frequencies can be obtained as shown in Table 3.3.

As is seen, there are four repeated frequencies in the system, namely, $\omega_5, \omega_6, \omega_7$ and ω_8. The mean-square velocity for the left front tire is

$$R_{\dot{x}_1\dot{x}_1}(0) = \sum_{\alpha=1}^{8} v_{1\alpha}^2 R_{\dot{q}_\alpha\dot{q}_\alpha}(0) + \sum_{\substack{\alpha=1 \\ \alpha\neq\beta}}^{8} \sum_{\beta=1}^{8} v_{1\alpha} v_{1\beta} R_{\dot{q}_\alpha\dot{q}_\beta}(0) = 5.4792 \times 10^{-6} S_0. \tag{3.28}$$

Omission of the cross-correlation terms yields an approximate value

$$\tilde{R}_{\dot{x}_1\dot{x}_1}(0) = \sum_{\alpha=1}^{8} v_{1\alpha}^2 R_{\dot{q}_\alpha\dot{q}_\alpha}(0) = 1.3698 \times 10^{-6} S_0 \tag{3.29}$$

associated with error

$$\tilde{\eta}_1 = \frac{\tilde{R}_{\dot{x}_1\dot{x}_1}(0) - R_{\dot{x}_1\dot{x}_1}(0)}{R_{\dot{x}_1\dot{x}_1}(0)} \times 100\% = 75\%. \tag{3.30}$$

Now we consider a more realistic vehicle model. It is slightly modified by Zuo and Nayfeh (2003) and Zuo and Nayfeh (2007), with dampers are introduced to the tires. The parameters are shown in Table 3.4. The frequencies of the asymmetric vehicle system change to be the results shown in Table 3.5.

There are no repeated frequencies in the systems but the last four frequencies still stay very close. The mean velocity of the left front tire is

$$R_{\dot{x}_1\dot{x}_1}(0) = 4.4226 \times 10^{-4} S_0. \tag{3.31}$$

Omission of the cross-correlation terms yields an approximate value

$$\tilde{R}_{\dot{x}_1\dot{x}_1}(0) = 2.0674 \times 10^{-4} S_0 \tag{3.32}$$

associated with error

Table 3.4 Realistic asymmetrical vehicle model and corresponding parameters

Description	Symbol	Value
Unsprung mass (front/rear)	m_{wf} and m_{wr}	40/40 kg
Sprung mass	m_b	1375.9 kg
Sprung moment of inertia about x-/y-axis	J_x and J_y	484.4/2344.4 kg m^2
Mass of driver's seat	m_c	60 kg
Vertical stiffness of one tire (front/rear)	k_{tf} and k_{tr}	182087/182087 N/m
Suspension stiffness per wheel (front/rear)	k_{sf} and k_{sr}	20984.8/19121.7 N/m
Vertical stiffness of driver's seat	k_c	10507.1 N/m
Damping coefficient of one tire (front/rear)	c_{tf} and c_{tr}	1820.9/1820.9 Ns/m
Suspension damping per wheel (front/rear)	c_{sf} and c_{sr}	209.8/191.2 Ns/m
Damping coefficient of driver's seat	c_c	105.1 Ns/m
Distance between x-axis and driver's seat	l_x	0.24 m
Distance between y-axis and driver's seat	l_y	0.34 m
Distance between x-axis and front/rear tires	l_f and l_r	1.125/1.511 m
Distance between left and right tires	d	1.44 m

Table 3.5 Natural frequencies of the realistic vehicle model

Natural frequencies	$\omega_1^2(\text{rad}^2/\text{s}^2)$	$\omega_2^2(\text{rad}^2/\text{s}^2)$	$\omega_3^2(\text{rad}^2/\text{s}^2)$	$\omega_4^2(\text{rad}^2/\text{s}^2)$
Values	15.9004	47.6554	151.5081	262.7300
Natural frequencies	$\omega_5^2(\text{rad}^2/\text{s}^2)$	$\omega_6^2(\text{rad}^2/\text{s}^2)$	$\omega_7^2(\text{rad}^2/\text{s}^2)$	$\omega_8^2(\text{rad}^2/\text{s}^2)$
Values	5.0310×10^3	5.0478×10^3	5.0778×10^3	5.0951×10^3

$$\tilde{\eta}_1 = \left| \frac{\tilde{R}_{\dot{x}_1\dot{x}_1}(0) - R_{\dot{x}_1\dot{x}_1}(0)}{R_{\dot{x}_1\dot{x}_1}(0)} \right| \times 100\% = 53\% \tag{3.33}$$

which is still an extremely large error. The reason for this phenomenon is that when the frequencies lie together or close, the contribution of interaction of separate normal modes (the cross-correlation terms) is very large and cannot be neglect.

Note also that the neglect of cross-correlations may lead to nonunique values of the estimated mean-square responses as demonstrated in Appendix. Further details can be found in the paper by Long, Elishakoff, Jiang, Han, and Hashemi (2014).

3.4 Conclusion

It is demonstrated that the cross-correlations have significant effect on the evaluation of random vibration of vehicle models with repeated or close natural frequencies. Besides the large error due to the omission of cross-correlations, it is remarkable to note that the response estimated depends on the eigenvector associated with repeated frequencies. As demonstrated in the Appendix, the latter eigenvectors turn out to be not unique, which makes it difficult to determine which approximation is valid. Hence, the cross-correlations cannot be neglected in the analysis when there are repeated or closely spaced natural frequencies.

References

Crandall, S. H. (1993). Modal-Sum and image-sum procedures for estimating wide-band random response of structures. *Probabilistc Engineering Mechanics, 8*(3–4), 187–196.

Dahlberg, T. (1982). Modal cross-spectral terms may be important and an alternative method of analysis be preferable. *Journal of Sound Vibration, 84,* 503–508.

Elishakoff, I. (1977a). Random vibrations of orthotropic plates clamped or simply supported all round. *Acta Mechanica, 28*(1–4), 165–176.

Elishakoff, I. (1977b). On the Role of cross—correlations in the random vibrations of shells. *Journal of Sound and Vibration, 50*(2), 239–252.

Elishakoff, I. (1977c). Random vibrations of orthotropic plates with all edges clamped or simply supported. *Acta Mechanica, 28,* 165–176.

Elishakoff, I. (1977d). Flutter and random vibration in plates. In B. L. Clarkson (Ed.), *Stochastic problems in dynamics* (pp. 390–410). London: Pitman.

Elishakoff, I., Van Zanten, A., & Crandall, S. H. (1979). Random Axisymmetric Vibration of cylindrical shells. *Journal of Applied Mechanics, 46*(2), 417–423.

Elishakoff, I. (1983) *Probabilistic Theory of Structures.* New York: Wiley (Dover, 1999; World Scientific, 2017).

Elishakoff, I. (1986). A model elucidating significance of cross-correlations in random vibration analysis. In I. Elishakoff & R. H. Lyon (Eds.), *Random vibration—status and recent developments* (pp 101–112). Amsterdam: Elseiver.

Elishakoff, I. & Ducreux, B (2014). Dramstic effect of cross-correlations in random vibration of point driven spherically curved panel, *Archive of applied mechanics*, 81(2), 021017.

Friswell, M. (1996). The Derivatives of repeated eigenvalues and their associated eigenvectors. *ASME Journal of Vibrations Acoustics, 118,* 390–397.

Kim, H., & Yoon, Y. S. (1995). Semi-Active Suspension With preview using a frequency-shape performance index. *Vehicle System Dynamics, 24,* 759–780.

Long, X. Y., Elishakoff, I., Jiang, C., Han, X., & Hashemi, J. (2014). Notes on Random Vibration of a vehicle model and other discrete systems possessing repeated natural frequencies. *Archive in Applied Mechanics, 84,* 1091–1101.

Lubliner, E., & Elishakoff, I. (1986). Random Vibration of system with finitely many degrees of freedom and several coalescent natural frequencies. *International Journal of Engineering Science, 24*(4), 461–470.

Zuo, L., & Nayfeh, S. A. (2003). Structured H2 optimization of vehicle suspensions based on multi-wheel models. *Vehicle System Dynamics, 40,* 351–371.

Zuo, L., & Nayfeh, S. A. (2007). H_2 Optimal Control of disturbance-delayed systems with application to vehicle suspensions. *Vehicle System Dynamics, 45,* 233–247.

Chapter 4
Hybrid Optimization and Anti-optimization of a Stochastically Excited Beam

4.1 Introductory Comments

Linear random vibrations of continuous structures constitute a well-studied subject that is explained in several monographs and papers, see e.g. Bolotin (1984), Crandall (1979a, 1979b), Elishakoff (1999), Kreé and Soize (1983), Wedig and Riemer (1981), Yang (1985), and others. Therefore, the pertinent works are those which consider some remarkable effects that were not studied before. Such is a problem of stationary random vibrations of beams that have point supports between the ends of the beam. To the best of the author's knowledge, this problem has not been dealt within the rich literature on random vibrations. When point supports are at the ends of the beam, one obtains the beam simply supported at its ends. When both supports tend to the middle cross section, the beam is equivalent to two half-beams coinciding with clamped-free beams of half-length.

This study was inspired by the definitive paper of Niordson and Simmonds (1975) who considered the *deterministic* problem of evaluation of closely spaced natural frequencies in a beam with two supports inside the span of the beam. They demonstrated that when the supports tend to each other, the natural frequencies become very close. The phenomenon of closely spaced frequencies and the associated effect of cross-correlations in *random* vibrations was studied by Elishakoff (1999). Therefore, it appeared natural to extend the study of Niordson and Simmonds (1975) to the random vibration setting. Additional motivation is based on the observation that the beam with supports inside the span often appears in mechanical, civil, and aerospace engineering.

The problem is defined as follows: Given the probabilistic description of the loading, evaluate the mean-square displacement of the beam; an additional task is the determination of the location of the supports inside the beam's span so that the maximum mean-square value is minimal. We consider the random vibration of such

I. Elishakoff, *Dramatic Effect of Cross-Correlations in Random Vibrations of Discrete Systems, Beams, Plates, and Shells*,
https://doi.org/10.1007/978-3-030-40394-2_4

beams subjected to either of three loadings: (a) symmetric and (b) anti-symmetric with respect to the middle cross section, and (c) general loading that lacks symmetry properties.

Note that the hybrid optimization and anti-optimization of structures are studied in detail in the monograph by Elishakoff and Ohsaki (2010). They considered the uncertainty as represented by *unknown-but-bounded* set of variables. In this study, we concentrate on the *probabilistic* optimization and anti-optimization, apparently for the first time in literature. Specifically, the anti-optimization consists in searching for the worst possible scenarios under uncertainty, whereas the deterministic optimization aims at best solutions in some predetermined sense. The concept of anti-optimization was introduced by Elishakoff (1990). Combined optimization and anti-optimization mathematically consists in solving min–max problem of finding the least favorable response of the system. It was widely employed in the literature in the recent years by Adali, Lena, Devaut, and Chiaruttini (2002), Bu, Jiang, and Jiao (2012), Catallo (2004), Elishakoff, Haftka, and Fang (1994), Elishakoff, Kriegesmann, Rolfes, Hünne, and Kling (2012), De Faria (2002), Gangadharan, Nikolaidis, Lee, Haftka, and Burdisso (1999), Hlaváček, Chleboun, and Babuška (2004), Lee, Haftka, Griffin, Watson, and Sensmeier (1994), Lombardi and Haftka (1998), McWilliam (2001), Ohsaki, Zhang, and Elishakoff (2012), Qiu and Wang (2010), Zingales and Elishakoff (2000).

4.2 Basic Equations

Consider first the free vibrations of the beam in order to find its natural frequencies and mode shapes, for their subsequent use in normal-mode approach. The governing equation for the beam's displacement is

$$EI\frac{\partial^4 w}{\partial x^4} + \rho A\frac{\partial^2 w}{\partial t^2} = 0. \tag{4.1}$$

In this equation x is the axial coordinate, t is the time, $w\ (x, t)$ is the transverse displacement, E is the modulus of elasticity, I is the second moment of area, ρ is the mass density, and A is the cross-sectional area. We seek solution in the form

$$w(x, t) = W(x)\sin(\omega t) \tag{4.2}$$

where $W(x)$ is the mode shape and ω is the sought natural frequency. Substituting Eq. (4.2) in Eq. (4.1) and changing the coordinate variable to the non-dimensional one $\xi = x/L$, we get

$$\frac{d^4 W}{d\xi^4} - \lambda^4 W = 0 \tag{4.3}$$

where the parameter λ is defined as

$$\lambda = \sqrt[4]{\frac{\rho A L^4 \omega^2}{EI}}. \tag{4.4}$$

The general solution of Eq. (4.3) reads

$$W(x) = a\sin(\lambda\xi) + b\cos(\lambda\xi) + c\sinh(\lambda\xi) + d\cosh(\lambda\xi) \tag{4.5}$$

The two supports are placed at a distance ε symmetrically located with respect to the middle cross section of the beam as shown in Fig. 4.1. Thus the sought solution, i.e. mean-square response of the beam, will be symmetric with respect to the cross section $= 1/2$, if the load is symmetric. In such a particular case, we can deal with only half of the beam, and extend the solution symmetrically with respect to the middle cross section. Analogously, if the load is anti-symmetric, the response will likewise exhibit the anti-symmetric behavior. In that case too we can limit ourselves by considering the half-beam, with appropriate boundary conditions. When the load does not fall in either category, natural frequencies of both kinds will be excited.

The half-beam is represented by two parts as shown in Fig. 4.2: the first part is at the left of the support, and the second part lies at its right. They exhibit different behaviors, although in each part the solution is written in the form (4.5), the coefficients will differ.

The characteristic equation will be obtained by finding the solution for each part and matching them at $\xi = 1/2 - \varepsilon$. The displacement of each part is denoted as W_i with $i = 1, 2$, respectively.

The boundary conditions are as follows. At the free-end we have:

$$W_1''(0) = W_1'''(0) = 0. \tag{4.6}$$

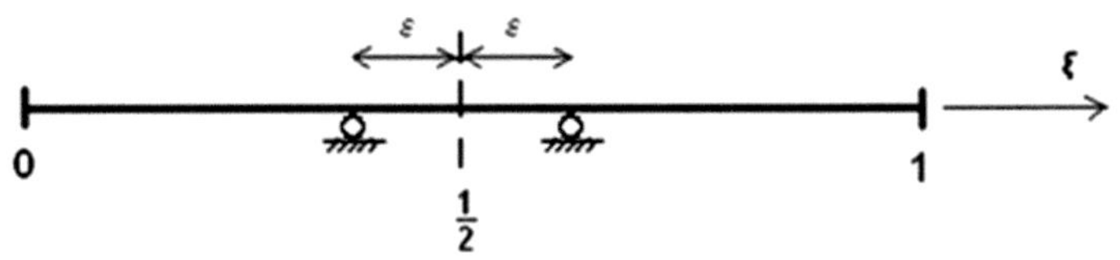

Fig. 4.1 Beam supported by two point supports

Fig. 4.2 Half-beam that ought to be considered for either symmetric or anti-symmetric cases

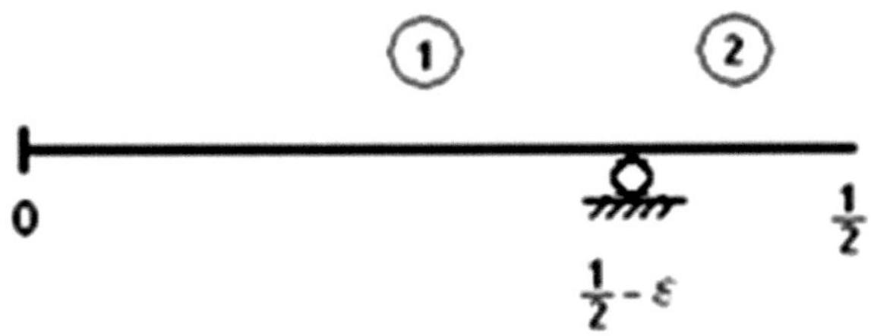

In the case of a symmetric load and behavior, at the other end of the half-beam we must impose that odd-numbered derivatives vanish:

$$W_2'\left(\frac{1}{2}\right) = W_2'''\left(\frac{1}{2}\right) = 0. \tag{4.7}$$

In other words, for the symmetric case, the boundary conditions at the right end of the half-beam correspond to those of the guided end. In the anti-symmetric case, at $\xi = 1/2$, the even-numbered derivatives are zero:

$$W_2\left(\frac{1}{2}\right) = W_2''\left(\frac{1}{2}\right) = 0 \tag{4.8}$$

which corresponds to a simply supported end. At the support point we demand that the displacements are zero:

$$W_1\left(\frac{1}{2} - \varepsilon\right) = W_2\left(\frac{1}{2} - \varepsilon\right) = 0 \tag{4.9}$$

Finally, at the same point, we have two continuity conditions:

$$W_1'\left(\frac{1}{2} - \varepsilon\right) = W_2'\left(\frac{1}{2} - \varepsilon\right) \tag{4.10}$$

and

$$W_1''\left(\frac{1}{2} - \varepsilon\right) = W_2''\left(\frac{1}{2} - \varepsilon\right). \tag{4.11}$$

Equation (4.10) signifies the continuity of the slope, whereas Eq. (4.11) indicates the continuity of the bending moment.

4.3 Evaluation of the Mode Shapes

Symmetric Mode Shapes

Using Eq. (4.5) and the boundary conditions, we find the expressions of the mode shapes of the beam for the symmetric case. The coefficients in Eq. (4.5) are supplied index 1 which indicates the first part. Combining Eqs. (4.5) and (4.6) results in $b_1 = d_1$ and $a_1 = c_1$. So we can rewrite Eq. (4.6) as follows:

$$W_1(x) = a_1[\sin(\lambda\xi) + \sinh(\lambda\xi)] + b_1[\cos(\lambda\xi) + \cosh(\lambda\xi)]. \tag{4.12}$$

Introducing Eq. (4.9) in the latter one, we obtain

$$W_1(x) = a_1[\sin(\lambda\xi) + \sinh(\lambda\xi)] + \alpha[\cos(\lambda\xi) + \cosh(\lambda\xi)] \tag{4.13}$$

where

$$\alpha = -\frac{\sin\left(\lambda\left(\frac{1}{2} - \varepsilon\right)\right) + \sinh\left(\lambda\left(\frac{1}{2} - \varepsilon\right)\right)}{\cos\left(\lambda\left(\frac{1}{2} - \varepsilon\right)\right) + \cosh\left(\lambda\left(\frac{1}{2} - \varepsilon\right)\right)}. \tag{4.14}$$

We further combine the derived forms of Eq. (4.5) with Eq. (4.7), this time the coefficients in Eq. (4.5) are marked with the index *2* since we now work on the second part of the half-beam. We add and subtract the equations obtained, which leads to two new expressions:

$$c_2 \cosh\left(\frac{\lambda}{2}\right) + d_2 \sinh\left(\frac{\lambda}{2}\right) = 0, \tag{4.15}$$

$$a_2 \cos\left(\frac{\lambda}{2}\right) - b_2 \sin\left(\frac{\lambda}{2}\right) = 0. \tag{4.16}$$

Satisfying the Eq. (4.9) in view of Eq. (4.15) and (4.16) we get:

$$\begin{gathered} a_2 \sin\left(\tfrac{\lambda}{2}\right)\cos(\lambda\varepsilon) + b_2 \cos\left(\tfrac{\lambda}{2}\right)\cos(\lambda\varepsilon) + c_2 \sinh\left(\tfrac{\lambda}{2}\right)\cosh(\lambda\varepsilon) + \\ d_2 \cosh\left(\tfrac{\lambda}{2}\right)\cosh(\lambda\varepsilon) = 0 \end{gathered}. \tag{4.17}$$

Finally, Eqs. (4.10) and (4.11) are simplified with the introduction of Eqs. (4.15) and (4.16) so we arrive at

$$\begin{aligned} a_1 M_1 = {} & a_2 \sin\left(\frac{\lambda}{2}\right)\sin(\lambda\varepsilon) + b_2 \cos\left(\frac{\lambda}{2}\right)\sin(\lambda\varepsilon) - c_2 \sinh\left(\frac{\lambda}{2}\right)\sinh(\lambda\varepsilon) \\ & - d_2 \cosh\left(\frac{\lambda}{2}\right)\sinh(\lambda\varepsilon) \end{aligned} \tag{4.18}$$

$$a_1 M_2 = -2a_2 \sin\left(\frac{\lambda}{2}\right)\cos(\lambda\varepsilon) - 2b_2 \cos\left(\frac{\lambda}{2}\right)\cos(\lambda\varepsilon) \tag{4.19}$$

where

$$\begin{aligned} M_1 = {} & \cos\left(\lambda\left(\frac{1}{2} - \varepsilon\right)\right) + \cosh\left(\lambda\left(\frac{1}{2} - \varepsilon\right)\right) \\ & + \alpha\left[\sinh\left(\lambda\left(\frac{1}{2} - \varepsilon\right)\right) - \sin\left(\lambda\left(\frac{1}{2} - \varepsilon\right)\right)\right] \end{aligned} \tag{4.20}$$

$$\begin{aligned} M_2 = {} & -\sin\left(\lambda\left(\frac{1}{2} - \varepsilon\right)\right) + \sinh\left(\lambda\left(\frac{1}{2} - \varepsilon\right)\right) + \alpha\left[\cosh\left(\lambda\left(\frac{1}{2} - \varepsilon\right)\right)\right. \\ & \left. -\cos\left(\lambda\left(\frac{1}{2} - \varepsilon\right)\right)\right] \end{aligned}. \tag{4.21}$$

Equations (4.15–4.19) represent a system of equations, where a_1, a_2, b_2, c_2 and d_2 are the five unknown constants. Thus, the following determinant has to be zero.

$$\begin{vmatrix} 0 & 0 & 0 & \cosh\left(\frac{\lambda}{2}\right) & \sinh\left(\frac{\lambda}{2}\right) \\ 0 & \cos\left(\frac{\lambda}{2}\right) & -\sin\left(\frac{\lambda}{2}\right) & 0 & 0 \\ 0 & \sin\left(\frac{\lambda}{2}\right)\cos(\lambda\varepsilon) & \cos\left(\frac{\lambda}{2}\right)\cos(\lambda\varepsilon) & \sinh\left(\frac{\lambda}{2}\right)\cosh(\lambda\varepsilon) & \cosh\left(\frac{\lambda}{2}\right)\cosh(\lambda\varepsilon) \\ M_1 & -\sin\left(\frac{\lambda}{2}\right)\sin(\lambda\varepsilon) & -\cos\left(\frac{\lambda}{2}\right)\sin(\lambda\varepsilon) & \sinh\left(\frac{\lambda}{2}\right)\sinh(\lambda\varepsilon) & \cosh\left(\frac{\lambda}{2}\right)\sinh(\lambda\varepsilon) \\ M_2 & 2\sin\left(\frac{\lambda}{2}\right)\cos(\lambda\varepsilon) & 2\cos\left(\frac{\lambda}{2}\right)\cos(\lambda\varepsilon) & 0 & 0 \end{vmatrix} = 0. \tag{4.22}$$

Instead of a direct evaluation of this determinant we simplify the procedure. Indeed, manipulating Eqs. (4.15–4.19), we obtain a_2, b_2, c_2 and d_2 expressed via a_1 as follows:

$$a_2 = -\frac{M_2}{2}\frac{\sin\left(\frac{\lambda}{2}\right)}{\cos(\lambda\varepsilon)}a_1,\; b_2 = -\frac{M_2}{2}\frac{\cos\left(\frac{\lambda}{2}\right)}{\cos(\lambda\varepsilon)}a_1,\; c_2 = -\frac{M_2}{2}\frac{\sinh\left(\frac{\lambda}{2}\right)}{\cosh(\lambda\varepsilon)}a_1,\; d_2 = \frac{M_2}{2}\frac{\cosh\left(\frac{\lambda}{2}\right)}{\cosh(\lambda\varepsilon)}a_1. \tag{4.23}$$

It must be emphasized that since $\cos(\lambda\varepsilon)$ appears in the denominator, the expression (4.23) is not valid when $\cos(\lambda\varepsilon) = 0$, i.e. when $\xi = 1/2$. This particular case will be dealt separately.

Let us concentrate on the case when $\xi \neq 1/2$. We introduce Eq. (4.23) in Eq. (4.18) and since a_1 differs from zero we obtain:

$$M_1 + \frac{M_2}{2}(\tan(\lambda\varepsilon) + \tanh(\lambda\varepsilon)) = 0. \tag{4.24}$$

It can be demonstrated that the evaluation of the determinant (4.22) leads to the expression (4.24). Having all the unknowns expressed through the single constant a_1 and letting the latter equal to unity, we write the mode shapes from Eqs. (4.14) to (4.6) as follows:

$$w_j^{(1)}(\xi) = \begin{cases} \sin(\lambda_j\xi) + \sinh(\lambda_j\xi) + \alpha_j\left[\cos(\lambda_j\xi) + \cosh(\lambda_j\xi)\right], & \text{for } 0 \le \xi \le \frac{1}{2} - \varepsilon \\ -\frac{M_{2j}}{2}\left[\frac{\sin\left(\frac{\lambda_j}{2}\right)\sin(\lambda_j\xi)}{\cos(\lambda_j\varepsilon)} + \frac{\cos\left(\frac{\lambda_j}{2}\right)\cos(\lambda_j\xi)}{\cos(\lambda_j\varepsilon)} + \frac{\sinh\left(\frac{\lambda_j}{2}\right)\sinh(\lambda_j\xi)}{\cosh(\lambda_j\varepsilon)} - \frac{\cosh\left(\frac{\lambda_j}{2}\right)\cosh(\lambda_j\xi)}{\cosh(\lambda_j\varepsilon)}\right], & \text{for } \frac{1}{2} - \varepsilon < \xi \le \frac{1}{2} \end{cases} \tag{4.25}$$

where M_{2j} is the value that is obtained from Eq. (4.21) when we substitute $\lambda = \lambda_j$ in it. The index (4.1) corresponds to the symmetric mode shapes. The λ_j are the values, solutions of Eq. (4.24), corresponding to each symmetric mode shape. They depend on the position of the support ε and are listed in Table 4.1.

For the special case $\varepsilon = 1/2$, the mode shapes are the ones of a beam of length L that is simply supported at its left end and guided at the other, namely,

Table 4.1 λ of the mode shapes depending on ε (symmetric case)

λ	ε										
	0	0.05	0.1	0.15	0.2	0.25	0.3	0.35	0.4	0.45	0.5
λ_1	3.750	3.798	3.933	4.148	4.426	4.682	4.683	4.348	3.902	3.489	3.142
λ_2	9.388	9.730	10.481	10.979	9.794	8.678	8.592	9.698	10.980	10.418	9.425
λ_3	15.710	16.483	17.165	14.394	14.728	16.719	16.683	15.050	15.440	17.179	15.708
λ_4	21.991	23.218	20.981	21.480	26.806	20.932	21.673	23.163	21.173	23.470	21.991
λ_5	28.274	29.824	26.983	29.666	35.134	29.323	27.692	28.516	28.491	28.678	28.274
λ_6	34.558	35.815	34.402	34.250	41.099	33.511	36.027	33.148	35.877	33.765	34.558
λ_7	40.841	40.390	41.888	39.698	46.188	41.888	39.271	41.353	41.832	39.921	40.841
λ_8	47.124	45.605	48.577	47.967	58.221	46.077	48.060	48.445	45.613	46.557	47.124
λ_9	53.407	52.002	52.360	54.288	66.548	54.454	53.149	51.864	52.351	53.338	53.407
λ_{10}	59.690	58.730	58.393	58.143	72.515	58.643	59.114	59.517	59.844	60.159	59.690

$$W_j^{(1)}(\xi) = \sin(\lambda_j \xi), \lambda_j = (2j - 1)\pi. \tag{4.26}$$

The expressions (4.25) and (4.26) pertain to the half-beam. Since we consider the symmetric case, the mode shapes of the entire beam are obtained from the values for $\xi \leq 1/2$ by symmetrically mirroring with respect to $\xi = 1/2$. Figure 4.3(a) portrays the first five symmetric mode shapes for $\varepsilon = 0.2$, bold vertical lines denoting the supports location. Figure 4.4 is associated with the case $\varepsilon = 0.4$ (Fig. 4.5).

4.4 Anti-symmetric Mode Shapes

Let us now determine the anti-symmetric mode shapes. The procedure is perfectly analogous to the one for the symmetric modes, the difference being in the use of Eq. (4.8) instead of Eq. (4.7) in the set of boundary conditions, in conjunction with Eq. (4.5). Thus, the displacement of the first part of the half-beam is given by the same function as one of the symmetric cases, i.e. Eq. (4.13). For the second part, the five equations obtained using the boundary conditions are

$$c_2 \sinh\left(\frac{\lambda}{2}\right) + d_2 \cosh\left(\frac{\lambda}{2}\right) = 0, \tag{4.27}$$

$$a_2 \sin\left(\frac{\lambda}{2}\right) + b_2 \cos\left(\frac{\lambda}{2}\right) = 0, \tag{4.28}$$

$$a_2 \cos\left(\frac{\lambda}{2}\right) \sin(\lambda\varepsilon) - b_2 \sin\left(\frac{\lambda}{2}\right) \sin(\lambda\varepsilon) + c_2 \cosh\left(\frac{\lambda}{2}\right) \sinh(\lambda\varepsilon) + d_2 \sinh\left(\frac{\lambda}{2}\right) \sinh(\lambda\varepsilon) = 0, \tag{4.29}$$

$$a_1 M_1 = a_2 \cos\left(\frac{\lambda}{2}\right) \cos(\lambda\varepsilon) - b_2 \sin\left(\frac{\lambda}{2}\right) \cos(\lambda\varepsilon) + c_2 \cosh\left(\frac{\lambda}{2}\right) \cosh(\lambda\varepsilon) + d_2 \sinh\left(\frac{\lambda}{2}\right) \cosh(\lambda\varepsilon), \tag{4.30}$$

$$a_1 M_2 = 2a_2 \cos\left(\frac{\lambda}{2}\right) \sin(\lambda\varepsilon) - 2b_2 \sin\left(\frac{\lambda}{2}\right) \sin(\lambda\varepsilon). \tag{4.31}$$

Once again, if $\xi \neq 1/2$, we simplify the system by expressing a_2, b_2, c_2, and d_2 via a_1, and we obtain the following frequency equation:

$$M_1 - \frac{M_2}{2}\left(\frac{1}{\tan(\lambda\varepsilon)} - \frac{1}{\tanh(\lambda\varepsilon)}\right) = 0 \tag{4.32}$$

which is equivalent to the determinant obtained from the matrix of the Eqs. (4.27)–(4.31). For $\varepsilon \neq 0$ and $\varepsilon \neq 1/2$ the values of λ found by solving Eq. (4.32) are listed in Table 4.2. In the case $\varepsilon = 0$, $\tan(\lambda\varepsilon) = 0$ so we cannot use Eq. (4.32) to find the natural frequencies. The anti-symmetric mode shapes are expressed by the following function:

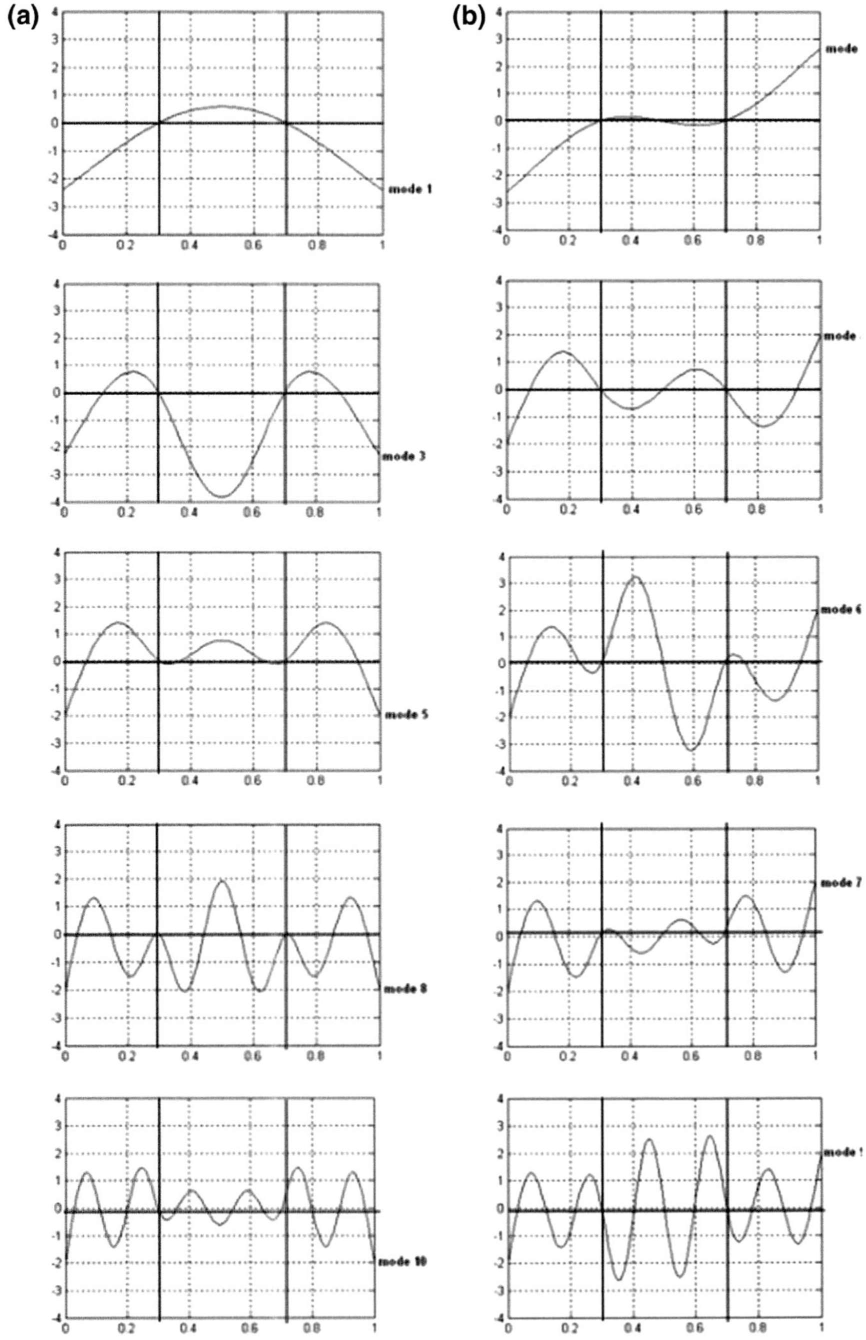

Fig. 4.3 The ten first mode shapes for $\varepsilon = 0.2$: **a** symmetric mode shapes; **b** anti-symmetric mode shapes (support locations are denoted by bold vertical lines)

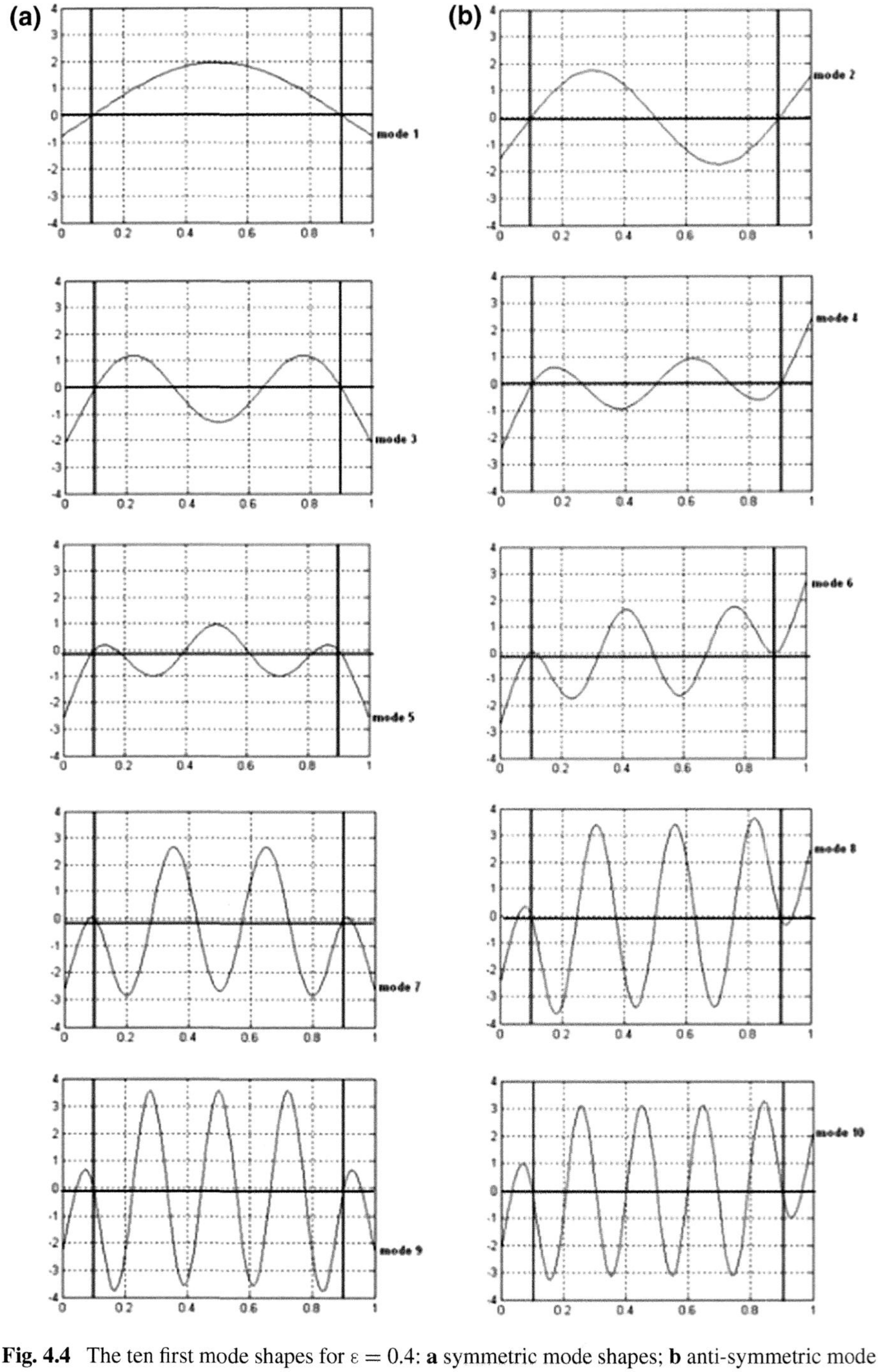

Fig. 4.4 The ten first mode shapes for $\varepsilon = 0.4$: **a** symmetric mode shapes; **b** anti-symmetric mode shapes (support locations are denoted by bold vertical lines)

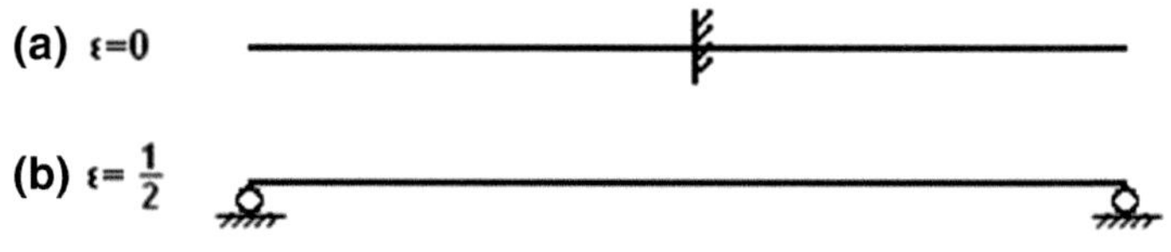

Fig. 4.5 Limiting cases: **a** corresponds to two C–F beams of length $L/2$ each; **b** corresponds to one beam of length L that is S–S at both ends

$$W_j^{(2)}(\xi) = \begin{cases} \sin(\lambda_j\xi) + \sinh(\lambda_j\xi) + \alpha_j\left[\cos(\lambda_j\xi) + \cosh(\lambda_j\xi)\right], & \text{for } 0 \le \xi \le \frac{1}{2} - \varepsilon \\ \frac{M_{2j}}{2}\left[\frac{\cos\left(\frac{\lambda_j}{2}\right)\sin(\lambda_j\xi)}{\sin(\lambda_j\varepsilon)} - \frac{\sin\left(\frac{\lambda_j}{2}\right)\cos(\lambda_j\xi)}{\sin(\lambda_j\varepsilon)} - \frac{\cosh\left(\frac{\lambda_j}{2}\right)\sinh(\lambda_j\xi)}{\sinh(\lambda_j\varepsilon)} + \frac{\sinh\left(\frac{\lambda_j}{2}\right)\cosh(\lambda_j\xi)}{\sinh(\lambda_j\varepsilon)}\right], & \text{for } \frac{1}{2} - \varepsilon < \xi \le \frac{1}{2} \end{cases} \tag{4.33}$$

The index (4.2) corresponds to the anti-symmetric mode shapes. If $\varepsilon = 1/2$, the mode shapes are the ones of a beam of length $L/2$ that is simply supported at its ends, namely,

$$W_j^{(1)}(\xi) = \sin(\lambda_j\xi),\ \lambda_j = 2j\pi. \tag{4.34}$$

The expressions of the anti-symmetric mode shapes for the entire beam are obtained from Eq. (4.33) to (4.34), with the values for $\xi \ge 1/2$ anti-symmetrically mirrored with respect to $\xi = 1/2$. Figure 4.3(b) portrays the first five anti-symmetric mode shapes for $\varepsilon = 0.2$, bold vertical lines denoting the supports location. Figure 4.4 is associated with the case $\varepsilon = 0.4$.

Comparison of Tables 4.1 and 4.2 leads to the following observations. Considering the first ten mode shapes, for all values of ε, except $\varepsilon = 0.2$ and $\varepsilon = 0.4$, the natural frequencies corresponding to the symmetric and anti-symmetric cases alternate one with another. It is remarkable that for $\varepsilon = 0.2$ between the third and fourth frequencies corresponding to the symmetric modes, i.e. in the interval [14.728, 26.806], there are two frequencies, namely 18.086 and 24.940, associated with the anti-symmetric modes. Likewise, for $\varepsilon = 0.4$, in the interval between the fifth and sixth natural frequencies corresponding to anti-symmetric modes, i.e. [32.218, 43.511], there are two natural frequencies 35.877 and 41.832 associated with the sixth and seventh symmetric mode shapes.

4.5 Limiting Cases

Consider now the limiting cases of the behavior. For $\varepsilon = 0$, the two supports tend to coincide with each other, and within Bernouilli–Euler theory we arrive at two clamped-free (C-F) beams of length $L/2$.

For the symmetric case, we substitute $\varepsilon = 0$ into Eq. (4.24) to arrive at

Table 4.2 λ of the mode shapes depending on ε (anti-symmetric case)

λ	ε										
	0	0.05	0.1	0.15	0.2	0.25	0.3	0.35	0.4	0.45	0.5
λ_1	3.750	4.025	4.360	4.779	5.317	6.024	6.934	7.771	7.647	6.966	6.283
λ_2	9.388	10.114	11.076	12.332	13.788	13.652	12.138	11.670	13.481	13.831	12.566
λ_3	15.710	16.971	18.677	20.398	18.086	17.749	20.299	19.089	17.915	20.419	18.850
λ_4	21.991	23.814	26.178	24.251	24.940	26.178	23.526	26.667	24.778	26.223	25.133
λ_5	28.274	30.675	32.869	30.573	32.593	30.371	32.350	30.035	32.218	31.093	31.416
λ_6	34.558	37.541	36.650	38.842	36.219	38.746	37.444	37.179	43.511	36.742	37.699
λ_7	40.841	44.400	42.685	44.290	45.232	42.935	43.406	45.373	48.755	43.209	43.982
λ_8	47.124	51.227	50.110	48.874	49.513	51.323	51.735	49.961	56.079	49.939	50.265
λ_9	53.407	57.961	57.596	57.063	56.354	55.501	54.978	55.406	63.595	56.747	56.549
λ_{10}	59.690	64.375	64.285	64.152	64.009	67.879	63.768	63.675	75.020	63.567	62.832

$$\cos\left(\frac{\lambda}{2}\right)\cosh\left(\frac{\lambda}{2}\right) = -1 \tag{4.35}$$

which corresponds to the characteristic equation of the C-F beam of length $L/2$. For this case the values of λ are well recorded; Rao (2007) and other authors list the values for $\lambda/2$ as 1.8751; 4.6941; 7.8547; 10.9956; etc. For $j \gg 1$ the values asymptotically tend to $\lambda_j \approx (j - 1/2)\pi$. But in the anti-symmetric case, it is not possible to directly introduce $\varepsilon = 0$ in Eq. (4.32) since denominator may take a zero value, i.e. $\tan(\lambda\varepsilon) = 0$. Instead, we make ε to tend to zero. This process leads to Eq. (4.35). Thus, we obtain the same values of λ as in the symmetric case. Consequently, if $\varepsilon = 0$ the natural frequencies of the two cases are coinciding. This corresponds to the conclusion of Niordson and Simmonds (1975). We verify it by calculating λ for $\varepsilon = 0.001$, $\varepsilon = 0.002$, and $\varepsilon = 0.003$, indeed symmetric and anti-symmetric natural frequencies are very close as shown in Table 4.3.

In case $\varepsilon = 1/2$, for the symmetric case instead of using Eq. (4.24), we proceed from the determinant (4.22) which reduces to

$$4\cos\left(\frac{\lambda}{2}\right)\cosh\left(\frac{\lambda}{2}\right) = 0. \tag{4.36}$$

Note that the $\cos(\lambda/2) = 0$ corresponds to the characteristic equation of a beam that is simply supported at one end and guided at the other one. In the anti-symmetric case, the determinant of the matrix stemming from the Eqs. (4.27) to (4.31) becomes

$$-4\sin\left(\frac{\lambda}{2}\right)\sinh\left(\frac{\lambda}{2}\right) = 0 \tag{4.37}$$

which corresponds to the characteristic equation of a beam that is simply supported at both ends.

Naturally, the combination of the symmetric and anti-symmetric mode shapes is equivalent to the entire spectrum of a beam of length L that is simply supported (S–S) at its ends:

$$\sin(\lambda) = 0 \tag{4.38}$$

whose solution is uniformly known as $\lambda_j = j\pi$.

4.6 Random Vibrations

The governing equation for a beam excited by random loading $q(x, t)$ reads

$$EI\frac{\partial^4 w}{\partial x^4} + c\frac{\partial w}{\partial t} + \rho A\frac{\partial^2 w}{\partial t^2} = q(x, t) \tag{4.39}$$

Table 4.3 Comparison of natural frequencies for small values of ε

ε	λ									
	λ_1		λ_2		λ_3		λ_4		λ_5	
	Sym	Anti-sym	Sym	Anti-sym	Sym	Anti-sym	Sym	Anti-sym	Sym	Anti-sym
0.001	3.750	3.755	9.388	9.401	15.710	15.731	21.992	22.021	28.276	28.312
0.002	3.750	3.760	9.389	9.413	15.711	15.752	21.995	22.050	28.280	28.351
0.003	3.750	3.765	9.390	9.426	15.714	15.773	21.999	22.080	28.288	28.389

where c is the damping coefficient. We utilize the normal-mode method. The load is expanded in terms of mode shapes $W_j(x)$ as follows:

$$q(x,t)=\sum_{j=1}^{\infty}q_j(t)W_j(x) \tag{4.40}$$

where W_j is the jth mode shape. From Eq. (4.40) we express $q_j(t)$ using the orthogonality property of the mode shapes as

$$q_j(t)=\nu_j^{-2}\int_0^L q(x,t)W_j(x)dx \tag{4.41}$$

In Eq. (4.41)

$$\nu_j^2=\int_0^L W_j^2(x)dx. \tag{4.42}$$

For the displacement a similar expansion is utilized:

$$w(x,t)=\sum_{j=1}^{\infty}w_j(t)W_j(x). \tag{4.43}$$

For stationary random vibrations we express $q_j(t)$ and $w_j(t)$ as Fourier integrals:

$$q_j(t)=\int_{-\infty}^{\infty}Q_j(\omega)e^{i\omega t}d\omega, \tag{4.44}$$

$$w_j(t)=\int_{-\infty}^{\infty}V_j(\omega)e^{i\omega t}d\omega. \tag{4.45}$$

Substitution of Eqs. (4.40) and (4.43) in Eq. (4.39), and using normal-mode properties, yields

$$V_j(\omega)=H_j(\omega)Q_j(\omega) \tag{4.46}$$

where

$$H_j(\omega)=\frac{1}{\rho A\left(\omega_j^2-\omega^2+2i\zeta_j\omega\omega_j\right)} \tag{4.47}$$

is the complex frequency response associated with the jth mode. In Eq. (4.47)

$$\zeta_j=\frac{c}{2\rho A\omega_j} \tag{4.48}$$

is the modal damping coefficient.

It must be pointed out that the autocorrelation function $R_q(x_1, x_2, \tau)$ of $q(x, t)$ should be specified in the outset in order to find the probabilistic characteristics of the response. The expression for $R_q(x_1, x_2, \tau)$ is expressed as follows:

$$R_q(x_1, x_2, \tau) = \sum_{j=1}^{\infty}\sum_{k=1}^{\infty} W_j(x_1)W_k(x_2)\int_{-\infty}^{\infty} S_{Q_jQ_k}(\omega)e^{i\omega\tau}d\omega \tag{4.49}$$

where $\tau = t_2 - t_1$. The cross-spectral densities of the generalized forces $S_{Q_jQ_k}(\omega)$ are given by Elishakoff (1999):

$$S_{Q_jQ_k}(\omega) = \nu_j^{-2}\nu_k^{-2}\int_0^L\int_0^L S_q(x_1, x_2, \omega)W_j(x_1)W_k(x_2)dx_1dx_2. \tag{4.50}$$

The expression $S_{Q_jQ_k}(\omega)$ allows us to evaluate the cross-spectral density $S_w(x_1, x_2, \omega)$ of the displacement:

$$S_w(x_1, x_2, \omega) = \sum_{j=1}^{\infty}\sum_{k=1}^{\infty} S_{Q_jQ_k}(\omega)H_j^*(\omega)H_k(\omega)W_j(x_1)W_k(x_2). \tag{4.51}$$

Finally, we obtain the sought mean-square value of the displacement by letting $x_1 = x_2 = x$ and $\tau = 0$ in Eq. (4.48):

$$d_w^2(x) = R_w(x, x, 0) = \int_{-\infty}^{\infty} S_w(x, x, \omega)\cdot d\omega. \tag{4.52}$$

For more details, the interested reader can consult monographs by Elishakoff (1999) and Krée and Soize (1983).

4.7 Symmetric Excitation

First we look for the response to the symmetric excitation. The loading is specified as

$$q(x, t) = Q(t)F(x) \tag{4.53}$$

where $F(x)$ is a symmetric function with respect to $\xi = 1/2$. For simplicity we let

$$F(x) = f_0. \tag{4.54}$$

The function $Q(t)$ is a stationary random function with a mean equaled to zero. Thus the autocorrelation function of the loading becomes

$$R_q(x_1, x_2, t_1, t_2) = f_0^2 E[Q(t_1)Q(t_2)] = f_0^2 R_q(t_2 - t_1). \tag{4.55}$$

We specify the time-wise autocorrelation function as a delta correlated random process

$$R_q(t_2 - t_1) = \delta(\tau). \tag{4.56}$$

Thus, the spectral density of $q(x, t)$ reads

$$S_q(x_1, x_2, \omega) = \frac{f_0^2}{2\pi}. \tag{4.57}$$

For the expression of the cross-spectral density of the generalized forces (4.50) in the non-dimensional coordinate system we obtain:

$$S_{Q_j Q_k}(\omega) = \frac{f_0^2}{2\pi} \nu_j^{-2} \nu_k^{-2} \int_0^1 W_j(\xi_1) d\xi_1 \int_0^1 W_k(\xi_2) d\xi_2 \tag{4.58}$$

We observe in Eq. (4.58) that if the mode shape is anti-symmetric the integrals would vanish. Therefore, only the symmetric mode shapes play role in this expression. This formula can be rewritten as follows:

$$S_{Q_j Q_k} = \frac{f_0^2}{2\pi} \frac{2\int_0^{\frac{1}{2}} W_j^{(1)}(\xi_1) d\xi_1 * 2\int_0^{\frac{1}{2}} W_k^{(1)}(\xi_2) d\xi_2}{2\int_0^{\frac{1}{2}} W_j^{(1)^2}(\xi_1) d\xi_1 * 2\int_0^{\frac{1}{2}} W_k^{(1)^2}(\xi_2) d\xi_2} \tag{4.59}$$

where $W_j^{(1)}$ is the symmetric mode shape defined in Eqs. (4.25) and (4.26). $S_{Q_j Q_k}$ does not depend on ω, so the mean-square value of the displacement (4.52) is expressed as follows:

$$d_w^2(\xi) = \sum_{j=1}^{\infty} \sum_{k=1}^{\infty} S_{Q_j Q_k} W_j^{(1)}(\xi) W_k^{(1)}(\xi) \int_{-\infty}^{\infty} H_j^*(\omega) H_k(\omega) d\omega. \tag{4.60}$$

The integral in Eq. (4.60) reads (Elishakoff, 1999, p. 410):

$$\int_{-\infty}^{\infty} H_j^*(\omega) H_k(\omega) d\omega = \frac{1}{(\rho A)^2} \int_{-\infty}^{\infty} \frac{d\omega}{\left(\omega_j^2 - \omega^2 - 2i\beta\omega\right)\left(\omega_k^2 - \omega^2 + 2i\beta\omega\right)} = \frac{1}{(\rho A)^2} \frac{4\beta\pi}{\left(\omega_j^2 - \omega_k^2\right)^2 + 2\beta\left(\omega_j^2 + \omega_k^2\right)} \tag{4.61}$$

where

$$\beta = \frac{c}{\rho A}. \tag{4.62}$$

Equation (4.61) is then rewritten using λ_j in Eq. (4.4) instead of the natural frequencies ω_j. Consequently, the mean-square value (4.60) becomes

$$d_w^2(\xi) = \frac{4\pi c L^8}{\rho A E^2 I^2} \sum_{j=1}^{\infty} \sum_{k=1}^{\infty} \frac{S_{Q_j Q_k} W_j^{(1)}(\xi) W_k^{(1)}(\xi)}{\left(\lambda_j^4 - \lambda_k^4\right)^2 + \eta\left(\lambda_j^4 + \lambda_k^4\right)} \quad (4.63)$$

where

$$\eta = \frac{2c^2 L^4}{\rho A E I} \quad (4.64)$$

is a non-dimensional coefficient. We evaluate the following non-dimensional mean-square displacement:

$$D_w^2(\xi) = \frac{\rho A E^2 I^2}{4\pi c L^8} d_w^2(\xi) = \sum_{j=1}^{\infty} \sum_{k=1}^{\infty} \frac{S_{Q_j Q_k} W_j^{(1)}(\xi) W_k^{(1)}(\xi)}{\left(\lambda_j^4 - \lambda_k^4\right)^2 + \eta\left(\lambda_j^4 + \lambda_k^4\right)} \quad (4.65)$$

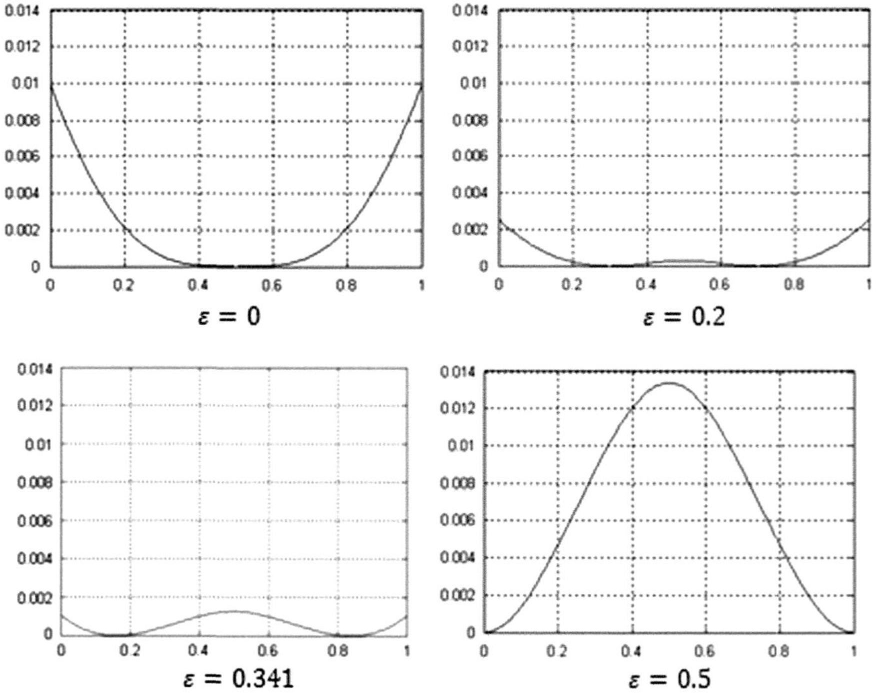

Fig. 4.6 $D_w^2(x)$ for various values of ε (symmetric case)

Table 4.4 Maximum values of $D_w^2(\xi)$ depending on ε (symmetric case)

ε	max $D_w^2(\xi)$	ε	max $D_w^2(\xi)$
0	0.009949	0.25	0.000545
0.001	0.009949	0.3	0.000701
0.05	0.009309	0.35	0.001650
0.1	0.007627	0.4	0.004499
0.15	0.005237	0.45	0.008464
0.2	0.002560	0.5	0.013250

for various values of ε, fixing $f_0 = 1$ N and $\eta = 0.1$. Figure 4.6 shows variations of $D_w^2(\xi)$ for several values of ε. In the context of anti-optimization, we are interested in location of maximum response. For $\varepsilon = 0$ the maximum mean-square displacement occurs at the ends of the beam as expected; analogous situation takes place up to $\varepsilon < 0.341$. Once ε exceeds this value, the maximum mean-square displacement occurs in the middle of the beam. These maxima are listed in Table 4.4. As designers, we are interested in optimization of the anti-optimized worst cases, i.e. in finding the location of the supports which corresponds to the minimum of the maximum mean-square displacement. This position is the optimum location under probabilistic uncertainty. The minimum of maxima is found when the supports are placed in the case $\varepsilon_1 = 0.27$ as evidenced in Fig. 4.7 (Fig. 4.8).

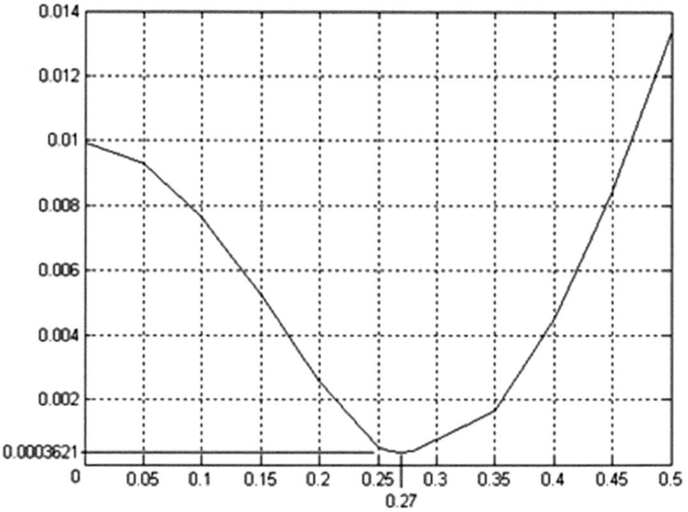

Fig. 4.7 Variation of the maximum value of $D_w^2(\xi)$ regarding to ε (symmetric case)

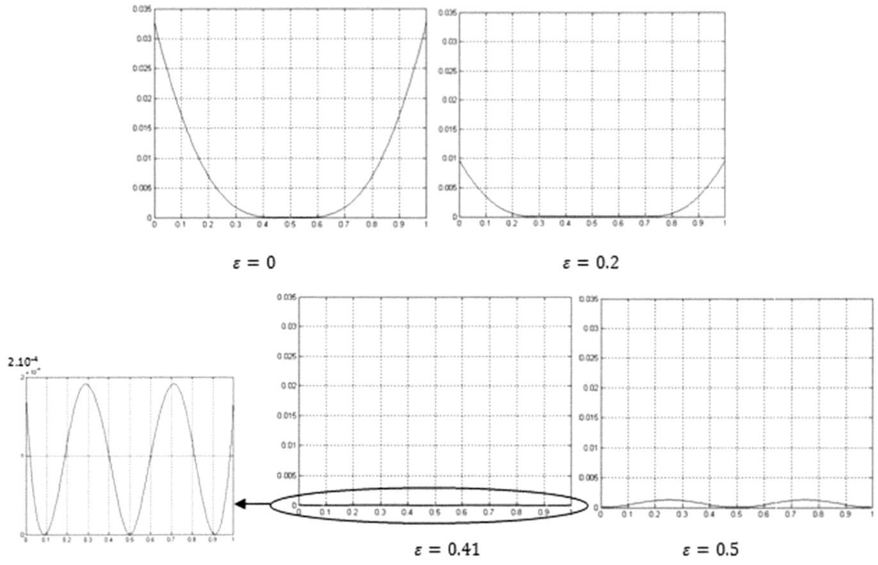

Fig. 4.8 $D_w^2(x)$ for various values of ε (anti-symmetric case)

4.8 Anti-symmetric Excitation

Let us deal now with the response to an anti-symmetric excitation. This time the loading is specified as in Eq. (4.53), where $F(x)$ is an anti-symmetric function with respect to the cross section $\xi = 1/2$. To simplify the analysis we use the following representation of the load:

$$F(x) = f_0 - \frac{2f_0}{L}x = f_0 - 2f_0\xi. \tag{4.66}$$

Thus, the autocorrelation function of the loading and its spectral density are

$$R_q(\xi_1, \xi_2, \tau) = (f_0 - 2f_0\xi_1)(f_0 - 2f_0\xi_2)\delta(\tau), \tag{4.67}$$

$$S_q(\xi_1, \xi_2, \omega) = (f_0 - 2f_0\xi_1)(f_0 - 2f_0\xi_2)\frac{1}{2\pi}. \tag{4.68}$$

The cross-spectral density of the generalized forces in Eq. (4.50) becomes

$$S_{Q_jQ_k}(\omega) = \nu_j^{-2}\nu_k^{-2}\int_0^1 (f_0 - 2f_0\xi_1)W_j(\xi_1)d\xi_1 \int_0^1 (f_0 - 2f_0\xi_2)W_k(\xi_2)d\xi_2. \tag{4.69}$$

We observe this time that only the anti-symmetric mode shapes are excited because for the symmetric ones the integrals vanish. Therefore, $S_{Q_jQ_k}$ can be put in the following form:

$$S_{Q_jQ_k} = \frac{2\int_0^{\frac{1}{2}}(f_0 - 2f_0\xi_1)W_j^{(2)}(\xi_1)d\xi_1 * 2\int_0^{\frac{1}{2}}(f_0 - 2f_0\xi_2)W_k^{(2)}(\xi_2)d\xi_2}{2\int_0^{\frac{1}{2}} W_j^{(2)^2}(\xi_1)d\xi_1 * 2\int_0^{\frac{1}{2}} W_k^{(2)^2}(\xi_2)d\xi_2}. \tag{4.70}$$

The non-dimensional mean-square displacement reads

$$D_w^2(\xi) = \frac{\rho A E^2 I^2}{4\pi c L^8} d_w^2(\xi) = \sum_{j=1}^{\infty}\sum_{k=1}^{\infty} \frac{S_{Q_jQ_k} W_j^{(2)}(\xi) W_k^{(2)}(\xi)}{\left(\lambda_j^4 - \lambda_k^4\right)^2 + \eta\left(\lambda_j^4 + \lambda_k^4\right)}. \tag{4.71}$$

We evaluate $D_w^2(\xi)$ for fixed values of $f_0 = 1$ N and $\eta = 0.1$. The variation of $D_w^2(\xi)$ is presented in Fig. 4.9 for various values of ε. The maximum mean-square displacement occurs at the ends of the beam in all the cases until the value $\varepsilon \leq 0.41$ is reached. Then the maximum appears in the vicinity of the middle of the half-beams: between $\xi = 0.29$ when $\varepsilon = 0.41$ and $\xi = 0.25$ when $\varepsilon = 0.5$; and symmetrically mirroring with respect to $\xi = 1/2$ for the second support. The maxima of these functions are listed in Table 4.5. The minimum value of the maxima is achieved when the supports are placed in the configuration $\varepsilon_2 = 0.41$ as seen in Fig. 4.9. One observes that contrary to the symmetric case this value matches the one where the position on the beam of the maximum mean-square displacement changes.

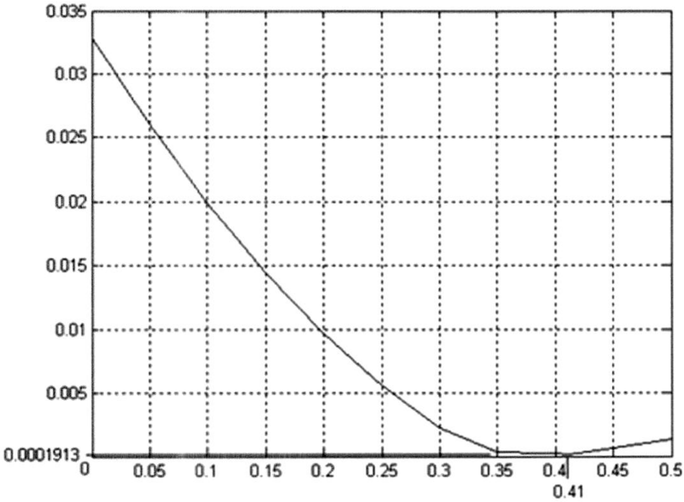

Fig. 4.9 Variation of the maximum value of $D_w^2(\xi)$ regarding ε (anti-symmetric case)

Table 4.5 Maximum values of $D_w^2(\xi)$ depending on ε (anti-symmetric case)

ε	max $D_w^2(\xi)$	ε	max $D_w^2(\xi)$
0	0.03274	0.25	0.005616
0.001	0.03260	0.3	0.002254
0.05	0.02598	0.35	0.000385
0.1	0.01989	0.4	0.000201
0.15	0.01446	0.45	0.000601
0.2	0.00969	0.5	0.001302

4.9 General Case: Non-symmetric Random Excitation

Now let us study the case of the general loading which lacks symmetric properties. We consider a loading as in Eq. (4.53) whose autocorrelation function is given as a white noise both in space and in time:

$$R_q(x_1, x_2, \tau) = \frac{R}{L}\delta(x_2 - x_1)\delta(\tau) = R\delta(\xi_2 - \xi_1)\delta(\tau) \tag{4.72}$$

where R is a positive constant. So the spectral density of the loading is

$$S_q(\xi_1, \xi_2, \omega) = \frac{R}{2\pi}\delta(\xi_2 - \xi_1). \tag{4.73}$$

Equation (4.50) permits to obtain the cross-spectral density of the generalized forces $S_{Q_jQ_k}(\omega)$ as

$$S_{Q_jQ_k}(\omega) = \nu_j^{-2}\nu_k^{-2}\frac{R}{2\pi}\int_0^1 W_j(\xi)W_k(\xi)d\xi = \frac{R}{2\pi\,\nu_j^2}\delta_{jk} \tag{4.74}$$

where δ_{jk} is the Kronecker's delta, because of the orthogonality of the normal modes $W_j(\xi)$ and $W_k(\xi)$. The mode shapes $W_j(\xi)$ correspond to the symmetric mode shapes $W_j^{(1)}(\xi)$ when j is odd, and to the anti-symmetric ones $W_j^{(2)}(\xi)$ when j is even, except cases $\varepsilon = 0.2$ and $\varepsilon = 0.4$ where the alternation is not valid. The mean-square displacement is obtained from Eq. (4.63) as follows:

$$d_w^2(\xi) = \frac{4\pi c L^8}{\rho A E^2 I^2}\sum_{j=1}^{\infty}\sum_{k=1}^{\infty}\frac{R}{2\pi}\frac{\delta_{jk}\nu_j^{-2}W_j(\xi)W_k(\xi)}{\left(\lambda_j^4 - \lambda_k^4\right)^2 + \eta\left(\lambda_j^4 + \lambda_k^4\right)} = \frac{RcL^8}{\rho A E^2 I^2}\sum_{j=1}^{\infty}\frac{W_j^2(\xi)}{\eta\nu_j^2\lambda_j^4}. \tag{4.75}$$

We simplify Eq. (4.75) utilizing η from Eq. (4.64) and we obtain

Table 4.6 Maximum values of $D_w^2(\xi)$ depending on ε (general case)

ε	max $D_w^2(\xi)$	ε	max $D_w^2(\xi)$
0	0.04166	0.25	0.01561
0.001	0.04158	0.3	0.01066
0.05	0.03711	0.35	0.00714
0.1	0.03199	0.4	0.01066
0.15	0.02653	0.45	0.01519
0.2	0.02098	0.5	0.02082

$$d_w^2(\xi) = \frac{RL^4}{2EIc}\sum_{j=1}^{\infty}\frac{W_j^2(\xi)}{\nu_j^2\lambda_j^4}. \tag{4.76}$$

We evaluate the following non-dimensional mean-square displacement:

$$D_w^2(\xi) = \frac{2EIc}{RL^4}d_w^2(\xi) = \sum_{j=1}^{\infty}\frac{W_j^2(\xi)}{\nu_j^2\lambda_j^4}. \tag{4.77}$$

The results of the evaluation of $D_w^2(\xi)$ for various values of ε are listed in Table 4.6. As in the symmetric case, for $\varepsilon = 0$ the maximum mean-square displacement occurs at the ends of the beam; analogous situation takes place for $\varepsilon < 0.345$; then the maximum mean-square displacement occurs at the middle of the beam. But as in the anti-symmetric case this value, denoted as ε_3 , corresponds to the supports configuration where the minimum of the maxima occurs, as observed in Fig. 4.10.

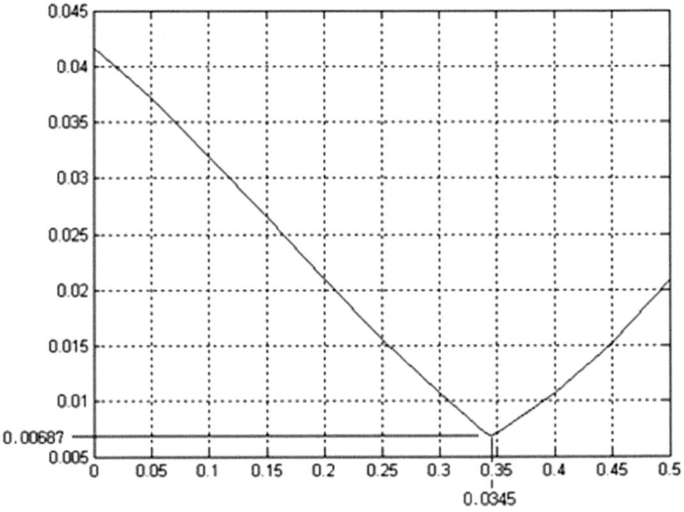

Fig. 4.10 Variation of the maximum value of $D_w^2(\xi)$ regarding ε (general case)

Moreover, it is remarkable to note an approximate relation between the optimum support locations of the 3 cases:

$$\varepsilon_3 \cong \frac{\varepsilon_1 + \varepsilon_2}{2}. \tag{4.78}$$

The authors are unaware of analogous situations in other optimization problems.

4.10 Point-Driven Excitation

Finally let us study the case of a point loading (Fig. 4.11). We consider a loading as in Eq. (4.53) with $F(x)$ expressed as follows:

$$F(\xi) = \delta(\xi - g) \tag{4.79}$$

where δ is the Dirac delta function and g is the non-dimensional position of the loading on the beam.

Thus, the autocorrelation function of the loading and its spectral density are, respectively:

$$R_q(\xi_1, \xi_2, \tau) = R\delta(\xi_1 - g)\delta(\xi_2 - g)\delta(\tau), \tag{4.80}$$

$$S_q(\xi_1, \xi_2, \omega) = \frac{R}{2\pi}\delta(\xi_1 - g)\delta(\xi_2 - g). \tag{4.81}$$

We express the cross-spectral density of the generalized forces from Eq. (4.50) as follows:

$$S_{Q_j Q_k}(\omega) = \frac{R}{2\pi \nu_j^2 \nu_k^2} \int_0^1 \delta(\xi_1 - g) W_j(\xi_1) d\xi_1 \int_0^1 \delta(\xi_2 - g) W_k(\xi_2) d\xi_2 = \frac{R W_j(g) W_k(g)}{2\pi \nu_j^2 \nu_k^2} \tag{4.82}$$

where $W_j(\xi)$ corresponds to the symmetric mode shapes $W_j^{(1)}(\xi)$ when j is odd, and to the anti-symmetric ones $W_j^{(2)}(\xi)$ when j is even, except cases $\varepsilon = 0.2$ and $\varepsilon = 0.4$ where the alternation is not valid. Regarding Eq. (4.63) we write the mean-square

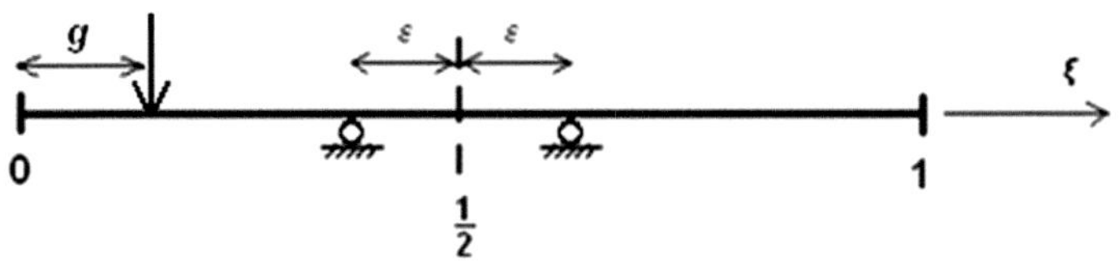

Fig. 4.11 Location of the point loading on the beam

Table 4.7 Maximum values of $D_w^2(\xi)$ depending on ε (point load case)

ε	max $D_w^2(\xi)$	ε	max $D_w^2(\xi)$
0	0.3600	0.25	0.0119
0.001	0.3497	0.3	0
0.05	0.1549	0.35	0.0058
0.1	0.1201	0.4	0.0200
0.15	0.0817	0.45	0.0436
0.2	0.0437	0.5	0.0733

displacement as follows:

$$d_w^2(\xi) = \frac{2cL^8R}{\rho AE^2I^2}\sum_{j=1}^{\infty}\sum_{k=1}^{\infty}\nu_j^{-2}\nu_k^{-2}\frac{W_j(g)W_k(g)W_j(\xi)W_k(\xi)}{\left(\lambda_j^4-\lambda_k^4\right)^2+\eta\left(\lambda_j^4+\lambda_k^4\right)}. \tag{4.83}$$

We fix $g = 0.2$ and $\eta = 0.1$ and evaluate the following non-dimensional mean-square displacement:

$$D_w^2(\xi) = \frac{\rho AE^2I^2}{2cL^8R}d_w^2(\xi) = \sum_{j=1}^{\infty}\sum_{k=1}^{\infty}\nu_j^{-2}\nu_k^{-2}\frac{W_j(g)W_k(g)W_j(\xi)W_k(\xi)}{\left(\lambda_j^4-\lambda_k^4\right)^2+\eta\left(\lambda_j^4+\lambda_k^4\right)}. \tag{4.84}$$

The results for $\max\left(D_w^2(\xi)\right)$ for various values of ε are listed in Table 4.7, and Fig. 4.12 shows the variation of the maximum mean-square displacement regarding ε, which exhibits the position of the optimum support location to minimize this value. The hybrid optimization and anti-optimization lead to conclusion that when $\varepsilon = 0.3$ the least maximum value is achieved, this result should have been expected because for $\varepsilon = 0.3$ the point load is applied on the support.

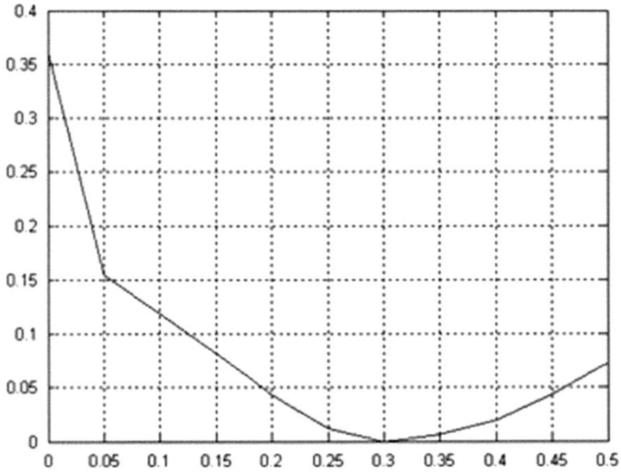

Fig. 4.12 Variation of the maximum value of $D_w^2(\xi)$ regarding ε (point load case)

This case permits to check the phenomenon of closely spaced frequencies and the associated effect of cross-correlations in *random* vibrations. Indeed, we established that for values of ε close to zero, the symmetric and anti-symmetric natural frequencies are tending to each other. At the value of ε equal to zero, the system may be viewed as two single beams of half-length with clamped-free ends, axis of symmetry passing through the clamped end.

Therefore, the more convenient representation of the non-dimensional displacement (4.84) as two sums is instructive:

$$D_w^2(\xi) = \sum_{j=1}^{\infty} A_{jj}(\xi) + \sum_{j=1}^{\infty}\sum_{k=1}^{\infty} A_{jk}(\xi)_{j\neq k} \tag{4.85}$$

where

$$A_{jk}(\xi) = \nu_j^{-2}\nu_k^{-2}\frac{W_j(g)W_k(g)W_j(\xi)W_k(\xi)}{\left(\lambda_j^4 - \lambda_k^4\right)^2 + \eta\left(\lambda_j^4 + \lambda_k^4\right)}. \tag{4.86}$$

The first sum in Eq. (4.85) corresponds to direct correlations, i.e. correlations between like modes. The second sum corresponds to cross-correlations, i.e. correlations between unlike modes. It is extremely interesting to evaluate the influence of the cross-correlations which are usually neglected in the literature. This neglect is unjustified for vanishingly small values of ε. For example, when $\varepsilon = 0.001$, we find $max\left(D_w^2(\xi)\right) = 0.3497$, and Eq. (4.85) leads to the following partial summations:

$$\sum_{j=1}^{\infty} A_{jj}(\xi) = 0.1796, \sum_{j=1}^{\infty}\sum_{k=1}^{\infty} A_{jk}(\xi)_{j\neq k} = 0.1701. \tag{4.87}$$

Cross-correlations contribute in this case 48.6% of the total response. For $\varepsilon \to 0$, the contribution will reach 50% since in the latter case the beam possesses two series of coinciding natural frequencies.

Thus, beam model by Niordson and Simmonds (1975) appears to be a generalization of the two-degree-of-freedom system discussed by Elishakoff (1999). The latter is composed of two masses each of value m and attached to wall with a spring constant k, whereas the spring constant between masses was taken as εk. When $\varepsilon \to 0$ the system reduces to two single-degree-of-freedom systems having coalescing natural frequencies equal to $\sqrt{k/m}$. Further details can be found in the paper by Elishakoff, Dujat, Lemaire, and Gadiot (2014).

4.11 Conclusion

In this chapter, we obtain the optimum location of supports to minimize the maximum mean-square displacement of beams subjected to various loadings. In the general non-symmetric case the optimum configuration of the supports to arrive at the maximal mean-square displacement which is as low as possible is $\varepsilon = 0.345$. This supports position corresponds approximately to the average value between the optimum locations associated with the symmetric and anti-symmetric cases, respectively.

With the excitation defined as a point load, we find a remarkable influence of the cross-correlation terms in random vibrations environment. When the two supports are very close, two series of natural frequencies exist and cross-correlation terms contribute as much as the direct correlations. Therefore, conventional neglect of cross-correlations is unjustified.

We conclude that varying the position of supports represents a viable means of reducing the maximum mean-square displacement providing the possibility of effective combination of optimization and anti-optimization.

References

Adali, S., Lena, F., Devaut, G., & Chiaruttini, V. (2002). Optimization of laminated composites under buckling uncertainties via anti-optimization. In *9th AIAA/ISSMO Symposium on Multidisciplinary Analysis and Optimization, AIAA Paper* 2002-5417, 4–6 Sept. 2002, Atlanta, Georgia.

Bolotin, V. V. (1984). *Random Vibrations of Elastic Systems*, The Hague, The Netherlands: Martinus.

Bu, J. L., Jiang, Z. G., & Jiao, S. H. (2012). Study of anti-optimization method to identify the parameters of the frequency correlation viscoelastic model for rubber. *Advanced Materials Research, 415–417*, 232–236.

Catallo, L. (2004). Genetic anti-optimization for reliability structural assessment of precast concrete structures. *Computers and Structures, 82*(13–14), 1053–1065.

Crandall, S. H. (1979a). Random vibration of one- and two-dimensional structures. In P. R. Krishnaiah (Ed.), *Applications of statistics* (pp. 1–82) (Proceeding of Symposium Held At Wright State University, Dayton, Ohio, 1976), Academic Press.

Crandall, S. H. (1979b). Random Vibration of vehicles and structures. In *Proceedings of the Seventh Canadian Congress of Applied Mechanics* (pp. 1–12), Sherbrooke.

De Faria, R. (2002). Buckling Optimization and anti-optimization of composite plates. *International Journal for Numerical Methods in Engineering, 53*, 719–732.

Elishakoff, I. (1990). An idea of the uncertainty triangle, (editorial). *Shock and Vibration Digest, 22*(10), 1.

Elishakoff, I. (1999). *Probabilistic Theory of Structures*, Mineola, N.Y: Dover publication.

Elishakoff, I., & Ohsaki, M. (2010). *Optimization and anti-optimization of structures under uncertainty*. London: Imperial College Press.

Elishakoff, I., Haftka, R. T., & Fang, J. (1994). Structural design under bounded uncertainty—Optimization with anti-optimization. *Computers and Structures, 53*(6), 1401–1405.

Elishakoff, I., Kriegesmann, B., Rolfes, R., Hünne, C., & Kling, A. (2012). Optimization and anti-optimization of buckling load for composite cylindrical Shells under uncertainties. *AIAA Journal, 50*(7), 1513–1534.

Elishakoff, I., Dujat, K., Lemaire, M., & Gadiot, G. (2014). Hybrid optimization and anti-optimization of a stochastically excited beam. *Journal of Applied Mechanics, 81*(2), article 021017.

Gangadharan, S., Nikolaidis, E., Lee, K., Haftka, R. T., & Burdisso, R. (1999).Anti-optimization for comparison of alternative structural models and damage detection. *AIAA Journal, 37*(7), 857–864.

Hlaváček, I., Chleboun, J., & Babuška, I. (2004). *Uncertain input data problems and the worst scenario method*. Amsterdam: Elsevier.

Krée, P., & Soize, C. (1983). *Mécanique Aléatoire*. Paris: Dunod (in French).

Lee, J., Haftka, R. T., Griffin, O. H., Jr., Watson, L. T., & Sensmeier, M. D. (1994). Detecting delamination in a composite beam using anti-optimization. *Structural Optimization, 8*(2/3), 93–100.

Lombardi, M., & Haftka, R. T. (1998). Anti-optimization technique for structural design under load uncertainties. *Computer Methods in Applied Mechanics and Engineering, 157*(1–2), 19–31.

McWilliam S. (2001). Anti-optimization of uncertain structures using interval analysis. *Computers and Structures, 79*, 421–430.

Niordson, F. I., & Simmonds, J. G. (1975). An improved Stodola method for computing close eigenvalues. In *Foundations of deformable media* (Galerkin 100th Birthday Volume), Nauka Publishers, Moscow, pp. 413–419 (in Russian).

Ohsaki, M., Zhang, J., & Elishakoff, I. (2012). Multi-objective Hybrid optimization and anti-optimization for force design of Tensegrity structures. *ASME Journal of Applied Mechanics*, article 021015.

Qiu Z. P., & Wang X. J. (2010). Structural Anti-optimization with interval design parameters, *Structural and Multidisciplinary Optimization, 41*, 397–406.

Rao, S .S. (2007). *Vibration of continuous systems* (pp. 333–335). New York: Wiley.

Wedig, W., & Riemer, M. (1981). Bauwerke unter Wellenlast—Tokussierungseffekte bei Poisson-erregung. In *VDI —Berichte* (pp. 201–207) (in German).

Yang C.Y., *Random vibrations of structures*, New York: Wiley-Interscience.

Zingales, M., & Elishakoff, I. (2000). Anti-optimization versus probability in an applied mechanics problem: Vector uncertainty. *ASME Journal of Applied Mechanics, 67*(3), 472–484.

Chapter 5
Random Vibration of a Point-Driven Two-Span Beam on an Elastic Foundation

5.1 Introductory Comments

Wide-band random vibration of point-driven elastic structures has been the object of considerable attention in recent years. Particularly, Crandall and Wittig (1972), Crandall (1973), Crandall (1974), Crandall (1977), Itao and Crandall (1978), Crandall (1979a, b), Crandall and Kulvets (1978), Crandall and Kulvets (1977), and Davies and Khandoker (1982) considered a large number of uniform structures subjected to a point force, with a time history in the form of a stationary wide-band random process. The distribution of the mean-square response velocity turned out to be surprisingly uniform, except for small zones or narrow bands of enhanced response. For example, if the force acts at $x = a$ on a string or a beam of length L, the zones of enhanced response are at $x = a$ and $x = L - a$. As the bandwidth of the excitation increases without limit, the relative enhancement tends to 3/2 asymptotically. This pattern was demonstrated experimentally and explained by normal-mode analysis from Crandall and Wittig (1972), Crandall (1973), and (1974) by the image-sum technique from Crandall (1977). The cross-correlations between unlike modes could generally be neglected, since there is little modal overlap due to the light damping and ample separation of the natural frequencies; this resulted in symmetrical distribution of the mean-square response velocity with respect to the midpoint of the string or the beam. This symmetry of the response distribution is also preserved in structures subjected to uncorrelated or fully correlated concentrated forces with the relative enhancement depending on the correlation and number of input force processes.

A structure where this symmetry is not preserved was considered by Elishakoff, Van Zanten and Crandall (1979) which investigated the wide-band random vibration of a uniform circular cylindrical shell excited by a ring load, uniform along the circumference and random in time. The response of this shell, although one-dimensional, turned out to differ both qualitatively and quantitatively from its counterparts for point-excited uniform strings and beams, because of large modal

I. Elishakoff, *Dramatic Effect of Cross-Correlations in Random Vibrations of Discrete Systems, Beams, Plates, and Shells*,
https://doi.org/10.1007/978-3-030-40394-2_5

overlaps at the low end of the frequency spectrum. The resulting modal cross-correlations caused a pronounced asymmetry of the distribution of mean-square response.

The present chapter considers another structure, namely, random vibration of a point-driven two-span beam on a linear elastic foundation.

The same example, without the elastic foundation, was considered previously by Crandall and Lotz (1971; Crandall, 1980) and by Elishakoff (1973) in different contexts. A remarkable feature of this study lies in the fact that a two-span beam on a linear elastic foundation reproduces the vibrational behavior of an axisymmetric cylindrical shell in that the natural frequencies are clustered together at the low end of the frequency spectrum. It is interesting to note that there is a significant difference between behaviors of the cylindrical shell and the two-span beam: in addition to the clustering natural frequencies, the spectrum comprises two groups of frequencies, the first corresponding to a simply supported beam of length L on an elastic foundation and the second being associated with a single-span beam of length L which is simply supported at the left end and clamped at the other.

5.2 Random Vibrations of a Two-Span Beam Without Elastic Foundation

In this section, a solution is presented for the stochastic problem associated with excitation of a symmetrically supported two-span beam without elastic foundation; using the method of generalized conditions. The field of external forces is assumed to be stationary-ergodic in time and, generally speaking, an arbitrary random function of the space coordinate. The cross-spectral densities of the generalized forces corresponding to different mode groups are determined, and integral probability parameters of the displacement field are estimated for the case of a random pressure field with a given correlation-function.

Consider a continuous two-span beam of uniform cross section, supported at its ends and midpoint and subjected to random pressure pulses with intensity $q(x, t)$, x denoting the space coordinate and t denoting time. All requirements for the problem to be considered in a statistically linear setting are assumed to be satisfied. The normal deflection $w(x, t)$ must satisfy the equation:

$$EI\frac{\partial^4 w}{\partial x^4} + 2\rho F\varepsilon\frac{\partial w}{\partial t} + \rho F\frac{\partial^2 w}{\partial t^2} = q(x, t), \left(0 \leqq x \leqq 2L\right) \tag{5.1}$$

where EI is the bending stiffness, F the cross-sectional area, ρ the density of the beam material, and ε the damping coefficient. Assigning local systems of coordinates $x_j = x - (j - 1)l,\ j = 1, 2$, Eq. (5.1) is supplemented by the boundary conditions:

$$w = \frac{\partial^2 w}{\partial x^4} = 0\,(x_1 = 0, x_2 = l)$$

$$w_- = w_+ = 0,\ \frac{\partial w_-}{\partial x} = \frac{\partial w_+}{\partial x},\ \frac{\partial^2 w_-}{\partial x^2} = \frac{\partial^2 w_+}{\partial x^2} \tag{5.2}$$

where

$$f_- = f(x)|_{x_1=l},\ f_+ = f(x)|_{x_2=0}.$$

The problem is formulated as follows: Given the probability behavior of the function $q(x,t)$, find that of the stress–strain field in the beam, dispensing with initial conditions. For the stochastic boundary value problem (5.1)–(5.2), we resort to the generalized coordinates method, which consists of expanding the given and sought functions in stochastic Fourier integrals and in series in terms of the modes of the undamped beam. For finding the frequencies and modes of the natural vibrations in Eq. (5.1), we assume $q(x,t) \equiv 0$, $\varepsilon \equiv 0$, and $w(x,t) = X(x)e^{i\omega t}$, where ω is the frequency.

Integration of the ordinary differential equation for $X(x)$

$$EI\frac{\mathrm{d}^4X}{\mathrm{d}x^4} - \rho F\omega^2 X = 0\ (0 \leqq x \leqq 2l)$$

derived under these assumptions from (5.1) yields

$$X(x) = A_1\cos kx_1 + A_2\sin kx_1 + A_3\sinh kx_1 + A_4\cosh kx_1,$$

$$\text{for}\ 0 \leqq x_1 \leqq l$$

$$X(x) = B_1\cos kx_2 + B_2\sin kx_2 + B_3\sinh kx_2 + B_4\cosh kx_2,$$

$$\text{for}\ 0 \leqq x_2 \leqq l$$

where A_j and B_j are constants of integration, $k^4 = \rho F\omega^2(EI)^{-1}$.

Equation (5.2) yields the following boundary conditions for $X(x)$:

$$X = \frac{\mathrm{d}^2X}{\mathrm{d}x^2} = 0\ (x_1 = 0, x_2 = l)$$

$$X_- = X_+ = 0,\ \frac{\mathrm{d}X_-}{\mathrm{d}x} = \frac{\mathrm{d}X_+}{\mathrm{d}x},\ \frac{\mathrm{d}^2X_-}{\mathrm{d}x^2} = \frac{\mathrm{d}^2X_+}{\mathrm{d}x^2},$$

which, satisfied subject to the additional requirement of nontrivility, yield in turn the frequency equation:

$$sS(sC - cS) = 0 \tag{5.3}$$

where

$$s = \sin kl,\ c = \cos kl,\ S = \sinh kl,\ C = \cosh kl.$$

The natural frequencies fall into two groups, the first corresponding to an interrupted beam ($s = 0$) and the second to a two-span beam with intermediate restraint ($sC - cS = 0$). The expressions for the corresponding modes have the form

$$\begin{aligned}
&X_{1\alpha}(x_1) = \sin\frac{\alpha\pi x_1}{l},\ X_{1\alpha}(x_2) = (-1)^\alpha \sin\frac{\alpha\pi x_2}{l}\\
&X_{2\alpha}(x_1) = \sin k_\alpha x_1 - \frac{s_\alpha}{S_\alpha}\sinh k_\alpha x_1,\ X_{2\alpha}(x_2) =\\
&-s_\alpha(\cosh k_\alpha x_2 - \cos k_\alpha x_2) + c_\alpha(\sinh k_\alpha x_2 - \sin k_\alpha x_2)
\end{aligned} \tag{5.4}$$

where

$$s_\alpha = \sin k_\alpha l,\ c_\alpha = \cos k_\alpha l,\ S_\alpha = \sinh k_\alpha l,\ C_\alpha = \cosh k_\alpha l\ k_\alpha^4 = \rho F \omega_\alpha^2 (EJ)^{-1}.$$

By Eq. (5.4), $X_{2\alpha}(x_1) \equiv X_{2\alpha}(x_2)|_{x_2=l-x_1}$; hence, $\int_0^{2l} X_{1\alpha}(x)X_{2\beta}(x)\mathrm{d}x = 0$ for any α and β. Bolotin's asymptotic method yields, for $\alpha \gg 1$ and $k_\alpha l \sim 1$,

$$\begin{aligned}
&X_{1\alpha}(x_1) = \sin\frac{\alpha\pi x_1}{l},\ X_{1\alpha}(x_2) = (-1)^\alpha \sin\frac{\alpha\pi x_2}{l}\\
&X_{2\alpha}(x_1) = \sin k_\alpha x_1 - 2^{-\frac{1}{2}}(-1)^\alpha e^{k_\alpha(x_1-l)}\\
&X_{2\alpha}(x_2) = \left(\cos k_\alpha x_2 - \sin k_\alpha x_2 - e^{-k_\alpha x_2}\right)(-1)^\alpha 2^{-\frac{1}{2}}.
\end{aligned} \tag{5.5}$$

Now we turn to the stochastic boundary value problem. Since the mathematical expectation of the displacement field is found deterministically, we hereafter confine ourselves to centered functions. Expanding the external load in series in terms of the modes of the undamped system, and using stochastic Fourier time-integrals, we have

$$\begin{aligned}
q(x,t) = &\sum_{\alpha=1}^{\infty}\int_{-\infty}^{\infty} Q_{1\alpha}(\omega)X_{1\alpha}(x)e^{i\omega t}\mathrm{d}\omega\\
&+\sum_{\beta=1}^{\infty}\int_{-\infty}^{\infty} Q_{2\beta}(\omega)X_{2\beta}(x)e^{i\omega t}\mathrm{d}\omega
\end{aligned} \tag{5.6}$$

where $Q_{j\alpha}(\omega)$ are generalized random complex functions of the frequency. By the assumption of the stationarity of $q(x,t)$, we have the following conditions for statistical orthogonality of $Q_{j\alpha}(\omega)$:

$$Q_{j\alpha}^*(\omega)Q_{k\beta}(\omega') = S_{j\alpha,k\beta}(\omega)\delta(\omega - \omega'),\ (j,k = 1,2) \tag{5.7}$$

with the asterisk denoting the complex conjugate, the triangular brackets denoting the expected value, and $\delta(x)$ denoting Dirac's delta function. $S_{1\alpha,1\beta}$ and $S_{2\alpha,2\beta}$ will be referred to as the cross-spectral densities of the generalized forces, corresponding to the first and second mode groups, respectively; $S_{1\alpha2\beta}$ and $S_{2\alpha1\beta}$ will be referred to as the cross-spectral densities corresponding to mixed groups.

For the space-time correlation-function of $q(x, t)$,

$$K_q\left(x, x', \tau\right) = \left\langle q^*(x, t) q\left(x', t + \tau\right)\right\rangle$$

we have, bearing in mind Eq. (5.7)

$$K_q(x, x, \tau) = \sum_{j,k} \sum_{\alpha\beta} \int_{-\infty}^{\infty} S_{j\alpha,k\beta}(\omega) X^*_{j\alpha}(x) X_{k\beta}\left(x'\right) e^{i\omega\tau} \mathrm{d}\omega. \tag{5.8}$$

As the probability characteristics of the external load are known, so is the Fourier time-transform of the correlation function:

$$S_q\left(x, x', \omega\right) = \frac{1}{2\pi} \int_{-\infty}^{\infty} K_q\left(x, x', \tau\right) e^{-i\omega\tau} \mathrm{d}\tau$$

which, applied to both sides of (5.8), yields:

$$S_q\left(x, x', \omega\right) = \sum_{j,k} \sum_{\alpha\beta} S_{j\alpha,j\beta}(\omega) X^*_{j\alpha}(x) X_{k\beta}\left(x'\right).$$

In view of the orthogonality referred to earlier, the sought expression for $S_{j\alpha,k\beta}(\omega)$ is obtainable from the above, namely,

$$S_{ja,kb}(\omega) = \frac{1}{\nu^2_{j\alpha} \nu^2_{k\beta}} \int_0^{2l} \int_0^{2l} S_q\left(x, x', \omega\right) X_{j\alpha}(x) X^*_{k\beta}\left(x'\right) \mathrm{d}x \mathrm{d}x' \tag{5.9}$$

where

$$\nu^2_{j\alpha} = \int_0^{2l} X_{j\alpha}(x) X^*_{j\alpha}(x) \mathrm{d}x.$$

Expanding $w(x, t)$ in a series in the same manner as (5.6) and recalling (5.1), we obtain expressions for the various probability parameters characterizing the stochastic behavior of the beam in terms of linear operators of $w(x, t)$. For the spectrum of space correlations of the displacement field we have

$$S_w(x, x', \omega) = \sum_{j,k} \sum_{\alpha,\beta} \frac{X^*_{j\alpha}(x)}{F_{j\alpha}(-i\omega)} \frac{X_{k\beta}(x')}{F_{k\beta}(i\omega)} S_{j\alpha,k\beta}(\omega) \tag{5.10}$$

the dispersion of the latter,

$$D_w(x, t) = \langle \omega^*(x, t)\omega(x, t) \rangle,$$

at a point with coordinate x, being

$$\begin{aligned} D_w(x, t) &= \int_{-\infty}^{\infty} S_w(x, x, \omega)\mathrm{d}\omega \\ &= \sum_{j,k} \sum_{\alpha,\beta} X^*_{j\alpha}(x) X_{k\beta}(x) \int_{-\infty}^{\infty} \frac{S_{j\alpha,k\beta}(\omega)\mathrm{d}\omega}{F_{j\alpha}(-i\omega)F_{k\beta}(i\omega)} \\ F_{j\alpha}(i\omega) &= \rho F\left(\omega^2_{j\alpha} - \omega^2 + 2i\omega\varepsilon\right) \\ \omega^2_{1\alpha} &= \frac{\alpha^4\pi^4}{l^4} \frac{EJ}{\rho F}, \quad \omega^2_{2\beta} = k^4_\beta \frac{EJ}{\rho F}, \\ &\left(k_1 l = 3.927,\ k_2 l = 7.069,\ k_\beta l = \beta + \frac{1}{4},\ \beta \geqq 3\right) \end{aligned} \tag{5.11}$$

$\omega_{1\alpha}, \omega_{2\alpha}$ being the natural frequencies corresponding to the first and second mode groups, respectively. Determination of the cross-spectral densities of the generalized forces for a given correlation-function $K_q(x, x', \tau)$ is facilitated by the asymptotic approach (Elishakoff, 1971a, b, c).
Example: Given a beam by an external load with a given correlation-function $K_q(x, x', \tau) = \delta(x - x')\delta(\tau)$. The integrals in Eq. (5.10) are obtainable as exact values. Integration of $D_w(x, t)$ over the length of the beam yields the average:

$$\begin{aligned} d &= \frac{1}{2l} \int_0^{2l} D_w(x, t)\mathrm{d}x \\ &= \frac{1}{8\varepsilon l^2(\rho F)^2} \left[\sum_{\alpha=1}^{\infty} \frac{1}{\omega^2_{1\alpha}} + \sum_{\beta=1}^{\infty} \frac{S_\beta}{S^2_\beta - s^2_\beta} \cdot \frac{1}{\omega^2_2} \right]. \end{aligned} \tag{5.12}$$

In applying Eq. (5.12), the first two modes of each mode group may be used with sufficient accuracy, the contribution of the second group to the integral probability parameter d being about 30%.

For the correlation-function is presented in product form

$$K_q(x, x', \tau) = K_{1q}(\tau)K_{2q}(x, x')$$

and the spectral density

$$S_{1q}(\omega) = \frac{1}{2\pi}\int_{-\infty}^{\infty} K_{1q}(\tau)e^{-i\omega t}\mathrm{d}\tau$$

is a rational fractional function, the integral in Eq. (5.11) is obtainable in exact terms; otherwise, the asymptotic methods proposed for negligible damping (Powell, 1958; Lin, 1967; Bolotin and Elishakoff, 1971) may be resorted to. In either, Eq. (5.12) retains its structure. The displacement dispersion (averaged over the length of the beam) is formulated as the sum of two series, representing the responses of a beam, respectively, simply supported and restrained at one end and supported at the other.

5.3 Formulation of the Problem

Consider a uniform two-span beam on a linear elastic foundation as shown in Fig. 5.1.

The beam is simply supported at its ends and additional support is placed at the midpoint. It is subjected to random distributed load $q(x, t) = P(t)\delta(x - a)$, x denoting the space coordinate and t time. All requirements for the problem to be considered in a probabilistically linear setting are assumed to be satisfied. The deflection $w(x, t)$ must satisfy the equation

$$EI\frac{\partial^4 w}{\partial x^4} + c\frac{\partial w}{\partial t} + \rho A\frac{\partial^2 w}{\partial t^2} + \chi w = q(x, t), \quad (-L \le x \le L) \tag{5.13}$$

where EI is the bending stiffness, A the cross-sectional area, ρ the density of the beam material, c the damping coefficient, and χ the stiffness of the elastic foundation. Equation (5.13) is supplemented by the boundary conditions

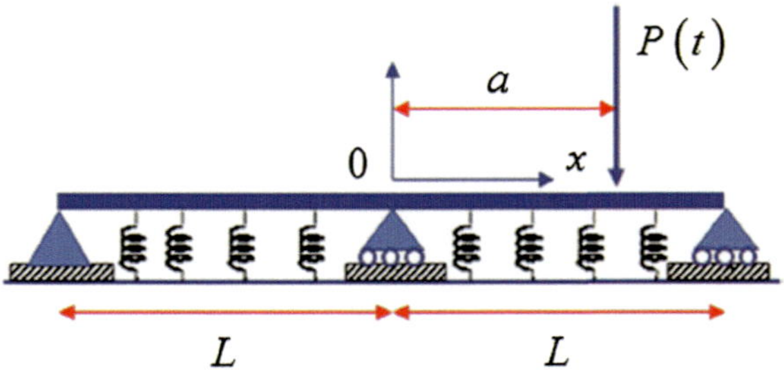

Fig. 5.1 Two-span beam on elastic foundation

$$w = \frac{\partial^2 w}{\partial x^2} = 0; \quad x = -L; x = L \tag{5.14}$$

and the continuity conditions

$$w_- = w_+ = 0, \quad \frac{\partial w_-}{\partial x} = \frac{\partial w_+}{\partial x}, \quad \frac{\partial^2 w_-}{\partial x^2} = \frac{\partial^2 w_+}{\partial x^2} \tag{5.15}$$

where $w_- = w(x)|_{x=-0}$ and $w_+ = w(x)|_{x=+0}$.

We consider the response due to a concentrated force acting at the section $x = a$, as shown in Fig. 5.1. For a stationary random loading the space-time correlation-function of the excitation reads

$$E[q(x_1 t_1)q(x_2, t_2)] = \delta(x_1 - a)\delta(x_2 - a)R_p(t_2 - t_1) \tag{5.16}$$

where $R_p(\tau)$ is the autocorrelation-function of the point loading P, supposed to be specified. In particular, the spectral density of the point force, defined by the Wiener–Khinchine formula

$$S_p(\omega) = \frac{1}{2\pi} \int_{-\infty}^{+\infty} R_p(\tau)^{-ikx} \mathrm{d}\tau \tag{5.17}$$

is given as the following band-limited white noise spectrum, along with the associated autocorrelation

$$S_P(\omega) = \begin{cases} S_0 & |\omega| < \omega_c \\ 0 & \omega_c < |\omega| \end{cases}, \quad R_p(\tau) = 2S_0 \frac{\sin \omega_c \tau}{\tau}. \tag{5.18}$$

The problem is formulated as follows: Given the spectral density of the point force $S_p(\omega)$, find the mean square of the transverse velocity $v(x, t) = \partial w / \partial t$.

With the normal-mode approachby Lin (1967), Crandall and Mark (1963), and Mercer (1965) in mind, we first find the natural frequencies and modes of the structure in question.

5.4 Natural Frequencies and Modes of the Two-Span Beam

For the above purpose, we assume in Eq. (5.1), $q(x, t) \equiv 0$, $c \equiv 0$, and $w(x, t) = X(x)e^{i\omega\tau}$.

Equation (5.13) becomes

$$\frac{\mathrm{d}^4 X}{\mathrm{d}x^4} + \frac{1}{EI}(\chi - \rho A\omega^2)X = 0 \tag{5.19}$$

It is seen from Eq. (5.19) that three different cases have to be considered

$$\text{(i)}\ \rho\omega^2 < \chi, \tag{5.20}$$

$$\text{(ii)}\ \rho\omega^2 = \chi, \tag{5.21}$$

$$\text{(iii)}\ \rho\omega^2 > \chi. \tag{5.22}$$

For the interval (5.20), the integration of Eq. (5.13) yields

$$X(x) = (A_1 \sin\gamma x + A_2 \cos\gamma x)e^{x} + (A \sin\gamma x + A_4 \cos\gamma x)e^{-xx} \quad -L \le x \le 0, \tag{5.23}$$

$$X(x) = (B_1 \sin\gamma x + B_2 \cos\gamma x)e^{x} + (B_3 \sin\gamma x + B_4 \cos\gamma x)e^{-ix} \quad 0 \le x \le L, \tag{5.24}$$

$$\gamma^4 = (\chi - \rho A\omega^2)/EI. \tag{5.25}$$

Bearing in mind the boundary conditions (5.14) and continuity conditions (5.15) we obtain the following frequency equation:

$$(\cos 2\gamma L - \cosh 2\gamma L)(\sin 2\gamma L - \sinh 2\gamma L) = 0. \tag{5.26}$$

We note that the vanishing of the first factor is associated with a simply supported beam of length L on elastic foundation with stipulation of condition (5.22), whereas vanishing of the second factor is associated with a similar beam simply supported at one end and clamped at the other under the same condition. Equation (5.26) has no real roots, hence there are no natural frequencies in the interval (5.20).

In the case exemplified in Eq. (5.21), $X(x)$ takes the form

$$X(x) = A_1 + A_2 x + A_3 x^2 + A_4 x^3 \quad -L \le x \le 0, \tag{5.27}$$

$$X(x) = B_1 + B_2 x + B_3 x^2 + B_4 x^3 \quad 0 \le x \le L. \tag{5.28}$$

Equations (5.27) and (5.28) are incompatible with the boundary and continuity conditions (5.14) and (5.15), hence, the frequency spectrum does not contain the frequency $\omega = \sqrt{\chi/\rho}$.

Finally, for the interval given in Eq. (5.22), integration of Eq. (5.13) yields

$$X(x) = A_1 \sin\vartheta x + A_2 \cos\vartheta x + A_3 \sinh\vartheta x + A_4 \cosh\vartheta x \quad -L \le x \le 0, \tag{5.29}$$

$$X(x) = B_1 \sin\vartheta x + B_2 \cos\vartheta x + B_3 \sinh\vartheta x + B_4 \cosh\vartheta x \quad 0 \le x \le L. \tag{5.30}$$

$$\vartheta^4 = \left(\rho A\omega^2 - \chi\right)/EI. \tag{5.31}$$

Bearing in mind the boundary conditions (5.14) and continuity conditions (5.15), we obtain the following frequency equation:

$$\sin\vartheta L(\sin\vartheta L\cosh\vartheta L - \cos\vartheta L\sinh\vartheta L) = 0. \tag{5.32}$$

The natural frequencies fall into two groups, the first corresponding to a simply supported beam as an elastic foundation

$$\sin\vartheta L = 0 \tag{5.33}$$

and the second to a two-span beam with intermediate clamping

$$\sin\vartheta L\cosh\vartheta L - \cos\vartheta L\sinh\vartheta L = 0 \tag{5.34}$$

(see Fig. 5.2). The mode shapes of the first group are anti-symmetric with respect to the midpoint of the beam, whereas those of the second group are symmetric. Note that Eq. (5.24) coincides formally with Eq. (5.15) by Elishakoff (1973), where random vibrations of a two-span beam without an elastic foundation was considered.

The anti-symmetric mode shapes are

$$X_{1a} = (-1)^a \sin\frac{\alpha\pi x}{L} \quad \alpha = 1, 2, 3, \ldots \quad (-L \leq x \leq L) \tag{5.35}$$

with natural frequencies

$$\omega_{1\alpha}^2 = \frac{\chi}{\rho A} + \frac{EI}{\rho A}\left(\frac{\alpha\pi}{L}\right)^4 \tag{5.36}$$

whereas the symmetric mode shapes read

$$X_{2\alpha}(x) = -\sin\vartheta_\alpha L(\cosh\vartheta_\alpha x - \cos\vartheta_\alpha x) + \cos\vartheta_\alpha L(\sinh\vartheta_\alpha x - \sin\vartheta_\alpha x) \tag{5.37}$$

$$\alpha = 1, 2, 3\ldots \quad (-L \leq x \leq L)$$

with natural frequencies

$$\omega_{2\alpha}^2 = \frac{\chi}{\rho A} + \frac{EI}{\rho A L^4}(\vartheta_\alpha L)^4 \tag{5.38}$$

$$\vartheta L = 3.92660, \quad \vartheta_2 L = 7.06858, \quad \vartheta_3 L = 10.2102$$
$$\vartheta_A L = 13.3518, \quad \vartheta_S L = 16.4934$$
$$\vartheta_\alpha L = (4\alpha + 1)\pi/4, \alpha > 5$$

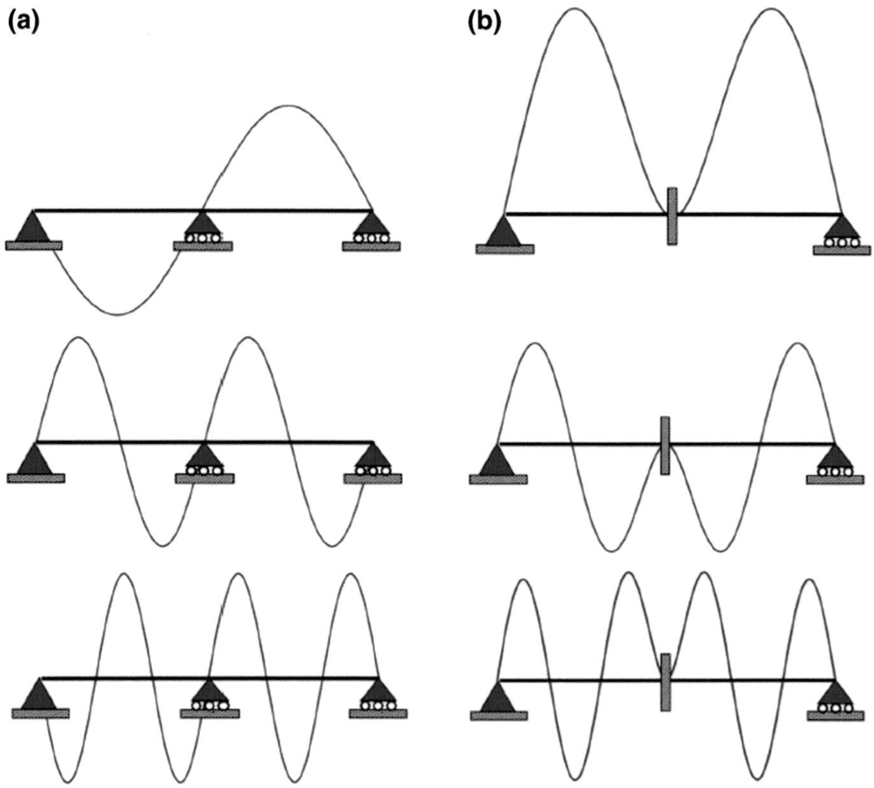

Fig. 5.2 **a** First group of modes, anti-symmetric with respect to the midpoint; **b** second group of modes, symmetric with respect to the midpoint

$\vartheta_\alpha L$ being the αth root of the transcendental Eq. (5.34).

For $a > 5$, the normal modes can be approximated with sufficient accuracy by Elishakoff (1973)

$$X_{2\alpha}(x) = \frac{\sqrt{2}}{2}(-1)^\alpha\left(\cos\vartheta_\alpha x - \sin\vartheta_\alpha x - e^{-\vartheta_\alpha x}\right). \tag{5.39}$$

We now normalize the mode shapes as follows:

$$\psi_{i\alpha}(x) = \frac{\sqrt{2L}X_{i\alpha}(x)}{\sqrt{\int_{-L}^{L} X_{i\alpha}^2(x)\mathrm{d}x}} \quad (i = 1, 2; \quad \alpha = 1, 2, 3, \ldots). \tag{5.40}$$

The normalized mode shapes have the following orthogonality property:

$$\int_{-L}^{L} \rho A \psi_{1\alpha}(x)\psi_{1\beta}(x)\mathrm{d}x = 2\rho A L \delta_{\alpha\beta}, \tag{5.41a}$$

$$\int_{-L}^{L} \rho A \psi_{2\alpha}(x)\psi_{2\beta}(x)\mathrm{d}x = 2\rho A L \delta_{\alpha\beta}, \tag{5.41b}$$

$$\int_{-L}^{L} \rho A \psi_{1\alpha}(x)\psi_{2\beta}(x)\mathrm{d}x = 0, \tag{5.41c}$$

$\delta_{\alpha\beta}$ denoting the Kronecker delta.

5.5 Random Vibration Response

We now proceed to the mean-square velocity response of a two-span beam. The external load is represented as follows:

$$-q(x,t) = \sum_{\alpha=1}^{\infty} \int_{-\infty}^{\infty} Q_{\alpha}(\omega)\psi_{1\alpha}(x)e^{i\omega t}\mathrm{d}\omega + \sum_{\beta=1-\infty}^{\infty} \int_{-\infty}^{\infty} Q_{2\beta}(\omega)\psi_{2\beta}(x)e^{i\omega t}\mathrm{d}\omega \tag{5.42}$$

where $Q_{j\alpha}(\omega)$ are generalized random complex functions of frequency, j denoting the number of the frequency group ($j = 1, 2$), and α or β the number of the components within a given group ($\alpha, \beta = 1, 2, \ldots$). Since $q(x,t)$ is a weakly stationary function of time, we have the following condition for statistical orthogonality of $Q_{j\alpha}(\omega)$

$$E\left[Q_{ja}^{*}(\omega)Q_{k\beta}(\omega')\right] = S_{j\alpha,k\beta}(\omega)\delta(\omega - \omega') \tag{5.43}$$

with the asterisk denoting the complex conjugate, $E[\bullet]$ the expected value, and $\delta(\bullet)$ the Dirac's Delta function. Following Elishakoff (1973), $S_{1\alpha 1\beta}$ and $S_{2\alpha,2\beta}$ are referred to here as the cross-spectral densities of the generalized forces corresponding to the first and second mode groups, respectively; $S_{1\alpha,2\beta}$ and $S_{2\alpha,1\beta}$ as the cross-spectral densities corresponding to mixed groups. Space-time correlation-function of $q(x,t)$ is then written in the following form:

$$E\left[q^{*}(x_1,t)q(x_2,t+\tau)\right] = R_1(x_1,x_2,\tau) + R_2(x_1,x_2,\tau) + R_{12}(x_1,x_2,\tau) \tag{5.44}$$

where

$$R_1(x_1,x_2,\tau)=\sum_{\alpha=1}^{\infty}\sum_{\beta=1}^{\infty}\psi_{1\alpha}(x_1)\psi_{1\beta}(x_2)\int_{-\infty}^{+\infty}S_{1\alpha_1\beta}(\omega)e^{i\omega\tau}\mathrm{d}\omega, \tag{5.45}$$

$$R_2(x_1,x_2,\tau)=\sum_{\alpha=1}^{\infty}\sum_{\beta=1}^{\infty}\psi_{2\alpha}(x_1)\psi_{2\beta}(x_2)\int_{-\infty}^{+\infty}S_{2\alpha,2\beta}(\omega)e^{i\omega\tau}\mathrm{d}\omega, \tag{5.46}$$

$$R_{12}(x_1,x_2,\tau)=\sum_{\alpha=1}^{\infty}\sum_{\beta=1}^{\infty}\left[\begin{array}{l}\psi_{1\alpha}(x_1)\psi_{2\beta}(x_2)\int_{-\infty}^{+\infty}S_{1\alpha,2\beta}(\omega)e^{i\omega\tau}\mathrm{d}\omega+\\ +\psi_{2\alpha}(x_1)\psi_{1\beta}(x_2)\int_{-\infty}^{+\infty}S_{2\alpha1\beta}(\omega)e^{i\omega\tau}\mathrm{d}\omega\end{array}\right]. \tag{5.47}$$

Equations (5.16) and (5.44)–(5.47) yield

$$S_{j\alpha,k\beta}(\omega)=\frac{\psi_{j\alpha}(\omega)}{2L}\frac{\psi_{i\beta}(\omega)}{2L}S_P(\omega). \tag{5.48}$$

We expand the displacement $w(x, t)$ in the same manner as in Eq. (5.42):

$$w(x,t)=\sum_{\alpha=1-\infty}^{\infty}\int_{-\infty}^{\infty}W_{1\alpha}(\omega)\psi_{1\alpha}(x)e^{i\omega t}\mathrm{d}\omega+\sum_{\beta=1}^{\infty}\int_{-\infty}^{\infty}W_{2\beta}(\omega)\psi_{2\beta}(x)e^{i\omega t}\mathrm{d}\omega. \tag{5.49}$$

Substituting Eqs. (5.42) and (5.49) in Eq. (5.13), we obtain

$$W_{j\alpha}(\omega)=\frac{Q_{j\alpha}}{\rho A\left(\omega_{j\alpha}^2+i\varepsilon\omega-\omega^2\right)},\qquad \varepsilon=\frac{c}{\rho A} \tag{5.50}$$

ε being the modal bandwidth. The mean square of the transverse displacement takes the form

$$E\left[w^2(x,t)\right]=S_1(x)+S_2(x)+S_{12}(x) \tag{5.51}$$

where

$$S_1(x)=\sum_{\alpha=1}^{\infty}\sum_{\beta=1}^{\infty}\psi_{1\alpha}(x)\psi_{1\beta}(x)\psi_{1\alpha}(a)\psi_{1\beta}(a)\Phi_{1\alpha1\beta}, \tag{5.52}$$

$$S_2(x)=\sum_{\alpha=1}^{\infty}\sum_{\beta=1}^{\infty}\psi_{2\alpha}(x)\psi_{2\beta}(x)\psi_{2\alpha}(a)\psi_{2\beta}(a)\Phi_{2\alpha2\beta}, \tag{5.53}$$

$$S_{12}(x) = \sum_{\alpha=1}^{\infty}\sum_{\beta=1}^{\infty}\left[\psi_{1\alpha}(x)\psi_{2\beta}(x)\psi_{1\alpha}(a)\psi_{1\beta}(a)\Phi_{1\alpha 2\beta}\right. \\ \left. + \psi_{2\alpha}(x)\psi_{1\beta}(x)\psi_{2\alpha}(a)\psi_{1\beta}(a)\Phi_{2\alpha 1\beta}\right] \quad (5.54)$$

In Eqs. (5.52–5.54)

$$\Phi_{j\alpha,k\beta}(\omega) = \int_{-\infty}^{\infty} H_{j\alpha}(-\omega)H_{k\beta}(\omega)S_p(\omega)\mathrm{d}\omega \quad (5.55)$$

where $H_{j\alpha}(\omega)$ is the frequency response function for the αth mode in the jth group

$$H_{j\alpha}(\omega) = \frac{1}{2\rho AL\left(\omega_{j\alpha}^2 + i\varepsilon\omega - \omega^2\right)}. \quad (5.56)$$

In complete analogy, we obtain the expression for the mean square of the transverse velocity $v(x,t) = \partial w/\partial t$:

$$E\left[v^2(x,t)\right] = T_1(x) + T_2(x) + T_{12}(x) \quad (5.57)$$

where $T_1(x)$, $T_2(x)$, and $T_{12}(x)$ are defined by the following expressions:

$$T_1(x) = \sum_{\alpha=1}^{\infty}\sum_{\beta=1}^{\infty}\psi_{1\alpha}(x)\psi_{1\beta}(x)\psi_{1\alpha}(a)\psi_{1\beta}(a)\Phi'_{1\alpha 1\beta} \quad (5.58)$$

$$T_2(x) = \sum_{\alpha=1}^{\infty}\sum_{\beta=1}^{\infty}\psi_{2\alpha}(x)\psi_{2\beta}(x)\psi_{2\alpha}(a)\psi_{2\beta}(a)\Phi'_{2\alpha 2\beta} \quad (5.59)$$

$$T_{12}(x) = \sum_{\alpha=1}^{\infty}\sum_{\beta=1}^{\infty}\left[\psi_{1\alpha}(x)\psi_{2\beta}(x)\psi_{1\alpha}(a)\psi_{2\beta}(a)\Phi'_{1\alpha 2\beta}\right. \\ \left. + \psi_{2\alpha}(x)\psi_{1\beta}(x)\psi_{2\alpha}(a)\psi_{1\beta}(a)\Phi'_{2\alpha 1\beta}\right] \quad (5.60)$$

with

$$\Phi'_{j\alpha,k\beta}(\omega) = \int_{-\infty}^{\infty} \omega^2 H_{j\alpha}(-\omega)H_{k\beta}(\omega)S_{\mathrm{p}}(\omega)\mathrm{d}\omega. \quad (5.61)$$

Bearing in mind the expression (5.18) for the spectral density of point loading, we obtain for $\Phi'_{j\alpha,k\beta}(\omega)$ the following expression:

$$\Phi'_{j\alpha,k\beta}(\omega_c) = \frac{S_0}{4\rho^2 A^2 L^2} \int\limits_{-\omega_c}^{\omega_j} \frac{\omega^2 \mathrm{d}\omega}{\left(\omega_{j\alpha}^2 - i\omega\varepsilon - \omega^2\right)\left(\omega_{k\beta}^2 + i\omega\varepsilon - \omega^2\right)}. \tag{5.62}$$

The final result, obtained after integration in Eq. (5.62), reads

$$\Phi'_{j\alpha,k\beta}(\omega_c) = \frac{S_0}{4\rho^2 A^2 L^2} \chi'\left(\omega_{j\alpha}, \omega_{k\beta}; \omega_c\right) \tag{5.63}$$

where

$$\chi'\left(\omega_{j\alpha'}\omega_{k\beta}; \omega_c\right) = \frac{F'\left(\omega_{j\alpha'}\omega_{k\beta}; \omega_c\right) + F'\left(\omega_{k\beta}, \omega_{j\alpha}; \omega_c\right)}{\left(\omega_{j\alpha}^2 - \omega_{k\beta}^2\right)^2 + 2\varepsilon^2\left(\omega_{j\alpha}^2 + \omega_{k\beta}^2\right)} \tag{5.64}$$

and

$$\begin{aligned} F'\left(\omega_{ja}, \omega_{k,\beta}; \omega_c\right) = &\frac{\omega_{j\alpha}^2\left(\omega_{j\alpha}^2 - \omega_{j\beta}^2\right) + \varepsilon^2\left(\omega_{j\alpha}^2 + \omega_{j\beta}^2\right)/2}{2\left(\omega_{j\alpha}^2 - \varepsilon^2/4\right)^{1/2}} \\ &\times \ln \frac{\omega_c^2 + \omega_{j\alpha}^2 - 2\omega_c\left(\omega j_{j\alpha}^2 - \varepsilon^2/4\right)^{1/2}}{\omega_c^2 + \omega_{j\alpha}^2 + 2\omega_c\left(\omega_{j\alpha}^2 - \varepsilon^2/4\right)^{1/2}} \\ &+ \varepsilon\left(\omega_{j\alpha}^2 + \omega_{k\beta}^2\right)\left\{\tan^{-1}\left[\frac{\omega_c - \left(\omega_{j\alpha}^2 - \varepsilon^2/4\right)^{1/2}}{\varepsilon/2}\right]\right. \\ &\left. + \tan^{-1}\left[\frac{\omega_c + \left(\omega_{j\alpha}^2 - \varepsilon^2/4\right)^{1/2}}{\varepsilon/2}\right]\right\}. \end{aligned} \tag{5.65}$$

Note that this expression is also obtained from formula (5.29) by Elishakoff et al. (1979) formal substitution of $\omega_{j\alpha} \to \omega_j, \omega_\beta \to \omega_k$.

It is interesting to correlate the random vibration response of the substructures, in particular, simply supported beams on an elastic foundation, with that of the original structure—a two-span beam—on an elastic foundation. For the simply supported beam the normal modes are normalized as follows:

$$\psi_\alpha(x) = \frac{\sqrt{L} X_{1\alpha}(x)}{\sqrt{\sum\limits_0^L X_{1\alpha}^2 \mathrm{d}x}} = \sqrt{2} \sin \frac{\alpha \pi x}{L} \tag{5.66}$$

and the mean-square expanse reads

$$E\left[v^2\right]_1 = \sum_{a=1}^{\infty}\sum_{\beta=1}^{\infty}\frac{S_0}{A^2L^2}\psi_\alpha(x)\psi_\beta(x)\psi_\alpha(a)\psi_\beta(a)\chi'\left(\omega_{1a},\omega_{1\beta};\omega_c\right) \tag{5.67}$$

with $\chi'\left(\omega_{1\alpha},\omega_\beta;\omega_c\right)$ as for Eq. (5.64). Comparison of Eqs. (5.67) and (5.58) yields

$$E\left[v^2\right]_1 = 4T_1(x) \tag{5.68}$$

Note also that Eq. (5.67) is similar to Eq. (5.34) by Elishakoff et al. (1979), where axisymmetric random response of a cylindrical shall be considered. Instead of shell thickness, Eq. (5.36) contains the beam cross-sectional area; the expression for the natural frequencies are also similar. For a beam with rectangular cross section, for example, with width b and depth h, Eq. (5.36) yields

$$\omega_{1\alpha}^2 = \frac{E}{\rho L^2}\left[\frac{L^2}{AE} + \frac{h^2}{L^2}\frac{(\alpha\pi)^4}{12}\right]. \tag{5.69}$$

It is seen that the above is analogues to Eq. (5.18) by Elishakoff et al. (1979); in particular, the natural frequencies of a beam on an elastic foundation with stiffness modulus χ, correspond to those of axisymmetric vibration of a cylindrical shell with radius R, such that formally

$$R^2 = AE/\chi. \tag{5.70}$$

The difference thus lies in the factor $1 - v^2$, being ν the Poisson's ratio, in the shell modulus, absent in the beam case. For a simply supported clamped beam, the normal modes are normalized as follows:

$$\psi_\alpha(x) = \frac{\sqrt{L}X_{2\alpha}(x)}{\sqrt{\int_0^L X_{2\alpha}^2\mathrm{d}x}}, \tag{5.71}$$

and the mean-square response is

$$E\left[v^2\right]_2 = \sum_{\alpha=1}^{\infty}\sum_{\beta=1}^{\infty}\frac{S_0}{A^2L^2}\psi_\alpha(x)\psi_\beta(x)\psi_\alpha(a)\psi_\beta(a)\chi'\left(\omega_{2\alpha},\omega_{2\beta};\omega_c\right) \tag{5.72}$$

with $\chi'\left(\omega_{2\alpha},\omega_{2\beta};\omega_c\right)$ as for Eq. (5.64). Comparison of Eqs. (5.72) and (5.58) yields

$$E\left[v^2\right]_2 = 4T_2(x). \tag{5.73}$$

The formulas of this section can be converted to dimensionless form via the following notations:

$$\xi = \frac{x}{L}; \quad \eta = \frac{a}{L}; \quad \Omega = \frac{\omega L}{(E/\rho)^{1/2}}; \quad \Omega_c = \frac{\omega_c L}{(E/\rho)^{1/2}}; \quad B = \frac{\varepsilon L}{(E/\rho)^{1/2}}.$$

$$d_v^2 = E\left[v^2\right]\rho^2 A^2 L(E/\rho)^{1/2}/S_0; \quad K = \chi L^2/AE \tag{5.74}$$

The dimensionless mean-square velocity d_v^2 takes the form

$$d_v^2 = \frac{1}{4}\sum_{j=1}^{\infty}\sum_{a=1}^{\infty}\sum_{k=1}^{2}\sum_{k=1}^{\infty}\psi_{j\alpha}(\xi)\psi_{k\beta}(\xi)\psi_{j\alpha}(\eta)\psi_{k\beta}(\eta)I_{j\alpha,k\beta}(\Omega_e) \tag{5.75}$$

where the dimensionless integral equals

$$I_{j\alpha,k\beta}(\Omega_c) = \frac{\rho^2 A^2 L(E/\rho)^{1/2}}{S_0}\Phi'_{j\alpha,k\beta}(\omega_c) = \int_{-\infty}^{\infty}\frac{\Omega^2}{\left(\Omega_{j\alpha}^2 - i\Omega B - \Omega^2\right)\left(\Omega_{kj}^2 - i\Omega B - \Omega^2\right)}\mathrm{d}\Omega. \tag{5.76}$$

Note that d_v^2 can also be written as follows:

$$d_v^2 = s_1 + s_2 + s_3 + s_4 + s_5 \tag{5.77}$$

where

$$s_1 = \frac{1}{4}\sum_{\alpha=1}^{\infty}\psi_{1\alpha}^2(\xi)\psi_{1\alpha}^2(\eta)I_{1\alpha,1\alpha}(\Omega_c)$$

$$s_2 = \frac{1}{4}\sum_{\alpha=1}^{\infty}\psi_{2\alpha}^2(\xi)\psi_{2\alpha}^2(\eta)I_{2\alpha,2\alpha}(\Omega_c)$$

$$s_3 = \frac{1}{4}\sum_{\substack{\alpha=1\\ \alpha\neq\beta}}^{\infty}\sum_{\beta=1}^{\infty}\psi_{1\alpha}(\xi)\psi_{1\beta}(\xi)\psi_{1\alpha}(\eta)\psi_{1\beta}(\eta)I_{1\alpha1\beta}(\Omega_c)$$

$$s_4 = \frac{1}{4}\sum_{\substack{a=1\\ a\neq\beta}}^{\infty}\sum_{\beta=1}^{\infty}\psi_{2\alpha}(\xi)\psi_{2\beta}(\xi)\psi_{2\alpha}(\eta)\psi_{2\beta}(\eta)I_{2\alpha,2\beta}(\Omega_c)$$

$$s_5 = \frac{1}{4}\sum_{a=1}^{\infty}\sum_{\beta=1}^{\infty}\left[\psi_{1\alpha}(\xi)\psi_{1\beta}(\xi)\psi_{2\alpha}(\eta)\psi_{2\beta}(\eta)I_{1\alpha,2\beta}(\Omega_c)\right.$$
$$\left. + \psi_{2\alpha}(\xi)\psi_{2\beta}(\xi)\psi_{1\alpha}(\eta)\psi_{1\beta}(\eta)I_{2\alpha,1\beta}(\Omega_c)\right] \tag{5.78}$$

S_1 and S_2 cover the autocorrelation between modes of the first and second groups, respectively; S_3 and S_4 represent correlations between different modes of the same group, while S_5 represents inter-group correlations.

Note also that

$$\begin{aligned} s_1 + s_3 &= T_1(x)\rho^2 A^2 L(E/\rho)^{1/2}/S_0 \\ s_2 + s_4 &= T_2(x)\rho^2 A^2 L(E/\rho)^{1/2}/S_0 \ . \\ s_5 &= T_{12}(x)\rho^2 A^2 L(E/\rho)^{1/2}/S_0 \end{aligned} \tag{5.79}$$

For a simply supported beam, the non-dimensional square velocity is obtained from Eq. (5.67)

$$\begin{aligned} d_{v(1)}^2 &= E\left[v^2\right]_1 \rho^2 A^2 L(E/\rho)^{1/2}/S_0 \\ &= \sum_{a=1}^{\infty}\sum_{\beta=1}^{\infty} \psi_\alpha(\xi)\psi_\beta(\xi)\psi_\alpha(\eta)\psi_\beta(\eta) I_{1\alpha 1\beta}(\Omega_c) = 4(s_1 + s_3) \end{aligned} \tag{5.80}$$

with $d_{1(1)} = 4s_1$ and $d_{2(1)} = 4s_3$.

For a single-span simply supported clamped beam the non-dimensional square velocity is obtained from Eq. (5.72)

$$\begin{aligned} d_{v(2)}^2 &= E\left[v^2\right]_2 \rho^2 A^2 L(E/\rho)^{1/2}/S_0 \\ &= \sum_{g=1}^{\infty}\sum_{\theta=1}^{\infty} \psi_\alpha(\zeta)\psi_\beta(\xi)\psi_\alpha(\eta)\psi_\beta(\eta) I_{2\alpha,2\beta}(\Omega_c) = 4(s_2 + s_4) \end{aligned} \tag{5.81}$$

with $d_{1(2)} = 4s_2$ and $d_{2(2)} = 4s_4$.

5.6 Numerical Results and Discussion

In the calculations which follow, the excitation is located at $\eta = a/L = 0.3$. The dimensionless modal bandwidth B is taken as 0.02. The dimensionless foundation modulus K is set at 4, which, in view of the analogy given by Eq. (5.70), corresponds to a shell with $R/L = 0.5$, the very case considered by Elishakoff et al. (1979). Three geometries of a two-span beam on an elastic foundation is considered, with h/L ratio set at 0.01, 0.001, and 0.003.

The fourth case is a two-span beam without an elastic foundation (corresponding to $R/L \to \infty$ by Elishakoff et al. (1979) with $h/L = 0.01$. In all cases, the dimensionless excitation cutoff frequency was set at $\Omega_c = 2\pi$. The number N_c of modes in the excitation band is defined such that Ω_c lies between N_c^{th} and $(N_t + 1)^{th}$ natural frequencies. For the cases with $h/L = 0.01$ (both $K = 4$ and $K = 0$), $N_c = 28$ (14 modes of the first group by Elishakoff et al. (1979) and 14 modes of the second group); the number of modes in the band is $N_c = 88$ for $h/L = 0.001$ and $N_c = 162$ for $h/L = 0.0003$; in the latter two cases again, equal numbers of modes of the first and second groups appear within the excitation band.

To illustrate the behavior of the spectrum of natural frequencies Ω_j(odd j's corresponds to the natural frequencies of the first group, even ones to those of the second group), the local overlap r_j, defined by Elishakoff et al. (1979) is introduced

$$r_j = \frac{B}{\Omega_j - \Omega_{j-1}} \quad j = 2, 3, \ldots . \tag{5.82}$$

The integrals I_{jk} for the two-span beam on the elastic foundation with modulus $K = 4$ and $h/L = 0.01$ are tabulated in Tables 5.1 and 5.2. Table 5.1 shows the clustering of the natural frequencies and the large local overlaps for the first eight natural frequencies and for the tenth natural frequency (note that in the single-span shell counterpart by Elishakoff et al. (1979) large overlaps occurred only for the three natural frequencies). Table 5.2 shows the behavior of I_{jk} up to $j = k = 32$. Comparison of I_{jk} with odd j and k with I_{jk} values of the single-span shell by Elishakoff et al. (1979) shows that the present I_{jk} values are a little higher (I_{13}, e.g. in present work, equals 153.188, whereas the corresponding single-span shell value is 152.471 by Elishakoff et al. (1979)) due to the fact that natural frequencies of the two-span beam are closer due to absence of the factor $1 - \nu^2$.

The mean-square velocity distribution over the length of the single-span beam is shown in Figs. 5.3, 5.4, 5.5, 5.6, 5.7 and 5.8. Figures 5.3, 5.4 and 5.5 are associated with beam simply supported at its ends, and Figs. 5.6, 5.7 and 5.8 with simply—supported—clamped beams.

In each case, the dimensionless mean-square velocity d_v^2 and its two components $d_{v(1)}^2(\xi)$ and $d_{y(2)}^2(\xi)$ are plotted as functions of the dimensionless axial coordinate $\xi = x/L$.

For the simply supported beam, the autocorrelation sum $d_{v(1)}^2(\xi)$ is non—negative and symmetric about the midpoint; the cross-correlation sum $d_{y(2)}^2(\xi)$ has no simple symmetry properties but its space average is zero.

For the clamped—simply supported beam neither $d_1(\xi)$ nor $d_2(\xi)$ is symmetric about the midpoint, as expected. For the beams without an elastic foundation (Figs. 5.3 and 5.6) the contribution of $d_2(\xi)$ is negligible and the distribution of d_v^2 is essentially that of $d_1(\zeta)$.

Introduction of the elastic foundation, however, increases the contribution of cross-correlation sums (Figs. 5.4, 5.5, 5.7 and 5.8). This sum introduces a substantial amount of asymmetry in the distribution of the mean-square velocity for the simply supported beam (Figs. 5.4 and 5.5). As is seen from Figs. 5.5 and 5.8, in some instances the cross-correlation sum $d_2(\xi)$ dominates the response at the drive point $\xi = \eta$; intensified response can also be observed at the image point $\xi = 1 - \eta$ (Fig. 5.5).

In the case of symmetric load ($\eta = 0.5$) for a simply supported beam (Fig. 5.9) both the components associated with modal autocorrelations and modal cross-correlations

Table 5.1 Values of natural frequencies Ω_j and local modal overlap ratios r_j ($K = 4$ and $h/L = 0.01$)

j	Ωj	r_j
1	2.0002	
2	2.0005	68.385
3	2.00324	7.2755
4	2.00519	10.2568
5	2.01637	1.78947
6	2.02251	3.2558
7	2.05129	0.694935
8	2.06515	1.44365
9	2.12305	0.3454
10	2.14864	0.781397
11	2.24767	0.201969
12	2.2888	0.486218
13	2.43906	0.133109
14	2.49854	0.336232
15	2.70645	0.0961927
16	2.78575	0.252223
17	3.05382	0.074606
18	3.15321	0.201241
19	3.48101	0.0610126
20	3.60002	0.168057
21	3.98556	0.0518743
22	4.12341	0.145085
23	4.56424	0.045369
24	4.72011	0.128313
25	5.21384	0.0405075
26	5.38699	0.115509
27	5.9316	0.0367235
28	6.12141	0.105368
29	6.71524	0.0336798
30	6.92122	0.0970952
31	7.56296	0.0311655
32	7.78473	0.090184

are symmetric with respect to the midpoint, apart from the presence of the elastic foundation.

For a clamped—simply supported beam (Fig. 5.10) no simple symmetry for both component sums is recognizable even if the point load is applied in the midpoint.

Table 5.2 Values of I_{jk} $(j \le k)$ for the two-span beam on elastic foundation with $K = 4$, $h/L = 0.01$ and $\Omega_c = 2\pi$ $(N_c = 28)$ (odd j's and k's correspond to the eigenfrequencies of the first group, even ones to those of the second group): (a) $j = 1, 2, \ldots, 11$; $k = 1, 2, \ldots, 11$; (b) $j = 1, 2, \ldots, 11$; $k = 12, 13, \ldots, 22$; (c) $j = 1, 2, \ldots, 11$; $k = 23, 24, \ldots, 32$; (d) $j = 12, 13, \ldots, 22$; $k = 12, 13, \ldots, 22$; (e) $j = 12, 13, \ldots, 22$; $k = 23, 24, \ldots, 32$; (f) $j = 23, 24, \ldots, 32$; $k = 23, 24, \ldots, 32$

j	*k*										
	1	2	3	4	5	6	7	8	9	10	11
(a)											
1	156.738	156.704	153.188	147.587	94.657	69.644	20.533	13.267	3.716	2.460	0.677
2		156.738	153.820	148.521	96.022	70.671	20.741	13.380	3.735	2.471	0.680
3			156.737	155.258	109.447	81.118	22.856	14.507	3.919	2.576	0.703
4				156.738	119.357	89.421	24.452	15.390	4.057	2.655	0.719
5					156.734	143.194	38.453	22.268	4.993	3.170	0.823
6						156.737	50.813	27.993	5.639	3.512	0.887
7							156.724	105.808	10.982	6.020	1.270
8								156.736	16.402	8.182	1.521
9									156.734	59.207	3.600
10										156.734	5.812
11											156.731
j	*k*										
	12	13	14	15	16	17	18	19	20	21	22
(b)											
1	0.408	−0.020	−0.094	−0.225	−0.251	−0.302	−0.314	−0.342	−0.350	−0.373	−0.382
2	0.410	−0.019	−0.094	−0.225	−0.251	−0.302	−0.314	−0.342	−0.350	−0.373	−0.382

(continued)

Table 5.2 (continued)

j	k										
	12	13	14	15	16	17	18	19	20	21	22
(b)											
3	0.424	−0.015	−0.091	−0.224	−0.250	−0.302	−0.314	−0.342	−0.350	−0.373	−0.382
4	0.435	−0.012	−0.089	−0.223	−0.249	−0.302	−0.314	−0.342	−0.350	−0.373	−0.382
5	0.499	0.005	−0.077	−0.219	−0.247	−0.301	−0.313	−0.342	−0.350	−0.373	−0.382
6	0.538	0.016	−0.070	−0.217	−0.245	−0.300	−0.312	−0.342	−0.350	−0.373	−0.382
7	0.763	0.071	−0.034	−0.205	−0.237	−0.297	−0.310	−0.341	−0.349	−0.374	−0.382
8	0.902	0.102	−0.014	−0.199	−0.232	−0.295	−0.309	−0.341	−0.349	−0.374	−0.382
9	1.909	0.279	0.096	−0.168	−0.211	−0.288	−0.30+	−0.339	0.348	−0.373	−0.382
10	2.790	0.394	0.163	−0.151	−0.200	−0.284	−0.301	−0.338	−0.347	−0.373	−0.382
1 1	29.687	1.398	0.642	−0.057	−0.140	−0.266	−0.289	−0.334	−0.345	−0.373	−0.382

j	k									
	23	24	25	26	27	28	29	30	31	32
(c)										
1	0.410	−0.422	−0.470	−0.493	−0.627	−0.749	−0.522	−0.446	−0.312	−0.283
2	−0.410	−0.422	−0.470	−0.493	−0.627	−0.749	−0.522	−0.446	−0.312	−0.283
3	−0.410	−0.422	−0.470	−0.493	−0.628	−0.749	−0.522	−0.446	−0.312	−0.283
4	−0.410	−0.422	−0.470	−0.493	−0.628	−0.750	−0.522	−0.446	−0.312	−0.283

(continued)

Table 5.2 (continued)

j	k									
	23	24	25	26	27	28	29	30	31	32
(c)										
5	−0.410	−0.422	−0.470	−0.493	−0.628	−0.750	−0.523	−0.446	−0.312	−0.283
6	−0.410	−0.422	−0.470	−0.493	−0.628	−0.750	−0.523	−0.446	−0.312	−0.283
7	−0.411	−0.423	−0.471	−0.494	−0.629	−0.752	−0.523	−0.446	−0.312	−0.283
8	−0.411	−0.423	−0.471	−0.495	−0.630	−0.752	−0.523	−0.447	−0.312	−0.283
9	−0.412	−0.424	−0.473	−0.496	−0.632	−0.755	−0.525	−0.447	−0.312	−0.283
10	−0.412	−0.424	−0.473	−0.497	−0.633	−0.758	−0.525	−0.447	−0.312	−0.282
11	−0.413	−0.425	−0.475	−0.499	−0.637	−0.762	−0.527	−0.448	−0.312	−0.282

j	k										
	12	13	14	15	16	17	18	19	20	21	22
(d)											
12	156.729	2.385	1.064	0.004	−0.103	−0.256	−0.282	−0.332	−0.343	−0.373	−0.382
13		156.724	15.600	0.516	0.161	−0.200	−0.246	−0.321	−0.336	−0.372	−0.382
14			156.722	1.081	0.397	−0.164	−0.224	−0.315	−0.332	−0.371	−0.382
15				156.714	9.030	0.147	−0.060	−0.280	−0.310	−0.390	−0.380
16					156.711	0.496	0.088	−0.257	−0.296	−0.363	−0.378
17						156.697	5.730	−0.050	−0.188	−0.342	−0.366

(continued)

Table 5.2 (continued)

j	*k*										
	12	13	14	15	16	17	18	19	20	21	22
(d)											
18							156.692	0.186	−0.087	−0.327	−0.358
19								156.671	3.902	−0.183	−0.285
20									156.662	−0.013	−0.213
21										156.656	2.777
22											156.609

j	*k*									
	23	24	25	26	27	28	29	30	31	32
(e)										
12	−0.414	−0.426	−0.476	−0.500	−0.639	−0.764	−0.528	−0.449	−0.312	−0.282
13	−0.454	−0.428	−0.480	−0.504	−0.645	−0.773	−0.531	−0.450	−0.311	−0.281
14	−0.416	−0.429	−0.481	−0.506	−0.648	−0.777	−0.532	−0.451	−0.311	−0.280
15	−0.418	−0.433	−0.487	−0.513	−0.659	−0.792	−0.537	−0.453	−0.310	−0.279
16	−0.419	−0.434	−0.489	−0.515	−0.663	−0.798	−0.539	−0.454	−0.309	−0.278
17	−0.420	−0.437	−0.498	−0.525	−0.680	−0.821	−0.546	−0.457	−0.307	−0.275
18	−0.420	−0.438	−0.501	−0.529	−0.687	−0.830	−0.548	−0.458	−0.306	−0.273

(continued)

Table 5.2 (continued)

j	k									
	23	24	25	26	27	28	29	30	31	32
(e)										
19	−0.412	−0.437	−0.512	−0.543	−0.713	−0.866	−0.558	−0.462	−0.301	−0.268
20	−0.403	−0.434	−0.516	−0.548	−0.723	−0.881	−0.562	−0.463	−0.299	−0.265
21	−0.305	−0.390	−0.525	−0.566	−0.764	−0.939	−0.574	−0.465	−0.289	−0.254
22	−0.179	−0.340	−0.526	−0.571	−0.781	−0.964	−0.578	−0.465	−0.284	−0.248
(f)										
23	156.541	1.990	−0.476	−0.568	−0.846	−1.064	−0.588	−0.458	−0.260	−0.223
24		156.508	−0.389	−0.543	−0.874	−1.109	−0.590	−0.452	−0.248	−0.2 II
25			156.341	1.283	−0.960	−1 301	−0.571	−0.403	−0.182	−0.146
26				156.241	−0.959	−1.389	−0.544	−0.365	−0.142	−0.108
27					155.318	−0.685	−0.132	0.038	0. 150	0.152
28						153.598	0.572	0.583	0.448	0.406
29							0.865	0.672	0.401	0.351
30								0.527	0.319	0.280
31									0.197	0.174
32										0.154

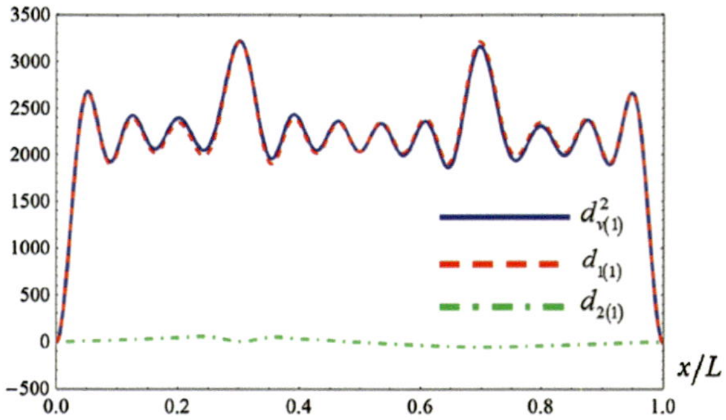

Fig. 5.3 Dimensionless mean-square velocity $d^2_{v(1)}$ and component sums $d_{1(1)}$ and $d_{2(1)}$ for single-span simply supported beam without elastic foundation ($\eta = 0.3$, $h/L = 0.01$, $\Omega_c = 2\pi$, $K = 4$, $N_c = 14$)

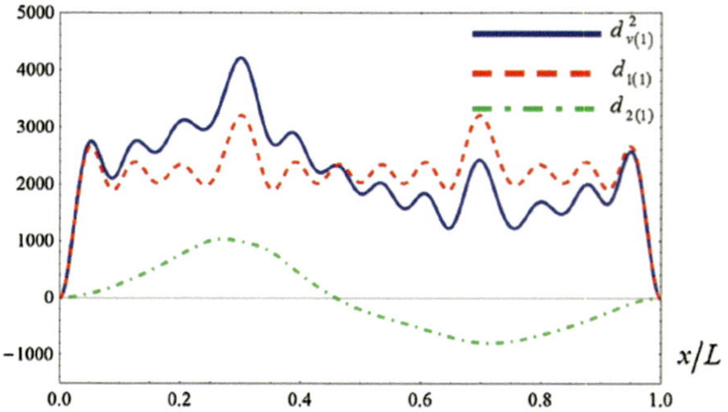

Fig. 5.4 Dimensionless mean-square velocity $d^2_{v(1)}$ and component sums $d_{1(1)}$ and $d_{2(1)}$ for single-span simply supported beam on elastic foundation ($\eta = 0.3$, $h/L = 0.01$, $\Omega_c = 2\pi$, $K = 4$, $N_c = 14$)

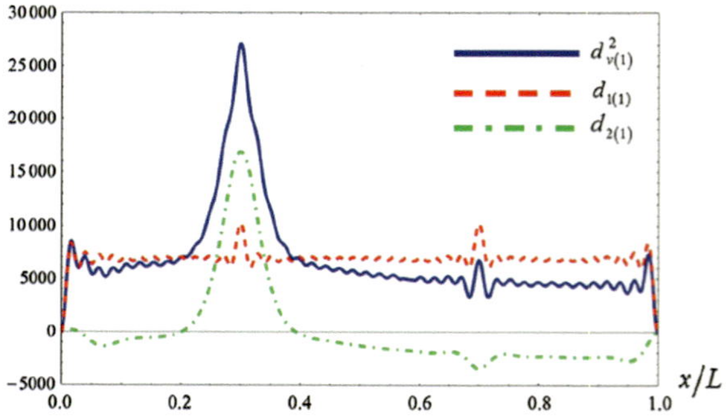

Fig. 5.5 Dimensionless mean-square velocity $d^2_{v(1)}$ and component sums $d_{1(1)}$ and $d_{2(1)}$ for single-span simply supported beam on elastic foundation ($\eta = 0.3$, $\Omega_c = 2\pi$, $K = 4$, $N_c = 44$)

Figures 5.11, 5.12 and 5.13 are associated with the two-span beams; Fig. 5.11 shows the distribution of the mean-square response velocity for the two-span beam without elastic foundation.

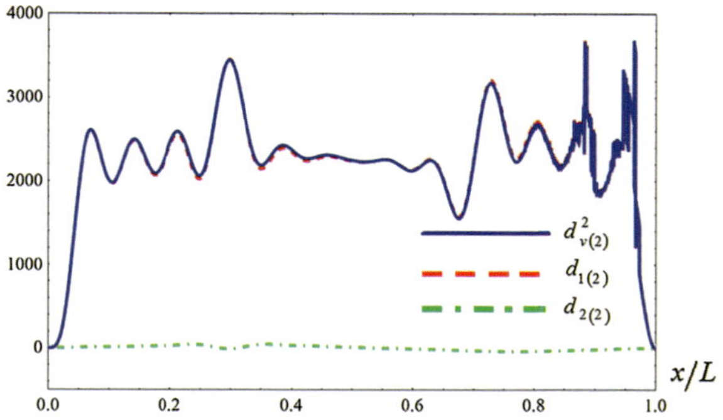

Fig. 5.6 Dimensionless mean-square velocity $d^2_{v(2)}$ and component sums $d_{1(2)}$ and $d_{2(2)}$ for single-span clamped simply supported beam without elastic foundation ($\eta = 0.3$, $h/L = 0.01$, $\Omega_c = 2\pi$, $N_c = 14$)

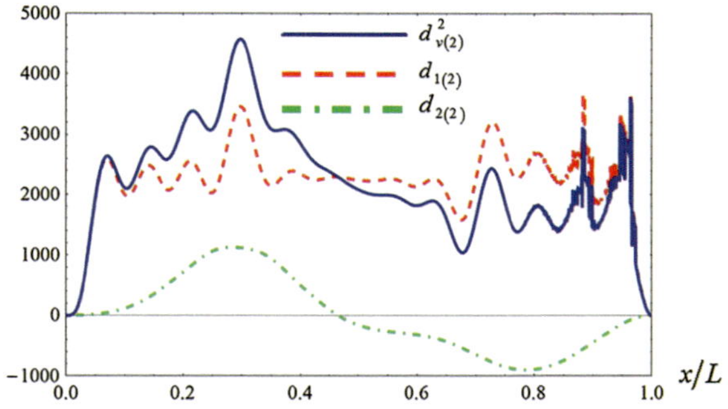

Fig. 5.7 Dimensionless mean-square velocity $d^2_{v(2)}$ and component sums $d_{1(2)}$ and $d_{2(2)}$ for single-span clamped simply supported beam on elastic foundation ($\eta = 0.3$, $h/L = 0.01$, $\Omega_c = 2\pi$, $K = 4$, $N_c = 14$)

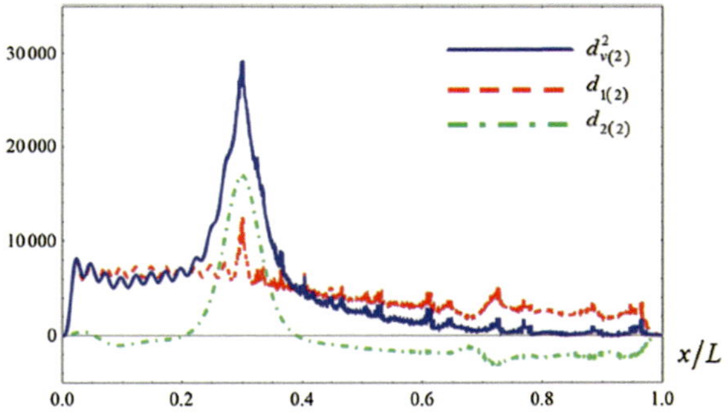

Fig. 5.8 Dimensionless mean-square velocity $d^2_{v(2)}$ and component sums $d_{1(2)}$ and $d_{2(2)}$ for single-span clamped simply supported beam on elastic foundation ($\eta = 0.3$, $h/L = 0.001$, $\Omega_c = 2\pi$, $K = 4$, $N_c = 44$)

The mean-square velocity is not symmetrically distributed about the midpoint of the structure ($\xi = 0$), due to the anti-symmetry property of the component s_5 which takes into account the inter-group correlation. Curve 1 is associated with d_v^2, and curve 2 with the autocorrelation sums ($S_1 + S_2$). Relative contribution of the cross-correlation sums ($S_3 + S_4 + S_5$) increases from almost zero percent for the single-span beams to about 9.45% at the driving section, the maximum difference being about 13% at section $\xi = 0.4$.

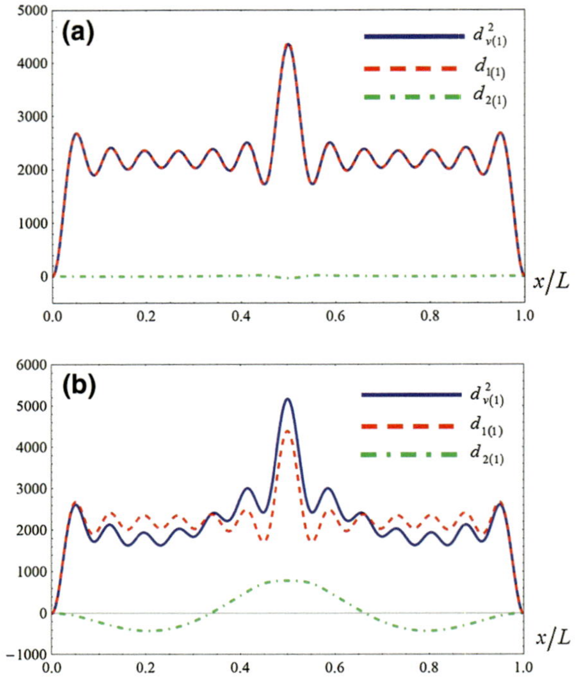

Fig. 5.9 Dimensionless mean-square velocity $d^2_{v(1)}$ and component sums $d_{1(1)}$ and $d_{2(1)}$ for single-span simply supported beam: **a** without elastic foundation ($\eta = 0.5$, $h/L = 0.01$, $\Omega_c = 2\pi$, $N_c = 14$); **b** on elastic foundation ($\eta = 0.5$, $h/L = 0.01$, $\Omega_c = 2\pi$, $K = 4$, $N_c = 14$)

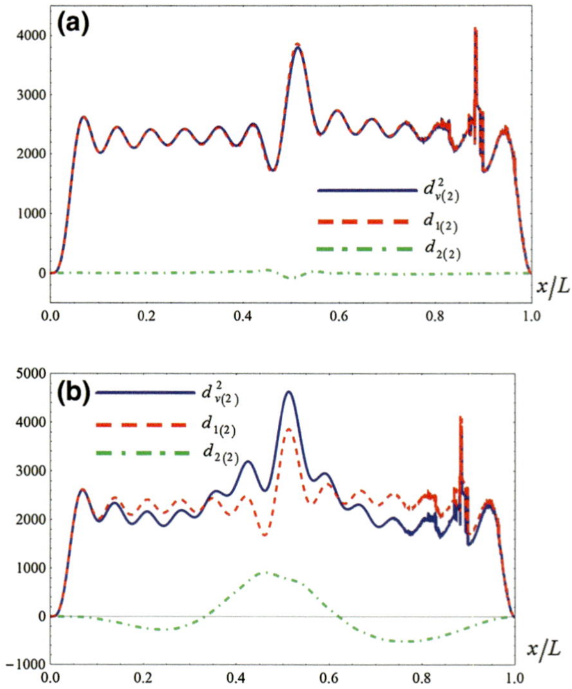

Fig. 5.10 Dimensionless mean-square velocity $d^2_{v(2)}$ and component sums $d_{1(2)}$ and $d_{2(2)}$ for single-span clamped simply supported beam: **a** without elastic foundation ($\eta = 0.5$, $h/L = 0.01$, $\Omega_c = 2\pi$, $K = 0$, $N_c = 14$); **b** on elastic foundation ($\eta = 0.5$, $h/L = 0.01$, $\Omega_c = 2\pi$, $K = 4$, $N_c = 14$)

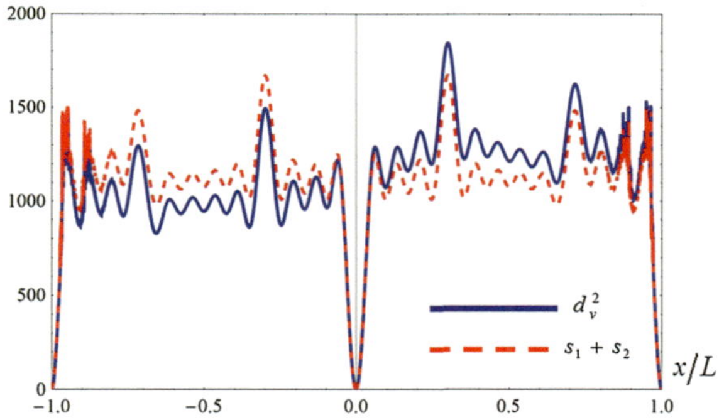

Fig. 5.11 Dimensionless mean-square velocity d_v^2 and its approximation, taking into account the correlation between like modes only, for two-span beam without elastic foundation ($\eta = 0.3$, $h/L = 0.01$ $\Omega_c = 2\pi$, $K = 0$, $N_c = 28$)

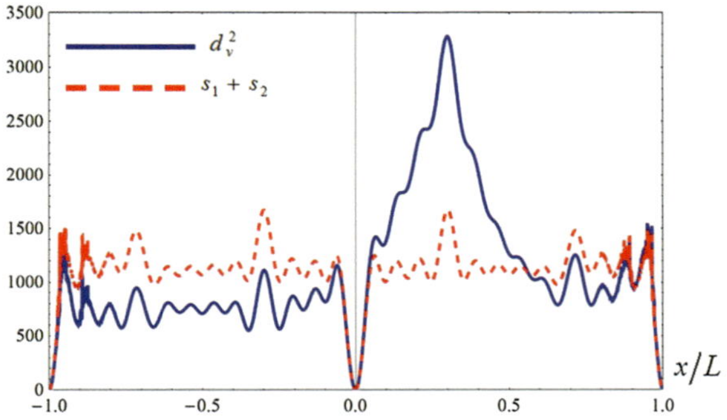

Fig. 5.12 Dimensionless mean-square velocity d_v^2 and its approximation, taking into account the correlation between like modes only, for two-span beam on elastic foundation ($\eta = 0.3$, $h/L = 0.01$, $\Omega_c = 2\pi$, $K = 4$, $N_c = 28$)

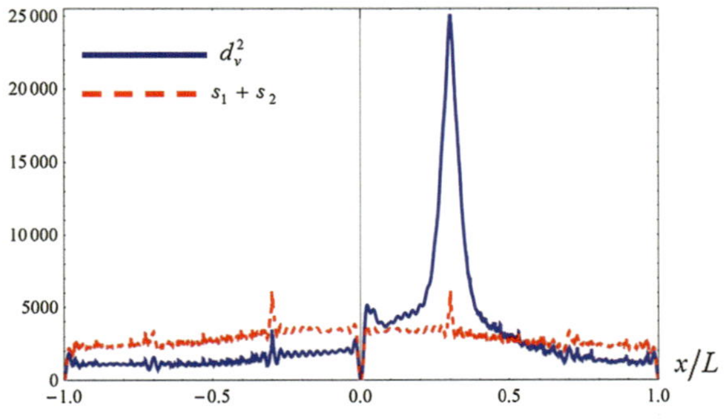

Fig. 5.13 Dimensionless mean-square velocity d_v^2 and its approximation, taking into account the correlation between like modes only, for two-span beam on elastic foundation ($\eta = 0.3$, $h/L = 0.001$, $\Omega_c = 2\pi$, $K = 4$, $N_c = 88$)

This is in agreement with the finding of Mercer (1965). He considered random vibrations of the nine-span beam simply supported at its ends, excited by boundary layer or jet noise.

The natural frequencies of the first band (nine natural frequencies range over close intervals from 500 to 612 Hz) were taken into account (for the two-span beam such an approach will imply use of the first two modes, X_{11} and X_{21} in the analysis).

This increase of the contribution of the cross-correlation terms stems from the clustering of natural frequencies (for the same h/L values, the two-span beam has approximately twice as much natural frequencies within the excitation band as the single-span beam).

Introduction of the elastic foundation further increases the role of cross-correlations (Figs. 5.12 and 5.13).

We observe that the contribution of the cross-correlation components increases to 49.2% at the driving section with non-dimensional parameter $H = 0.01$ and number of modes N_c in excitation band equals 28 until an increment of 77.7% in the same section for a structure with $H = 0.001$ and $N_c = 88$.

Figures 5.14 and 5.15 show the mean-square velocity response for the driving section located at $\eta = 0.5$ and at $\eta = 0.7$.

Their examination also readily shows the drastic increase in the contribution of cross-correlations in the overall response for increasing N_c.

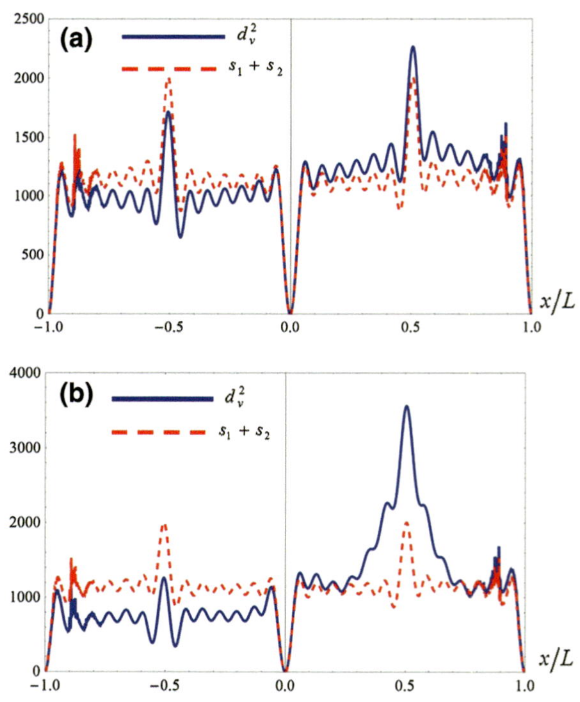

Fig. 5.14 Dimensionless mean-square velocity d_v^2 and its approximation, taking into account the correlation between like modes only, for two-span beam: **a** without elastic foundation ($\eta = 0.5$, $h/L = 0.01$, $\Omega_c = 2\pi$, $K = 0$, $N_c = 28$); **b** on elastic foundation ($\eta = 0.5$, $h/L = 0.01$, $\Omega_c = 2\pi$, $K = 4$, $N_c = 28$)

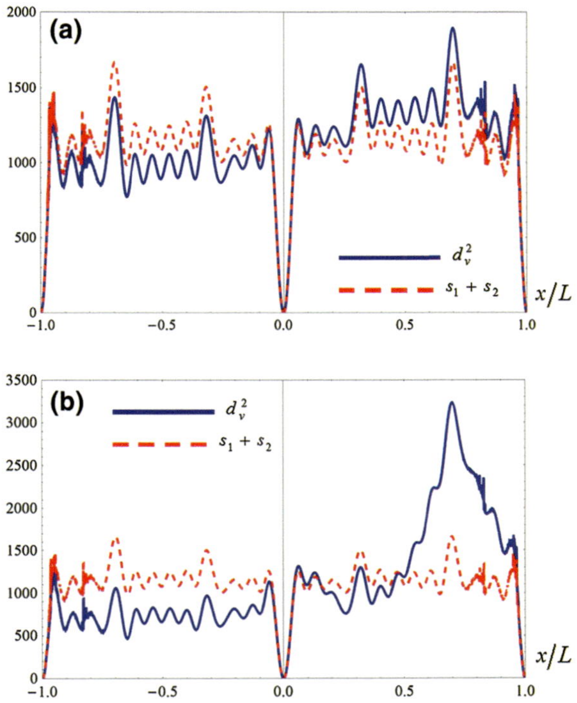

Fig. 5.15 Dimensionless mean-square velocity d_v^2 and its approximation, taking into account the correlation between like modes only, for two-span beam: **a** without elastic foundation ($\eta = 0.7$, $h/L = 0.01$, $\Omega_c = 2\pi$, $K = 0$, $N_c = 28$) **b** on elastic foundation ($\eta = 0.7$, $h/L = 0.01$, $\Omega_c = 2\pi$, $K = 4$, $N_c = 28$)

The contributions to the drive-section mean-square response for the single-span and two-span beams both without and with an elastic foundation are summarized in Tables 5.3, 5.4 and 5.5. For the driving-section located at $\eta = 0.3$, as the h/L parameter decreases from 0.01 to 0.003 the ratio of the true response to the prediction based on the autocorrelation sum alone, increases from 1.002 to 1.477 for the single-span beam without an elastic foundation, whereas in the presence of the elastic foundation this ratio changes from 1.31 to 4.3; of the same order of magnitude are these ratios for the clamped simply supported beams (Table 5.4). The situation rather changes for the two-span beam: for the absence of the elastic foundation the ratio $d_v^2/(s_1 + s_2)$ equals 2.69 ($h/L = 0.0003$), as against 9.6 for the same beam on an elastic foundation: thus the prediction based only on the correlations between the like modes of vibration constitutes only about 12% of true mean-square response. Further details can be found in the paper by Elishakoff and Santoro (2014).

Table 5.3 Components of driving-point response for simply supported beams with $K = 0$ or 4 ($\Omega_c = 2\pi$)

η	K	h/L	N_c	$d_{1(1)}$	$d_{2(1)}$	$d^2_{v(1)}$	$d^2_v/d_{1(1)}$
0.3	0	0.01	14	3212.5	7.1	3219.6	1.002
		0.001	44	10264.1	1814.3	12078.4	1.177
		0.0003	81	19071.6	9098.6	28170.2	1.477
	4	0.01	14	3211.43	999.98	4211.41	1.314
		0.001	44	10261	16836.5	27097.5	2.641
		0.0003	81	19064.4	62352.8	81417.1	4.217
0.5	0	0.01	14	4387.67	−36.184	4351.5	0.992
		0.001	44	13789.6	1104.13	14893.7	1.08
		0.0003	81	25696.4	6769.8	32466.3	1.263
	4	0.01	14	4386.4	782.32	5168.7	1.19
		0.001	44	13785.2	15336.9	29122.1	2.11
		0.0003	81	25685.3	58773	84458.4	3.29
0.7	0	0.01	14	3212.5	7.1	3219.6	1.002
		0.001	44	10264.1	1814.3	12078.4	1.177
		0.0003	81	19071.6	9098.58	28170.2	1.477
	4	0.01	14	3211.43	999.98	4211.4	1.311
		0.001	44	10261	16836.5	27097.5	2.641
		0.0003	81	19064.4	62352.8	81417.1	4.27

5.7 Conclusion

The effect of cross-correlation between the modes within each group, as well as between the modes belonging to different groups was shown to be significant for the two-span beams. The often reported symmetric distribution in the mean-square response by Crandall and his coworkers was shown to be lacking in the two-span beam. It can be conjectured that analogous results can be expected for multi-span beams and plates that form the often used elements in mechanical, ocean, and aerospace engineering.

Table 5.4 Components of driving-point response for clamped simply supported beams with $K = 0$ or 4 ($\Omega_c = 2\pi$)

η	K	h/L	N_c	$d_{1(1)}$	$d_{2(1)}$	$d^2_{v(2)}$	$d^2_v/d_{1(2)}$
0.3	0	0.01	14	3452.14	−10.89	3441.25	0.997
		0.001	44	12370.1	1888.92	14259	1.153
		0.0003	81	12375.1	10125.7	22500.8	1.818
	4	0.01	14	3451.14	1 116.27	4567.41	1.323
		0.001	44	12366.4	16798.1	29164.5	2.358
		0.0003	81	12373.2	63080.8	75454	6.098
0.5	0	0.01	14	3511.87	−33.39	3478.49	0.99
		0.001	44	6315.06	1921.61	8236.66	1.304
		0.0003	81	6315.19	9075.04	15390.2	2.437
	4	0.01	14	3510.56	844.517	4355.07	1.056
		0.001	44	6314.07	16906.8	23220.8	3.678
		0.0003	81	6314.24	56653.2	62967.5	9.97
0.7	0	0.01	14	3428.34	−15.08	3413.28	0.997
		0.001	44	4003.46	2106.74	6110.2	1.526
		0.0003	81	4003.47	8273.03	12276.5	3.066
	4	0.01	14	3424.71	921.645	4346.96	1.269
		0.001	44	4002.85	16562.7	20565.6	5.138
		0.0003	81	4002.87	36434.4	40437.3	10.10

Table 5.5 Components of driving-point response for two-span beams with $K = 0$ or 4 ($\Omega_c = 2\pi$)

η	K	h/L	N_c	s_1	s_2	s_3	s_4	s_5	d_v^2	$d_v^2/(s_1+s_2)$
0.3	0	0.01	28	806.74	863.1	1.8	−2.7	174.889	1844.56	1.1041
		0.001	88	2566.02	3088.71	453.6	470.69	2728.53	9307.52	1.65
		0.0003	162	4767.9	3089.96	2274.64	2547.16	8502.23	21181.9	2.696
	4	0.01	28	802.9	862.8	250	279.23	1082.9	3277.84	1.97
		0.001	88	2565.24	3087.81	4209.13	4198.4	11165.7	25225.3	4.5
		0.0003	162	4766.1	3089.48	15588.2	15824	36018.4	75286.9	9.6
0.5	0	0.01	28	1100.58	879.85	−9.046	−8.3	269.9	2232.99	1.13
		0.001	88	3447.4	1505.06	276.033	481.35	2840.51	8550.35	1.7
		0.0003	162	6424.11	1505.09	1692.45	2257.52	7215.21	19094.3	2.4
	4	0.01	28	1096.6	IS78.59	195.6	211.64	140.52	3522.94	1.78
		0.001	88	3446.3	1504.8	3834.22	4227.4	11047.7	24060.4	4.9
		0.0003	162	6421.33	1504.86	14694.1	13964.1	31619.6	68203.1	8.6
0.7	0	0.01	28	806.7	860.34	1.8	−3.7	229.6	1894.72	1.13
		0.001	88	2566.02	998	453.6	527.8	2494.64	7040.16	1.98
		0.0003	162	4767.9	998	2274.64	2062.74	6067.82	16171.7	2.8
	4	0.01	28	802.9	858.4	250	232.2	1089.08	3232.49	1.95
		0.001	88	2565.2	997.78	4209.13	4139.65	10272.1	22169	6.22
		0.0003	162	4766.1	997.78	15588.2	8911.45	24065.7	54409.2	9.43

References

Bolotin, V. V., & Elishakoff, I. (1971). Random vibrations of elastic shells containing an acoustic medium. *Mechanics of Solids, 6,* 99–107.

Crandall, S. H. (1973). Wide-band random vibrations of point-driven strings and plates. *Mekhanika Tvergodo Tela, 2,* 37–46. (in Russian).

Crandall S. H. (1974). Wide-band random vibration of structures. In *Proceedings of the 7th U.S. National Congress of Applied Mechanics* (pp. 131–138). New York: ASME Press.

Crandall S. H. (1977). Structured response patterns due to wide—Band random excitation. In B. L. Clarkson (Ed.), *Stochastic problems in dynamics* (pp. 366–389). London: Pitman.

Crandall, S. H. (1979a) Random vibration of one- and two-dimensional structures. In Paruchuri, R. Krishnaiah (Eds.), *Applications of statistics* (Proc. Symp. held at Wright State University, Dayton, Ohio, 1976) (pp. 1 – 82). Academic Press.

Crandall, S. H. (1979b). Random vibration of vehicles and structures. In *Proceedings of the Seventh Canadian Congress of Applied Mechanics*, Sherbrooke (pp. 1 – 12).

Crandall S. H. (1980). Private communication, 7 January.

Crandall, S. H., & Kulvets, A. P. (1978). Vibrations of homogeneous beams and plates under correlated stationary random forces. *Mashinovedenie, 1,* 16–21. (in Russian).

Crandall S. H., & Kulvets A. P. (1979). Random vibration of mechanisms on plates. In *Proceedings of the Fifth World Congress on Theory of Machines and Mechanisms* (pp. 1568–1571). New York: ASME Press.

Crandall, S. H., & Lotz, R. (1971). On the coupling loss factor in statistical energy analysis. *Journal of the Acoustical Society of America, 49*(1B), 352–356.

Crandall, S. H., & Mark, W. D. (1963). *Random vibration in mechanical systems.* New York: Academic Press.

Crandall, S. H., & Wittig, L. E. (1972). Chladni's patterns for random vibration of a plate. In G. Herrmann & N. Perrone (Eds.), *Dynamic response of structures* (pp. 35–71). New York: Pergamon Press.

Davies, H. G., & Khandoker, S. I. (1982). Random point excitation of coupled beams. *Journal of Sound and Vibration, 84*(4), 557–562.

Elishakoff, I. (1971a). On evaluation of spectral densities of external forces for plates and shells. In V. V. Bolotin & A. Chyras (Eds.), *Problems of reliability in structural mechanics* (pp. 37–44). Vilnius, Luthuania. (in Russian).

Elishakoff, I. (1971b). On the determination of spectral densities of external forces of plates and shells. In V. V. Bolotin, A. A. Čyras, (Eds.), *Reliability problems in structural mechanics* (pp. 36–44). Vilnius.

Elishakoff, I. (1971c). Vibration fields in cylindrical shells, excited by random forces. Ph.D. thesis, Department of Dynamics and Strength of Machines, Moscow Power Engineering Institute and State University (in Russian).

Elishakoff, I. (1973). Random vibrations of a two-span beams. *Israel Journal of Technology, 11*(5), 317–320.

Elishakoff, I., Van Zanten, A. Th., & Crandall, S. H. (1979). Wide-band random axisymmetric vibration of cylindrical shells. *Journal of Applied Mechanics, 46*(2), 417–423.

Elishakoff, I., & Santoro, R. (2014). Random vibration of a point-driven two-span beam on an elastic foundation. *Archive of Applied Mechanics, 84,* 355–374.

Itao, K. & Crandall, H. (1978). Wide-band random vibration of circular plates. *Journal of Engineering Design,100*(4), 690-695.

Lin Y. K. (1967). *Probabilistic theory of structural dynamics*, Chap. 7 (pp. 203–252). New York: McGraw-Hill.

Mercer, C. A. (1965). Response of a multi-supported beam to a random pressure field. *Journal of Sound and Vibration, 2,* 293–306.

Powell, A. (1958). On the response of structures to random pressures and to jet noise in particular. In S. H. Crandall (Ed.), *Random vibration*, Chap. 8 (pp. 187–229). New York: Wiley.

Chapter 6
Random Vibrations of Orthotropic Plates

6.1 Introductory Comments

Investigations dealing with random vibrations of beams, plates, and shells are numerous (the references in monographs of Crandall and Mark (1963), Lin (1967), Bolotin (1979), and Parkus (1969) contain an appropriate bibliography). Early investigators of this problem considered only a hypothetical field (white noise) having a constant power spectral density, and adopted the simplifying assumption that the structure is infinitely large. This model proved equivalent to an integral approximation of the space-averaged spectral density in the high frequency range, where the boundary conditions practically have no influence on the probabilistic behavior of the structure (see Bolotin, 1963; Elishakoff & Efimtsoff, 1972; Elishakoff, 1975a, 1975b; Palmov, 1965; Powell, 1958; Ribner, 1966; Strawderman, 1969a, 1969b). However, even though it yields a satisfactory estimate of the average structural response, it is impracticable in problems of structural integrity, where an integral probabilistic characterization is inadequate (Dowell, 1971).

The finite-system model also attracted considerable attention in the literature. Most of these works are confined to simply supported plates (Clarkson, 1968; Crocker, 1969; Dyer, 1959; Strawderman, 1969a, 1969b; Wagner & Bhat, 1970)—an assumption which greatly simplifies the analysis, permitting the solutions to be expressed in closed forms. However, although these analytical results give better agreement with experiment than those obtained with the infinite-plate assumption, they tend to overestimate the response of a realistic plate.

Since realistic boundary conditions of a plate are intermediate between all-round simple support and all-round clamping, the latter problem seems to be of importance. This chapter considers random vibrations of orthotropic plates under these conditions. An approximate analysis proposed coincides with the exact analytical approach when applied to a plate with at least one pair of opposite edges simply supported.

I. Elishakoff, *Dramatic Effect of Cross-Correlations in Random Vibrations of Discrete Systems, Beams, Plates, and Shells*,
https://doi.org/10.1007/978-3-030-40394-2_6

6.2 Analytical Formulation

We confine this discussion to orthotropic plates with a plane of elastic symmetry parallel to the middle plate. The correlation scale of the external normal load acting on the plate is assumed to be much larger, and the mean-square deviation of the displacements much smaller than the thickness of the plate, so that the problem may be considered in a statistically linear setting. In these circumstances, the equation for the force vibration of the plate is given by

$$D_x\left(\frac{\partial}{\partial t}\right)\frac{\partial^4 w}{\partial x^4}+2H\left(\frac{\partial}{\partial t}\right)\frac{\partial^4 w}{\partial x^2\partial y^2}+D_y\left(\frac{\partial}{\partial t}\right)\frac{\partial^4 w}{\partial y^4}+2\varrho\varepsilon\frac{\partial w}{\partial t}+\varrho\frac{\partial^2 w}{\partial t^2}=q(x,y,t) \quad (6.1)$$

where h is the thickness of the plate, $w(x, y, t)$—normal displacement of the plate, $q(x, y, t)$—the random pressure intensity, x and y—the space coordinates, and t—time. $D_x(\partial/\partial t)$, $D_y(\partial/\partial t)$, $H(\partial/\partial t)$ are viscoelastic operators of the time-differentiation parameter $\partial/\partial t$, corresponding to the flexural rigidities of the orthotropic plate, ϱ is the mass density.

The differential Eq. (6.1) is supplemented by one of the alternative sets of boundary conditions, corresponding to all-round clamping and all-round simple support:

$$w=\frac{\partial w}{\partial n}=0, \quad (6.2)$$

$$w=\Delta w=0, \quad (6.3)$$

respectively, where Δ denotes the Laplace operator on the middle surface and n denotes the direction normal to the contour.

The problem is formulated as follows: Given the probability behavior of the function $q(x, y, t)$, find that of the displacement and/or stress field in the plate, dispensing with the initial conditions.

For the stochastic boundary value problem (6.1)–(6.2), an approximate method will be presented, the first step of which consists (as in the normal-mode approach) in expanding the given and sought functions in stochastic Fourier integrals and in series in terms of the modes of the undamped plate. For finding the frequencies and mode shapes, we assume in Eq. (6.1) $q(x, y, t) \equiv 0$, $D(\partial/\partial t) = \text{const.}$, $D_y(\partial/\partial t) = \text{const}$, $H(\partial/\partial t)$, and $w(x, y, t) = W(x, y)e^{i\omega t}$, where ω is an eigenfrequency. Equation (6.1) becomes

$$D_x\frac{\partial^4 W}{\partial x^4}+2H\frac{\partial^4 W}{\partial x^2\partial y^2}+D_y\frac{\partial^4 W}{\partial y^4}-\varrho\omega^2 W=0. \quad (6.4)$$

The eigenvalues and eigenfrequencies of the system (6.3), (6.4) are

$$\omega_{mn}=\left(\frac{H}{\varrho}\right)^{1/2}\left[\alpha\left(\frac{m\pi}{a}\right)^4+2\left(\frac{m\pi}{a}\right)^2\left(\frac{n\pi}{b}\right)^1+\beta\left(\frac{n\pi}{b}\right)^4\right]^{1/2}, \quad (6.5)$$

$$\alpha = \frac{D_x}{H},\ \beta = \frac{D_y}{H},\ W_{mn} = X_m(x)Y_n(y),\ X_m(x) = \sin\frac{m\pi x}{a},$$
$$Y_n(y) = \sin\frac{n\pi y}{b} \quad (m, n = 1,\ 2,\ 3,\ \ldots) \tag{6.6}$$

where m and n are positive integers, representing the numbers of half-waves in the x- and y-directions, respectively, and a and b are the plate sides in the x- and y-directions, respectively.

An approximate method for determining the eigenfrequencies and eigenfunctions of the system (6.2), (6.4) was presented by Elishakoff (1974a, 1974b); it consisted in approximating the eigenfrequency in a form analogous to (6.5)

$$\omega_{mn} = \left(\frac{H}{\varrho}\right)^{1/2}\left[\alpha\left(\frac{p\pi}{a}\right)^4 + 2\left(\frac{p\pi}{a}\right)^2\left(\frac{q\pi}{b}\right)^2 + \beta\left(\frac{q\pi}{b}\right)^4\right]^{1/4} \tag{6.7}$$

(where p and q are analogous of the numbers of half-waves in the coordinate directions, $m = [p]$ and $n = [q]$ representing the actual values of these numbers), and solving two auxiliary problems of the Voight–Levy type. The relevant transcendental equations:

$$1 - c_1C_1 + \frac{1}{\alpha}\left(2\frac{p^2}{q^2}\frac{1}{\alpha\delta^2} + \frac{p^4}{q^4}\frac{1}{\delta^4}\right)^{-1/2} s_1S_1 = 0, \tag{6.8}$$

$$1 - c_2C_2 + \frac{1}{\beta}\left(2\frac{q^2}{p^2}\frac{\delta^2}{\beta} + \frac{q^4}{p^4}\delta^4\right)^{-1/2} s_2S_2 = 0, \tag{6.9}$$

where

$$c_1 = \cos p\pi,\ s_1 = \sin p\pi,\ C_1 = \cosh\theta\pi,\ S_1 = \sinh\theta\pi, \tag{6.10}$$

$$c_2 = \cos q\pi,\ s_2 = \sin q\pi,\ C_2 = \cosh\chi\pi,\ S_2 = \sinh\chi\pi,$$

$$\chi = \left(\frac{2}{\beta}\frac{p^2}{\delta^2} + q^2\right)^{1/2},\quad \theta = \left(\frac{2}{\alpha}q^2\delta^2 + p^2\right)^{1/2},\ \delta = \frac{a}{b}, \tag{6.11}$$

and the approximate mode shapes are

$$X_{[p][q]}(x) = \cosh\frac{\theta\pi x}{a} - \cos\frac{p\pi x}{a} - \frac{C_1 - c_1}{S_1 - \theta p^{-1}s_1}\left(\sinh\frac{\theta\pi x}{a} - \frac{\theta}{p}\sin\frac{p\pi x}{a}\right), \tag{6.12}$$

$$Y_{[p][q]}(y) = \cosh\frac{\chi\pi y}{b} - \cos\frac{q\pi y}{b} - \frac{C_2 - c_2}{S_2 - \theta q^{-1}s_2}\left(\sin\frac{\chi\pi y}{b} - \frac{\chi}{q}\sin\frac{q\pi y}{b}\right). \tag{6.13}$$

For a square plate, $\delta = 1$, the equations of the paper by Elishakoff (1974a, 1974b) are a particular case.

As was shown by Elishakoff (1974a, 1974b), the functions $X_{[p][q]}(x)$ and $Y_{[p][q]}(y)$ do not satisfy the differential Eq. (6.4), but do satisfy all the boundary conditions (6.2); following Elishakoff (1974a, 1974b), we use the product

$$\psi_{[q][p]}(x, y) = X_{[p][q]}(x) Y_{[p][q]}(y) \tag{6.14}$$

as an approximate mode shape.

The function $q(x, y, t)$ is assumed to be represented by following Fourier integral

$$q(x, y, t) = \int_{-\infty}^{\infty} Q(x, y, \omega) e^{i\omega t} d\omega, \tag{6.15}$$

where $Q(x, y, \omega)$ is a random complex-valued spectrum of $q(x, y, t)$. For the plate displacement $w(x, y, t)$, we assume likewise

$$w(x, y, t) = \int_{-\infty}^{\infty} W(x, y, \omega) e^{i\omega t} d\omega \tag{6.16}$$

with $W(x, y, \omega)$—the random complex-valued spectrum of $w(x, y, t)$.

Substitution of (6.15) and (6.16) in (6.1) yields

$$D_x(i\omega)\frac{\partial^4 W}{\partial x^4} + 2H(i\omega)\frac{\partial^4 W}{\partial x^2 \partial y^2} + D_y(i\omega)\frac{\partial^4 W}{\partial y^4} + 2\varrho\varepsilon i\omega W - \varrho\omega^2 W = Q, \tag{6.17}$$

where

$$D_x(i\omega) = e^{-i\omega t} D_x(\partial/\partial t) e^{i\omega t}, \quad D_y(i\omega) = e^{-i\omega t} D_y(\partial/\partial t) e^{i\omega t}, \tag{6.18}$$

$$H(i\omega) = e^{-i\omega t} H(\partial/\partial t) e^{i\omega t}.$$

Furthermore, we assume that

$$D_x(i\omega) = D_x(1 + i\eta), \; D_y(i\omega) = D_y(1 + i\eta), \; H(i\omega) = H(1 + i\eta), \varepsilon \equiv 0 \tag{6.19}$$

η being the damping factor (hysteresis damping).

In consequence, Eq. (6.17) becomes

$$(1 + i\eta)\left[D_x\frac{\partial^4 W}{\partial x^4} + 2H\frac{\partial^4 W}{\partial x^2 \partial y^2} + D_y\frac{\partial^4 W}{\partial y^4}\right] - \varrho\omega^2 W = Q. \tag{6.20}$$

The Fourier transforms $W(x, y, \omega)$ and $Q(x, y, \omega)$ are expressed as the following truncated series:

$$W(x, y, \omega) = \sum_{[p]=1}^{P} \sum_{[q]=1}^{R} W_{[p][q]}(\omega)\psi_{[p][q]}(x, y), \tag{6.21}$$

$$Q(x, y, \omega) = \sum_{[p]=1}^{P} \sum_{[q]=1}^{R} W_{[p][q]}(\omega)\psi_{[p][q]}(x, y). \tag{6.22}$$

We substitute first Eqs. (6.21) and (6.22) in (6.17), to obtain

$$\sum_{[p]=1}^{P} \sum_{[q]=1}^{R} \Bigg[W_{[p][q]}(1+i\eta)\Bigg(D_x \frac{d^4 X_{[p][q]}}{dx^4} Y_{[p][q]} + 2H \frac{d^2 X_{[p][q]}}{dx^2} \frac{d^2 Y_{[p][q]}}{dy^2} + D_y \frac{d^4 Y_{[p][q]}}{dy^4} X_{[p][q]} - \varrho\omega^2 X_{[p][q]} Y_{[p][q]} \Bigg) - Q_{[p][q]} X_{[p][q]} Y_{[p][q]} \Bigg] = 0. \tag{6.23}$$

Next, we apply Bubnov–Galerkin's method, namely, multiply both sides of Eq. (6.22) by $X_{[r][s]}Y_{[r][s]}$, and integrate over the surface of the plate. We obtain

$$\sum_{[p]=1}^{P} \sum_{[q]=1}^{R} A_{[p][q][r][s]} Q_{[p][q]} - B_{[p][q][r][s]} W_{[p][q]} = 0$$
$$[r] = 1, 2, \ldots, P; \quad [q] = 1, 2, \ldots, R, \tag{6.24}$$

where

$$A_{[p][q][r][s]} = \big(X_{[p][q]}, X_{[r][s]}\big)\big(Y_{[p][q]}, Y_{[s][r]}\big), \tag{6.25}$$

$$\begin{aligned} B_{[p][q][r][s]} =& (1+i\eta)[D_x \big(X^{IV}_{[p][q]}, X_{[p][q]}\big)\big(Y_{[p][q]}, Y_{[r][s]}\big) \\ &+ 2H \big(X^{II}_{[p][q]}, X_{[r][s]}\big)\big(Y^{II}_{[p][q]}, Y_{[r][s]}\big) \\ &+ D_y \big(Y^{IV}_{[p][q]}, Y_{[r][s]}\big)\big(X_{[p][q]}, X_{[r][s]}\big) \\ &- \varrho\omega^2 \big(X_{[p][q]}, X_{[r][s]}\big)\big(Y_{[p][q]}, Y_{[r][s]}\big), \end{aligned} \tag{6.26}$$

$$\big(X_{[p][q]}, X_{[r][s]}\big) = \int_0^a X_{[p][q]}(x) X_{[r][s]}(x) dx, \tag{6.27}$$

$$(Y_{[p][q]}, Y_{[r][s]} = \int_0^b Y_{[p][q]}(y) Y_{[r][s]}(y) dy. \tag{6.28}$$

(ϕ, ψ) is a scalar product, and the primes denote differentiation with respect to the appropriate variable.

For each fixed frequency ω, the set of Eq. (6.23) may be represented in the following matrix form

$$B\bar{W} = A\bar{Q} \tag{6.29}$$

where W and Q are column matrices:

$$[\bar{W}] = [W_{11}, W_{12}, \ldots, W_{1R}, W_{21}, W_{22}, \ldots, W_{2R}, \ldots, W_{P1}, W_{P2}, \ldots, W_{PR}]_{PR\times 1}, \tag{6.30}$$

$$[\bar{Q}] = [Q_{11}, Q_{12}, \ldots, Q_{1R}, Q_{21}, Q_{22}, \ldots, Q_{2R}, \ldots, Q_{P1}, Q_{P2}, \ldots, Q_{PR}]_{PR\times 1}, \tag{6.31}$$

and B and A are rectangular matrices.

The solution of Eq. (6.24) may be represented as

$$\bar{W}_i = \sum_{j=1}^{P\times R} \left(B^{-1}A\right)_{ij} \bar{Q}_j, \; i = 1, 2, \ldots, P \times R, \tag{6.32}$$

where $(A^{-1}B)_{ij}$ are elements of the matrix $B^{-1}A$, with i and j replacing, as is seen from Eqs. (6.30) and (6.31), the relevant pairs of indices, corresponding to the integer parts of the numbers of half-waves in the coordinate directions.

Equation (6.32) may be rewritten in the following form

$$W_{jk} = \sum_{\mu=1}^{P} \sum_{\nu=1}^{R} C_{\mu\nu jk} Q_{\mu\nu}, \quad j = [p], \quad k = [q]. \tag{6.33}$$

Further, we resort to the normal-mode approach. The spectral density of the plate displacement at any point x, y of the middle surface can be expressed as follows:

$$S_w(x, y, \omega) = \sum_{j,g,\mu,\xi}^{P} \sum_{k,h,\nu,\eta}^{R} C^*_{\mu\nu jk} C_{\xi\eta gh} S_{\mu\nu\xi\eta} \psi_{jk}(x, y) \psi_{gh}(x, y) \tag{6.34}$$

where $S_{\mu\nu\xi\eta}$ are the cross-spectral densities of the generalized forces.

By virtue of Eq. (6.22), we have for the cross-spectral density of external forces

$$S_q\left(x', y', x'', y'', \omega\right) = \sum_{\mu}^{P} \sum_{\nu}^{R} \sum_{\xi}^{P} \sum_{\eta}^{R} S_{\mu\nu\xi\eta}(\omega) \psi_{\mu\nu}\left(x', y'\right) \psi_{\xi\eta}\left(x'', y''\right). \tag{6.35}$$

We multiply both sides of this equation by $\psi_{jk}\left(x', y'\right)\psi_{gh}\left(x'', y''\right)$ and integrate over the surface of the plate. We obtain

$$D_{jkgh} = \sum_{\mu}\sum_{\nu}\sum_{\xi}\sum_{\eta} S_{\mu\nu\xi\eta} E_{jk\mu\nu} F_{gh\xi\beta} \tag{6.36}$$

where

$$D_{jkgh} = \int_0^a\int_0^b\int_0^a\int_0^b S_q\left(x', y'; x'', y''; \omega\right)\psi_{jk}\left(x', y'\right)\psi_{gh}\left(x'', y''\right)dx'dy'dx''dy'', \tag{6.37}$$

$$E_{jk\mu\nu} = \int_0^a\int_0^b \psi_{\mu\nu}\left(x', y'\right)\psi_{jk}\left(x', y'\right)dx', dy', \tag{6.38}$$

$$F_{gh\xi\eta} = \int_0^a\int_0^b \psi_{\xi\eta}\left(x'', y''\right)\psi_{gh}\left(x'', y''\right)dx''dy''.$$

Equation (6.36) permits determination of the cross-spectral densities of generalized forces $S_{\mu\nu\xi\eta}(\omega)$.

Let $S_q\left(x', y'; x'', y''; \omega\right)$ have the property

$$S_q\left(x', y; x'', y''; \omega\right) = \phi(\omega) S_x\left(x'' - x', \omega\right) S_y\left(y'' - y'; \omega\right). \tag{6.39}$$

This representation of the cross-spectral density of the pressure corresponds to Clarkson's (1968) representation. By virtue of Eqs. (6.37) and (6.38), we have

$$D_{jkgh} = \phi(\omega) D^I_{jkgh} D^{II}_{jkgh}, \tag{6.40}$$

$$D^I_{jkgh} = \int_0^a\int_0^a S_x\left(x'' - x', \omega\right) X_{jk}\left(x'\right) X_{gh}\left(x''\right) dx'dx'', \tag{6.41}$$

$$D^{II}_{jkgh} = \int_0^b\int_0^b S_y\left(y'' - y', \omega\right) Y_{jk}\left(y'\right) Y_{gh}\left(y''\right) dy'dy'', \tag{6.42}$$

$$E_{jk\mu\nu} = \int_b^a X_{\mu\nu}\left(x'\right) X_{jk}\left(x'\right) dx' \int_0^b Y_{\mu\nu}\left(y'\right) Y_{jk}\left(y'\right) dy', \tag{6.43}$$

$$F_{gh\xi\eta} = \int_0^a X_{gh}\left(x''\right) X_{\xi\eta}\left(x''\right) dx'' \int_0^b Y_{gh}\left(y''\right) Y_{\xi\eta}\left(y''\right) dy''. \tag{6.44}$$

Functions D^I_{jkgh} and D^{II}_{jkgh} can be interpreted as coefficients of the cross-spectral densities of the generalized forces in the x- and y-directions, respectively.

We can represent them as follows:

$$\mathrm{Re} D^I_{jkgh} = \int_0^a \mathrm{Re} S_x(\xi, \omega) M(\xi, \omega) d\xi, \tag{6.45}$$

$$\mathrm{Im} D_{jkgh} = \int_0^a \mathrm{Im} S_x(\xi, \omega) N(\xi, \omega) d\xi, \tag{6.46}$$

where

$$M(\xi,\omega)=\int_{\xi}^{a-\xi}\left[X_{jk}\left(\eta-\frac{\xi}{2}\right)X_{gh}\left(\eta+\frac{\xi}{2}\right)+X_{jk}\left(\eta+\frac{\xi}{2}\right)X_{gh}\left(\eta-\frac{\xi}{2}\right)\right]d\eta, \tag{6.47}$$

$$N(\xi,\omega)=\int_{\xi}^{a-\xi}[X_{jk}\left(\eta-\frac{\xi}{2}\right)X_{gh}\left(\eta+\frac{\xi}{2}\right)-X_{jk}\left(\eta+\frac{\xi}{2}\right)X_{gh}\left(\eta-\frac{\xi}{2}\right)d\eta. \tag{6.48}$$

Formulae for D^{II}_{jkgh} are obtained by replacing $a\to b$, $X\to Y$, $S_x\to S_y$.

Under weak damping, the local maxima of the spectral density of the displacements are close to the eigenfrequencies of the undamped structure. These are determined by the $B^{-1}A$ matrix singularity conditions (for $\eta\equiv 0$):

$$\det\left[B^{-1}A\right]=0. \tag{6.49}$$

This equation is also obtainable directly by applying Galerkin's procedure in determining the eigenfrequencies of an orthotropic plate, clamped all round, with the product $X_{[p][q]}Y_{[p][q]}(y)$ used as a comparing function.

Further simplification is possible by means of the following assumption:

$$B_{[p][q][r][s]}\ll B_{[p][q][p][q]},\quad A_{[p][q][r][s]}\ll A_{[p][q][p][q]} \tag{6.50}$$

for $p\neq r$ and $q\neq s$. This is equivalent to the assumption that the matrices A and B are nearly diagonal.

Equation (6.33) takes the form

$$W_{[p][q]}=\frac{A_{[p][q][p][q]}}{B_{[p][g][p][q]}}Q_{[p][q]} \tag{6.51}$$

and Eq. (6.49) is now given by

$$\mathrm{Re}B_{[p][q][p][q]}=0 \tag{6.52}$$

which may be interpreted as the frequency equation in Galerkin's procedure with only diagonal terms retained in the appropriate determinant.

The eigenfrequency (wave number relationship was postulated by Eq. (6.7)). Equation (6.52) then represents the second "improved" approximation of the eigenfrequency

$$\begin{aligned}\varrho\bar{\omega}^{-2}_{[p][q]}&=D_x\left(X^{IV}_{[p][q]},X_{[p][q]}\right)\left(X_{[p][q]},X_{[p][q]}\right)^{-1}\\&+2H\left(X''_{[p][q]},X_{[p][q]}\right)\left(Y''_{[p][q]},Y_{[p][q]}\right)A^{-1}_{[p][q][p][q]}\\&+D_y\left(Y^{IV}_{[p][q]},Y_{[p][q]}\right)\left(Y_{[p][q]},Y_{[p][q]}\right)^{-1}.\end{aligned} \tag{6.53}$$

Taking into account Eq. (6.11) Elishakoff's (1974a, 1974b) paper and its counterpart for the second eigenvalue problem, we arrive at the following final expression for the eigenfrequency in the second approximation

$$\bar{\omega}^2_{[p][q]} = \omega^2_{[p][q]} + 2\left(\frac{H}{\varrho}\right)\left[\left(\frac{p\pi}{a}\right)^2\left(\frac{q\pi}{a}\right)^2 + \frac{\left(X''_{[p][q]}, X_{[p][q]}\right)\left(Y''_{[p][q]}, Y_{[p][q]}\right)}{\left(X_{[p][q]}, X_{[p][q]}\right)\left(Y_{[p][q]}, Y_{[p][q]}\right)}\right.$$
$$\left. + \left(\frac{p\pi}{a}\right)^2\frac{\left(Y''_{[p][q]}, Y_{[p][q]}\right)}{\left(Y_{[p][q]}, Y_{[q][q]}\right)} + \left(\frac{q\pi}{b}\right)^2\frac{\left(X''_{[p][q]}, X_{[p][q]}\right)}{\left(X_{[p][q]}, X_{[p][p]}\right)}\right] \quad (6.54)$$

where $\omega_{[p][q]}$ is determined by Eq. (6.7).

Equation (6.29) takes the form

$$W_{[p][q]} = \frac{Q_{[p][q]}}{\varrho(1 + i\eta)\bar{\omega}_{[p][q]} - \varrho\omega^2}. \quad (6.55)$$

The formula (6.34) changes accordingly to

$$S_w(x, y, \omega) = \sum_{[p]}\sum_{[q]}\sum_{[r]}\sum_{[s]} S_{w_{[p][q]}w_{[r][s]}}(\omega)X_{[p][q]}Y_{[p][q]}X_{[r][s]}Y_{[r][s]} \quad (6.56)$$

where the cross-spectral densities of the displacements $S_{w_{[p][q]}w_{[r][s]}}$ are

$$S_{w_{[p][q]}w_{[r][s]}} = \frac{S_{[p][q][r][s]}}{\varrho^2\left[\bar{\omega}^2_{[p][q]} - \omega^2 + i\eta\bar{\omega}^2_{[p][q]}\right][\bar{\omega}^2_{[r][s]} - \omega^2 - i\eta\bar{\omega}^2_{[r][v]}}. \quad (6.57)$$

The quadruple summation in Eq. (6.56) can be carried out as follows:

$$S_w(x, y, \omega) = \sum_{[p]}\sum_{[q]} S_{w_{[p][q]}w_{[p][q]}}\psi^2_{[p][q]}(x, y)$$
$$+ \sum_{[p]}{}'\sum_{[q]}{}'\sum_{[r]}{}'\sum_{[s]}{}' S_{w_{p][q]}w_{[r][s]}}\psi_{[p][q]}\psi_{[r][s]}$$
$$+ \sum_{[p]}{}''\sum_{[q]}{}''\sum_{[r]}{}''\sum_{[s]}{}'' S_{w_{[p][q]}w_{[r][s]}}\psi_{[p][q]}\psi_{[r][v]}. \quad (6.58)$$

The first sum refers to the cross-correlations between like vibration modes; the second and third refer to those between unlike modes; namely, the contribution of groups of vibration modes $[p][q]$ and $[r][s]$ corresponding to the multiple eigenfrequencies $\omega_{[p][q]} = \omega_{[r][s]}$ and $\omega_{[p][q]} \neq \omega_{[r][s]}$, respectively, (in indication whereof the summation sign is supplemented by single and double primes). It is generally accepted that the cross-terms are very small except where the eigenfrequencies are very close together. The non-multiple eigenfrequencies are widely separated and if additionally the damping is weak, the coupling between the unlike modes is negligible; this is not the case with the second sum.

By virtue of Eq. (6.57), the mean-square value of the displacement $\sigma_w^2(x, y)$ is given by

$$\sigma_w^2(x, y) = \sum_{[p]}\sum_{[q]} \varrho^{-2}\psi_{[p][q]}^2 \int_{-\infty}^{\infty} \frac{S_{[p][q][p][q]}(\omega)d\omega}{\left(\bar{\omega}_{[p][q]}^2 - \omega^2\right)^2 + \eta^2\bar{\omega}_{[p][q]}^4}$$
$$+ \sum_{[p]}'\sum_{[q]}'\sum_{[r]}'\sum_{[s]}' \varrho^{-2}\psi_{[p][q]}\psi_{[r][s]} \int_{-\infty}^{\infty} \frac{S_{[p][g][r][s]}(\omega)d\omega}{\left((\bar{\omega})_{[[p][q]]} - \omega^2\right)^2 + \eta^2\bar{\omega}_{[p][q]}^{-4}}. \tag{6.59}$$

Next, we evaluate the integrals in Eq. (6.59), approximating for the strong inequality $\eta \ll 1$, by Laplace's asymptotic method. The principal contribution to the integral is determined by the value of the integrand function close to the values of ω, which maximizes it. The later can only be the minimum points $\omega = \bar{\omega}_{[p][q]}$ of the frequency function in the denominator if the cross-spectral densities of the generalized forces are smooth functions. Thus

$$\int_{-\infty}^{\infty} \frac{S_{[p][q][r][s]}d\omega}{\left(\bar{\omega}_{pq}^2 - \omega^2\right)^2 + \eta^2\bar{\omega}_{[p][q]}^2} = \frac{\pi\left[1 + \left(1 + \eta^2\right)^{1/2}\right]^{1/2}}{2^{1/2}\eta(1 + \eta)^{1/2}\bar{\omega}_{[p][q]}^3} S_{[p][q][r][s]}\left(\bar{\omega}_{[p][q]}\right). \tag{6.60}$$

Therefore, for $\eta \ll 1$, Eq. (6.59) becomes

$$\sigma_w^2(x, y) = \sum_{[p]}\sum_{[q]} \frac{\pi\psi_{[p][q]}^2 S_{[p][q][p][q]}\left(\bar{\omega}_{[p][q]}\right)}{\varrho^2\eta\bar{\omega}_{[p][q]}^3}$$
$$+ \sum_{[p]}'\sum_{[q]}'\sum_{[p]}'\sum_{[q]}' \frac{\pi\psi_{[p][q]}\psi_{[r][s]}}{\varrho^2\eta\bar{\omega}_{[p][q]}^3} S_{[p][q][r][s]}\left(\bar{\omega}_{[p][q]}\right). \tag{6.61}$$

Approximation Eq. (6.50) was used by Dimentberg (1961) with the mode shapes and eigenfrequencies according to Bolotin's method (1961), and by Chyu and Au-Yang (1972) with the mode shape and eigenfrequencies according to Warburton's method (1964).

6.3 Random Vibrations of Plates with Opposite Edges Simply Supported

Consider a plate with the $x = 0$, $x = a$ edges simply supported, and the $y = 0$, $y = b$ edges clamped. Instead of Eqs. (6.8) and (6.9), we have

$$1 - C_2 c_2 + \frac{1}{\beta}\left(2\frac{q^2}{p^2}\frac{\delta^2}{\beta} + \frac{q^4}{p^4}\delta^4\right)^{-1/2} S_2 s_2 = 0, \tag{6.62}$$

$$s_1 = 0. \tag{6.63}$$

The latter has a solution $p = m = 1, 2, 3, \ldots$ and Eq. (6.62) takes the form

$$1 - C_2 c_2 + \frac{1}{\beta}\left(2\frac{q^2}{m^2}\frac{\delta^2}{\beta} + \frac{q^4}{m^4}\delta^4\right)^{-1/2} S_2 s_2 = 0 \tag{6.64}$$

where χ is now defined as

$$\chi = \left(\frac{2}{\beta}\frac{m^2}{\delta^2} + q^2\right)^{1/2}. \tag{6.65}$$

The eigenfrequencies $\omega_{[p][q]}$ take the form

$$\omega_{[p][q]} = \left(\frac{H}{\varrho}\right)^{1/2}\left[\alpha\left(\frac{m\pi}{a}\right)^4 + 2\left(\frac{m\pi}{a}\right)^2\left(\frac{q\pi}{b}\right)^2 + \beta\left(\frac{q\pi}{b}\right)^4\right]^{1/2} \tag{6.66}$$

and the mode shapes are given by

$$W(x, y) = X_{m[q]}(x)Y_{m[q]}(y), \quad X_{m[q]} \equiv \frac{m\pi x}{a} \tag{6.67}$$

where $Y_{m[q]}$ is determined by Eq. (6.12) with χ as per (6.65). From the construction of the approximate solutions described in Sect. 6.2, it is clear that the above solutions are exact (see, for example, Huffington and Hoppmann (1959). Furthermore, the expression in parenthesis in Eq. (6.22) takes the form

$$\begin{aligned} D_x \frac{d^4 X_{m[q]}}{dx^4} Y_{m[q]} + 2H \frac{d^2 X_{m[q]}}{dx^2}\frac{d^2 Y_{m[q]}}{dy^2} + D_y \frac{d^4 Y_{m[q]}}{dy^4} X_{m[q]} \\ -\varrho\omega^2 X_{m[q]} Y_{m[q]} = \varrho\left(\omega^2_{m[q]} - \omega^2\right) X_{m[q]} Y_{m[q]} \end{aligned} \tag{6.68}$$

and, because of the orthogonality of the eigenfunctions

$$B_{[p][q][r][s]} = B_{m[q]m'[s]} = B_{m[q]}\delta_{mm'}\delta_{[q][s]}, \tag{6.69}$$

$$A_{[p][q][r][s]} = A_{m[v]m'[s]} = A_{m[q]}\delta_{mm'}\delta_{[q][s]}, \tag{6.70}$$

where $\delta_{mm'}$ and $\delta_{[q][s]}$ are Kronecker deltas. Equation (6.55) becomes

$$W_{m[q]} = \frac{Q_{m[q]}}{\varrho(1+i\eta)\omega_{m[q]}^2 - \varrho\omega^2}. \tag{6.71}$$

All the formulae (6.51)–(6.59) become exact for this case, and can be used for our analysis by replacing $[p] \to m$, $[r] \to m'$. Expression (6.41), for example, takes the form

$$S_{m[q]m'[s]} = \psi(\omega)\Phi_{mm'}\Psi_{[q][s]}. \tag{6.72}$$

Functions $\Phi_{mm'}$ and $\Psi_{[q][s]}$ may be interpreted as coefficients of the cross-spectral densities of simply-supported and clamped beam strips of length a and b, respectively.

Expression (6.47) for $m = m'$ reduces to Powell's formula (1958):

$$M(\xi,\omega) = (a-x)\cos\frac{m\pi x}{a} - \frac{a}{m\pi}\sin\frac{m\pi x}{a}. \quad N(\xi,\omega) = 0 \tag{6.73}$$

and for $m \neq m'$ to Lin's formula (1967):

$$M(\xi,\omega) = \frac{1-(-1)^{m+m_1+1}}{\pi\left(m_1 - m^2\right)}\left(m_1\sin\frac{m\pi x}{a} - m\sin\frac{m_1\pi x}{a}\right), \tag{6.74}$$

$$N(\xi,\omega) = \frac{1-(-1)^{m+m_1}}{\pi\left(m_1^2 - m^2\right)}\left(m_1\sin\frac{m\pi x}{a} + m\sin\frac{m_1\pi x}{a}\right). \tag{6.75}$$

Analogous of Powell's and Lin's formulae for the general vector differential equation were obtained by Elishakoff (1971a, 1971b, 1971c).

For the plate, simply supported at all edges, normal modes are given by Eq. (6.5). The probabilistic characteristics of displacements are obtainable from the formulae of the preceding section, with X_m replacing $X_{[p][q]}$ and Y_n replacing $Y_{[p][q]}$. The expression in parentheses in Eq. (6.54) is then identically zero, and $\bar{\omega}_{[p][q]} = \omega_{[p][q]}$, the latter being given by the exact expression (6.5) for the eigenfrequencies ω_{mn}. Moreover,

$$A_{mnm'n'} = A_{mnmn}\delta_{mm'}\delta_{nn'}, \quad A_{mnmn} = \frac{ab}{4}, \tag{6.76}$$

$$B_{mnm'n'} = B_{mnmn}\delta_{mm'}\delta_{nn'}. \tag{6.77}$$

Instead of Eq. (6.71), we have

$$W_{mn} = \frac{Q_{mn}}{\varrho(1 + i\eta)\omega_{mn}^2 - \varrho\omega^2}. \tag{6.78}$$

The subsequent steps are as before.

6.4 Conclusion

New analytical results for random vibrations of clamped orthotropic plates have been presented. We have used "Postulate" (6.7) and two Voight–Levý type solutions (6.12), (6.13) to approximate natural frequencies and mode shapes of vibrating clamped orthotropic plates. Galerkin's procedure, together with the expansion of the displacements and external forces, resulted in a set of linear equations for the cross-spectral densities of displacements. The essential advantage of the method presented is that it coincides with the normal-mode approach for the cases capable of exact solutions.

References

Bolotin, V. V. (1961). An asymptotic method for the study of the problem of eigenvalues for rectangular regions. In J. R. M. Dodds (Ed.), *Problems in continuum mechanics*, Volume dedicated to N.I. Muskhelishvili (pp. 56–58). Philadelpha, PA: Society of Industrial and Applied Mathematics.

Bolotin, V. V. (1963). On the density of the distribution of natural frequencies of thin elastic shells. *PMM: Journal of Applied Mathematics and Mechanics, 27*(2), 362–364 (in Russian).

Bolotin, V. V. (1984). *Random vibrations of elastic systems*. The Hague, The Netherlands: Martinus Nijhoff Publishers.

Chyu, W. J., & Au-Yang, M. K. (1972). Random response of rectangular panels to the pressure field beneath a turbulent boundary layer in subsonic flows. *NASA TND-6970*.

Clarkson, B. L. (1968). Stresses in skin panels subjected to random acoustic loading. *The Aeronautical Journal of the Royal Aeronautical Society, 72,* 1000–1010.

Crandall, S. H., & Mark, W. D. (1963). *Random vibration in mechanical systems*. New York: Academic Press.

Crocker, M. J. (1969). The response of a supersonic transport fuselage to boundary layer and to reverberant noise. *Journal of Sound and Vibration, 9,* 6–20.

Dimentberg, M. F. (1961). Forced vibrations of plates under load representing spatial-temporary random process. *Inzhenernyi Zhurnal, 1*(2), 97–105 (in Russian).

Dowell, E. H. (1971). Comments on "turbulent-induced plate vibrations: An evaluation of finite- and infinite-plate models". *The Journal of the Acoustical Society of America, 49,* 376.

Dyer, I. (1959). Response of plates to a decaying and connecting random pressure field. *Journal of the Acoustical Society of America, 31,* 922–928.

Elishakoff, I. (1971a). On evaluation of spectral densities of external forces for plates and shells. In V. V. Bolotin & A. Chyras (Eds.), *Problems of reliability in structural mechanics* (pp. 37–44). Vilnius, Luthuania (in Russian).

Elishakoff, I. (1971b) On the determination of spectral densities of external forces of plates and shells. In V. V. Bolotin & A. A. Čyras (Eds.), *Relability problems in structural mechanics* (pp. 36–44). Vilnius.

Elishakoff, I. (1971c). *Vibration fields in cylindrical shells, excited by random forces*. Ph.D. Thesis, Department of Dynamics and Strength of Machines, Moscow Power Engineering Institute and State University (in Russian).
Elishakoff, I. (1974a). Mean-square stability of elastic bodies in supersonic flow. *Journal of Sound and Vibration, 33,* 67–78.
Elishakoff, I. (1974b). Vibration analysis of clamped square orthotropic plates. *AIAA Journal, 12,* 921–924.
Elishakoff, I. (1975a). Distribution of natural frequencies in certain structural elements. *The Journal of the Acoustical Society of America, 57*(2), 361–369.
Elishakoff, I. (1975b). Turbulent flow excited vibration of a shallow cylindrical shell. *AIAA Journal, 13,* 1109–1118.
Elishakoff, I., & Efimtsov, B. (1972). On the vibrations of unbounded plates and the field of random forces. *Soviet Applied Mechanics, 8,* 903–906.
Huffington, J., Jr., & Hoppmann, N. H. (1959). On the transverse vibrations of rectangular orthotropic plates. *Journal of Applied Mechanics, 25,* 389–395.
Lin, Y. K. (1967). *Probabilistic theory of structural dynamics*, Chap. 7 (pp. 203–252). New York: McGraw-Hill.
Palmov, V. A. (1965). Thin plates under action of wide-band random loading. In *Proceedings of the Leningrad Politechnics Institute Named After M.I. Kalinin* (Vol. 252, pp. 97–106) (in Russian).
Parkus, H. (1969). *Random processes in mechanical sciences*. Vienna: Springer.
Powell, A. (1958). On the fatigue failure of structures due to vibrations excited by random pressure fields. *The Journal of the Acoustical Society of America, 30*(12), 1130–1135.
Ribner, H. S. (1966). Response of flexible panel to turbulent flow: Running wave versus modal density analysis. *The Journal of the Acoustical Society of America, 40*(3), 721–726.
Strawderman, W. A. (1969a). Turbulent-flow-excited vibration of a simply supported, recangular flate plate. *The Journal of the Acoustical Society of America, 45*, 177–192.
Strawderman, W. A. (1969b). Turbulent-induced plate vibrations: An evaulation of finite- and infinite-plate models. *The Journal of the Acoustical Society of America, 46*, 1294–1307.
Wagner, H., & Bhat, B. R. (1970). Linear response of an elastic plate to actual random load. *Ingenieur-Archiv, 39*(3), 149–158.
Warburton, G. B. (1964). *The dynamical behaviour of structures*. London: Pergamon Press (Second edition, Elsevier, 2014).

Chapter 7
Wide-Band Random Axisymmetric Vibration of Cylindrical Shells

7.1 Introductory Comments

When a uniform structure (string, beam, membrane, or plate) is excited by a point force whose time history is a stationary wide-band random process, the distribution of mean-square response velocity is surprisingly uniform except for small zones or narrow lanes in which there is enhanced response. The location of these regions depends on the position of the exciting force and on the symmetry of the structure. For example, if the force acts at $x = a$ on a beam or a string of length l the zones of enhanced response are at $x = a$ and $x = l - a$. As the bandwidth of the excitation increases without limit, the relative enhancement approaches 3/2 asymptotically. On a rectangular plate the lanes of enhanced response form a tic-tac-toe pattern with the drive point at one of the intersections. The relative enhancement in the lanes again approaches 3/2 while the relative enhancement at the drive point and its three image points approaches 9/4. There are also regions at the boundaries where the responses may be either enhanced or diminished depending on whether the boundary is free or fixed. These patterns of enhanced response were demonstrated experimentally and explained analytically by natural-mode analyses (Crandall and Wittig, 1972; Crandall, 1974; Itao and Crandall, 1978) and by image-sum techniques (Crandall, 1977). In the natural-mode analyses, it was generally possible to neglect the cross-correlations between modes because there was little modal overlap due to light damping and to ample separation between the natural frequencies. For uniform one-dimensional structures, the modal separation is easily predicted. For two-dimensional structures, the average modal separation can easily be predicted but the actual spacing is erratic. The accidental overlap of a few modes out of a hundred or more probably has very little effect on the resulting distribution. If, however, there is a systematic occurrence of repeated roots (which occurs in rectangular membranes and plates with certain aspect ratios), then it is necessary to account for the cross-correlation between these modes. For example, it is the cross-correlation between modes with repeated frequencies in a square plate which produces enhanced

I. Elishakoff, *Dramatic Effect of Cross-Correlations in Random Vibrations of Discrete Systems, Beams, Plates, and Shells*,
https://doi.org/10.1007/978-3-030-40394-2_7

response along a diagonal (Crandall, 1977), in addition to the usual tic-tac-toe pattern for rectangular plates, when the exciting force acts on the diagonal.

The relative importance of the modal cross-correlations in comparison with the autocorrelations in random vibration analysis has been the subject of investigation for 20 years. In 1958, Powell (1958a, 1958b) showed that the spatial average of the mean-square response over the surface of a panel was unaffected by the cross-correlations. In 1961, Bolotin (1961) gave a quantitative statement of the requirement that the modal overlap by small if the cross-correlation terms are to be neglected in calculating local responses. In an early calculation, Dimentberg (1961) used Bolotin's asymptotic method (Bolotin, 1961; Elishakoff, 1976) to evaluate the maximum RMS stress in a clamped plate and found a 15.8% drop when the cross-correlation terms due to repeated frequencies were omitted. Similar studies have been made for various structures under various random excitations. In some cases, the contribution of the cross-correlations has been negligible (Chyu and Au-Yang, 1972) or only a few percent (Schweiker and Davis, 1966). In the case of a multi-span beam with closely spaced natural frequencies, Mercer (1965) found discrepancies of up to 20% between calculations including cross-correlation terms and calculations which neglected them. In 1965, independently of Bolotin's work, Davis (1965) derived estimates of the errors incurred by neglecting cross-correlation terms in random vibration response calculations and indicated errors as large as 50% in the response spectral density. Recently, Elishakoff (1977a, 1977b, 1977c, 1977d) studied shallow spherical shell panels, rectangular in plan, where because of the shell action the natural frequencies are clustered together at the low-frequency end of the spectrum, and where there are, in addition, many repeated roots because of specially selected aspect ratios. The local error due to neglecting cross-correlation terms in the mean-square response was shown to exceed 250% when the excitation was a spatially uniform random pressure with uniform spectral density and 60 modes were included in the response.

In the present paper, we study an essentially one-dimensional problem similar to the point excitation of a string or a beam except that, since the structure is a shell, the natural frequencies are clustered and considerable modal overlap occurs even with light damping. A uniform circularly cylindrical shell is excited by a ring load which is uniform around the circumference and whose time history is a stationary random process. The stationary response is also uniform in the circumferential direction. Our object is to investigate how the strong modal overlap influences the distribution of mean-square response.

7.2 Random Vibration of a Thin Cylindrical Shell

Consider a uniform cylindrical shell simply supported at its ends $x = 0$ and $x = L$ as shown in Fig. 7.1. If the shell is subjected to a viscous damping force c per unit area per unit velocity, and excited by a transverse force $f(x, \theta, t)$ per unit area, the transverse displacement $w(x, \theta, t)$ satisfies the shallow shell equations

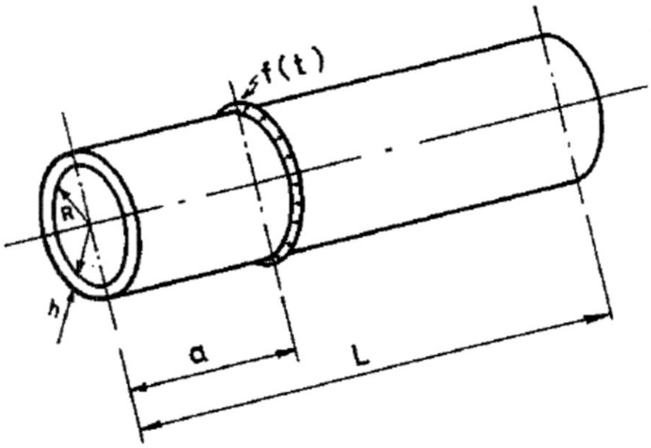

Fig. 7.1 Circular cylindrical shell driven by uniform ring loading at section $x = a$

$$D\Delta\Delta w - \frac{1}{R}\frac{\partial^2\chi}{\partial x^2} + c\frac{\partial w}{\partial t} + \rho h\frac{\partial^2 w}{\partial t^2} = f(x,\theta,t)$$
$$\Delta\Delta\chi + \frac{Eh}{R}\frac{\partial^2 w}{\partial x^2} = 0, \quad \Delta = \frac{\partial^2}{\partial x^2} + \frac{1}{R^2}\frac{\partial^2}{\partial\theta^2} \tag{7.1}$$

where $\chi(x,\theta,t)$ is the stress function, E is the tension modulus, D is the flexural rigidity, ρ is the mass density, h is the thickness, and R is the radius. We consider the circularly symmetric response $w(x,t)$ due to a uniformly distributed ring force located at $x = a$, as shown in Fig. 7.1. For a stationary random ring loading, the space-time correlation of the excitation is

$$E[f(x_1,\theta_1,t_1)f(x_2,\theta_2,t_2)] = \delta(x_1 - a)\delta(x_2 - a)R_f(t_2 - t_1) \tag{7.2}$$

where $R_f(\tau)$ is the autocorrelation function of the ring loading. The natural-mode formalism for the stationary response (1967) leads to the following expression for the mean square of the transverse *displacement*

$$E\left[w^2(x)\right] = \sum_{j=1}^{\infty}\sum_{k=1}^{\infty}\psi_j(x)\psi_k(x)\psi_j(a)\psi_k(a)\Phi_{jk} \tag{7.3}$$

where the axisymmetric natural modes

$$\psi_j(x) = \sqrt{2}\sin j\pi x/L \tag{7.4}$$

satisfy the eigenfunction equation

$$D\frac{d^4\psi_j}{dx^2} + \frac{Eh}{R^2}\psi_j = \rho h\omega_j^2\psi_j \tag{7.5}$$

with

$$\omega_j^2 = \frac{E}{\rho L^2}\left[\frac{L^2}{R^2} + \frac{h^2}{L^2}\frac{(j\pi)^4}{12(1-\nu^2)}\right] \tag{7.6}$$

and satisfy the orthogonality condition

$$\int_0^L \rho h \psi_j \psi_k dx = \rho h L \delta_{jk} \tag{7.7}$$

where δ_{jk} is the Kronecker delta. In (7.3) the symbol Φ_{jk} stands for the integral

$$\Phi_{jk} = \int_{-\infty}^{\infty} H_j(-\omega)H_k(\omega)d\omega \tag{7.8}$$

where $H_j(\omega)$ is the frequency response function for the jth mode

$$H_j(\omega) = \frac{1}{\rho h L\left(\omega_j^2 + i\beta\omega - \omega^2\right)}, \quad \beta = \frac{c}{\rho h}. \tag{7.9}$$

in terms of the modal bandwidth β, $S_f(\omega)$ is the spectral density of the ring loading defined by the Wiener–Khintchine relation

$$S_f(\omega) = \frac{1}{2\pi}\int_{-\infty}^{\infty} R_f(\tau)e^{-i\omega\tau}d\tau. \tag{7.10}$$

For the subsequent calculations, we take $S_f(\omega)$ to be the following band-limited white noise spectrum

$$S_f(\omega) = \begin{matrix} S_0, & |\omega| < \omega_c \\ 0, & \omega_c < |\omega| \end{matrix}. \tag{7.11}$$

A sum similar to (7.3) can be obtained for the mean square of the transverse *velocity* $v(x,t) = \partial w/\partial t$. The only difference is that the integral Φ_{jk} of (7.8) must be replaced by

$$\Phi'_{jk} = \int_{-\infty}^{\infty} \omega^2 H_j(-\omega)H_k(\omega)S_f(\omega)d\omega. \tag{7.12}$$

The integrals (7.8) and (7.12) are real and symmetric, in that $\Phi_{jk} = \Phi_{kj}$ and $\Phi'_{jk} = \Phi'_{kj}$. In preparation for the numerical studies which follow, we have obtained closed-form expressions for these integrals in the case where $H_j(\omega)$ is given by (7.9) and $S_f(\omega)$ is given by (7.11). By partial fraction expansion, the integrands can be reduced to elementary tabulated forms. For the displacement integral (7.8)

$$\Phi_{jk}(\omega_c) = \frac{S_0}{\rho^2 h^2 L^2} \int_{-\omega_c}^{\omega_c} \frac{d\omega}{\left(\omega_j^2 - i\omega\beta - \omega^2\right)\left(\omega_k^2 + i\omega\beta - \omega^2\right)}, \tag{7.13}$$

we obtain

$$\Phi_{jk}(\omega_c) = \frac{S_0}{\rho^2 h^2 L^2} \frac{F\left(\omega_j, \omega_k; \omega_c\right) + F\left(\omega_k, \omega_j; \omega_c\right)}{\left(\omega_j^2 - \omega_k^2\right)^2 + 2\beta^2\left(\omega_j^2 + \omega_k^2\right)} \tag{7.14}$$

where

$$\begin{aligned} F\left(\omega_j, \omega_k; \omega_c\right) = & \frac{\omega_j^2 - \omega_k^2 - \beta^2}{2\left(\omega_j^2 - \beta^2/4\right)^{1/2}} \\ & \times \ln \frac{\omega_c^2 + \omega_j^2 - 2\omega_c\left(\omega_j^2 - \beta^2/4\right)^{1/2}}{\omega_c^2 + \omega_j^2 + 2\omega_c\left(\omega_j^2 - \beta^2/4\right)^{1/2}}. \\ & + 2\beta \tan^{-1} \frac{\beta\omega_c}{\omega_j^2 - \omega_c^2} \end{aligned} \tag{7.15}$$

For the velocity integral (7.12), we obtain

$$\Phi'_{jk}(\omega_c) = \frac{S_0}{\rho^2 h^2 L^2} \frac{F'\left(\omega_j, \omega_k; \omega_c\right) + F'\left(\omega_k, \omega_j; \omega_c\right)}{\left(\omega_j^2 - \omega_k^2\right)^2 + 2\beta^2\left(\omega_j^2 + \omega_k^2\right)} \tag{7.16}$$

where

$$\begin{aligned} F'\left(\omega_j, \omega_k; \omega_c\right) = & \frac{\omega_j^2\left(\omega_j^2 - \omega_k^2\right) + \beta^2\left(\omega_j^2 + \omega_k^2\right)/2}{2\left(\omega_j^2 - \beta^2/4\right)^{1/2}} \\ & \times \ln \frac{\omega_c^2 + \omega_j^2 - 2\omega_c\left(\omega_j^2 - \beta^2/4\right)^{1/2}}{\omega_c^2 + \omega_j^2 + 2\omega_c\left(\omega_j^2 - \beta^2/4\right)^{1/2}}. \\ & + \beta\left(\omega_j^2 + \omega_k^2\right) \tan^{-1} \frac{\beta\omega_c}{\omega_j^2 - \omega_c^2} \end{aligned} \tag{7.17}$$

These results are apparently new. The displacement integral (7.14) reduces to Bolotin's result (1961)

$$\Phi_{jk}(\infty) = \frac{S_0}{\rho^2 h^2 L^2} \frac{4\pi\beta}{\left(\omega_j^2 - \omega_k^2\right)^2 + 2\beta^2\left(\omega_j^2 + \omega_k^2\right)} \tag{7.18}$$

when $\omega_c \to \infty$ and to Crandall and Mark's result (1963)

$$\Phi_{jj}(\omega_c) = \frac{1}{\rho^2 h^2 L^2 \beta \omega_j^2} \left[\frac{\beta}{4\left(\omega_j^2 - \beta^2/4\right)^{1/2}} \right. \times \ln \frac{\omega_c^2 + \omega_j^2 + 2\omega_c\left(\omega_j^2 - \beta^2/4\right)^{1/2}}{\omega_c^2 + \omega_j^2 - 2\omega_c\left(\omega_j^2 - \beta^2/4\right)^{1/2}} \left. + \tan^{-1} \frac{\beta\omega_c}{\omega_j^2 - \omega_c^2} \right] \tag{7.19}$$

when $k = j$.

The summations for mean-square displacement and velocity for the shell are formally the same as those for a uniform string or simply supported beam. The only differences lie in the spectra of natural frequencies associated with the various structures. For strings and beams the natural frequencies are well separated so that for light damping, the terms with $j \neq k$ (which involve modal cross-correlations) can be neglected in comparison with the terms with $j = k$ (which involve modal autocorrelations) in summations such as (7.3). For thin shells on the other hand, the natural frequencies ω_j are tightly clustered for low values of j. The average modal density $n(\omega)$ (number of modes per unit of circular frequency) obtained from (7.6)

$$n(\omega) = \begin{cases} 0, & \omega < \omega_1 \\ \frac{L}{2\pi}\left(\frac{\rho h}{D}\right)^{1/4} \frac{\omega}{\left[\omega^2 - E/\rho R^2\right]^{3/4}}, & \omega_1 \leq \omega \end{cases} \tag{7.20}$$

has a high peak at $\omega = \omega_1$ and gradually decreases with increasing ω. The local modal overlap ratio

$$r_j = \frac{\text{modal bandwidth}}{\text{modal separation}} = \frac{\beta}{\omega_j - \omega_{j-1}}, \quad j = 2, 3, \ldots \tag{7.21}$$

can be very much greater than unity for the lower frequency modes of a thin shell.

7.3 Description of the Calculations

To study the effects of modal overlap, we have evaluated the mean-square *velocity*

$$E\left[\upsilon^2(x)\right] = \sum_{j=1}^{\infty}\sum_{k=1}^{\infty} \psi_j(x)\psi_k(x)\psi_j(a)\psi_k(a)\Phi'_{jk}(\omega_c) \tag{7.22}$$

for several sets of shell parameters including the limiting case $R \to \infty$ where the shell becomes a flat plate excited by a line load at $x = a$. The response, in that case, is the same as that for a simply supported beam excited by a point load at $x = a$, except for the factor of $1 - \nu^2$ in the plate modulus as compared with the beam modulus. We call this case the beamlike limit. Our main results are computer calculated distributions of mean-square velocity for several shell configurations. These results are presented in dimensionless form.

The non-dimensional response location ξ and excitation location η are defined by

$$\xi = \frac{x}{L} \quad \eta = \frac{a}{L}. \tag{7.23}$$

The dimensionless frequency Ω and dimensionless modal bandwidth B are defined by

$$\Omega = \frac{\omega L}{(E/\rho)^{1/2}} \quad B = \frac{\beta L}{(E/\rho)^{1/2}}. \tag{7.24}$$

We consider the length L of the shell to be fixed and that the shell shape is defined by the radius ratio parameter R/L and the thickness ratio parameter h/L. To illustrate the relative importance of the cross-correlation terms in comparison with the autocorrelation terms, the dimensionless mean-square velocity

$$d_v^2 = E[v^2]\rho^2 h^2 L(E/\rho)^{1/2}/S_0 \tag{7.25}$$

is split into two sums

$$d_v^2 = s_1 + s_2 \tag{7.26}$$

where

$$\begin{aligned} s_1 &= 4\sum_{j=1}^{\infty} \sin^2 j\pi\xi \sin^2 j\pi\eta I_{jj}(\Omega_c) \\ s_2 &= 8\sum_{j=1}^{\infty}{}_{j<k}\sum_{k=1}^{\infty} \sin j\pi\xi \sin k\pi\xi \sin j\pi\eta \sin k\pi\eta I_{jk}(\Omega_c) \end{aligned} \tag{7.27}$$

and the dimensionless integrals are defined by

$$\begin{aligned} I_{jk}(\Omega_c) &= \frac{\rho^2 h^2 L(E/\rho)^{1/2}}{S_0}\Phi'_{jk}(\omega_c) \\ &= \int_{-\Omega_c}^{\Omega_c} \frac{\Omega^2 d\Omega}{\left[\Omega_j^2 - i\Omega B - \Omega^2\right]\left[\Omega_k^2 + i\Omega B - \Omega^2\right]}. \end{aligned} \tag{7.28}$$

In the calculations, the location of the excitation is fixed at $\eta = a/L = 0.3$. The dimensionless modal bandwidth is fixed at $B = 0.02$ and Poisson's ratio is fixed at $\nu = 0.3$. Four shell geometries are considered. In three cases, the radius ratio is fixed at $R/L = 0.5$ while the thickness ratio h/L takes the values 0.01, 0.001, and 0.0003. The fourth case is the beamlike limit in which $R/L \to \infty$ with $h/L = 0.01$. In all four cases, the dimensionless excitation cutoff frequency is taken as $\Omega_c = 2\pi$. In addition a wider excitation band with $\Omega_c = 20\pi$ is also considered for the beamlike case.

The number N_c of modes in the excitation band is defined as the largest value of j such that $\Omega_j < \Omega_c$. For the cases with $h/L = 0.01$ (both $R/L = 0.5$ and $R/L \to \infty$), the number of modes in the band with $\Omega_c = 2\pi$ is $N_c = 14$. For the same excitation, the number of modes in the band is $N_c = 44$ for $h/L = 0.001$, and $N_c = 81$ for $h/L = 0.0003$. For the beamlike case with $\Omega_c = 20\pi$, the number of modes in the excitation band is $N_c = 47$. For the shells with $R/L = 0.5$, the fundamental frequency Ω_1 is slightly greater than 2.0. This implies that the modal loss factors B/Ω_j are always smaller than 0.01 which represents light damping. In the beamlike case the fundamental frequency is much lower: $\Omega_1 = 0.029867$, so that the fundamental loss factor is 0.67, which represents heavy damping. Even in this case, most of the modes excited are lightly damped because the loss factor of higher modes decreases in inverse proportion to the frequency. Since the natural frequencies Ω_j in the beamlike case increase in proportion to the square of j, it turns out that only the first three modes have loss factors greater than 0.042.

To illustrate the behavior of the spectrum of natural frequencies Ω_j, the local modal overlaps r_j, and the integrals I_{jk}, some results for the shell with $R/L = 0.5$ and $h/L = 0.01$ are tabulated in Tables 7.1 and 7.2. In Table 7.1, note the clustering of the natural frequencies and the large local overlaps for the first three modes. Table 7.2 shows the behavior of I_{jk} up to $j = k = 18$. Note the large values I_{jk} for $j \neq k$ for small values of j and k where the local modal overlaps are large and the gradual decrease in the importance of the off-diagonal I_{jk} for larger values of j and k. Note the almost uniform behavior of I_{jj} up to $j = N_c = 14$, and the precipitous drop thereafter. Note the change in the behavior of the I_{jk} values in three regions: (*i*) when both i and j are less than N_c; (*ii*) when only one is greater than N_c; and (*iii*) when both are greater than N_c. In our evaluations of the sums s_1 and s_2 of (7.27), we truncated the sums at $j, k = N_c$ was compared with the corresponding distribution for upper limits of $j, k = N_c + 10$. In no case did the maximum discrepancy exceed 1.15%.

The I_{jk} values used here were computed from the closed-form expressions (7.16) and (7.17). A small damping approximation that is often made (e.g. in Crandall and Wittig, 1972; Crandall, 1974; and Itao and Crandall, 1978) in evaluating $I_{jj}(\Omega_c)$ is to use $I_{jj}(\infty)$ for $\Omega_j < \Omega_c$ and zero for $\Omega_c < \Omega_j$. Had this approximation been used, we would have had $I_{jj}(\infty) = \pi/B = 157.07963$ for $j = 1, \ \ldots, \ 14$. Comparing this with Table 7.2, we see that the discrepancy in I_{jj} varies from 0.22% at $j = 1$ to 3.67% at $j = 14$.

Table 7.1 Values of natural frequencies Ω_j and local modal overlap ratios r_j for shells with $\frac{R}{L} = 0.5$ and $\frac{h}{L} = 0.01$

j	Ω_j	r_j
1	2.000223	
2	2.003565	5.984
3	2.017983	1.387
4	2.056297	0.522
5	2.134834	0.255
6	2.270697	0.147
7	2.478256	0.096
8	2.766538	0.069
9	3.138881	0.054
10	3.594474	0.044
11	4.130391	0.037
12	4.743103	0.033
13	5.429283	0.029
14	6.186115	0.026
15	7.011330	0.024
16	7.903147	0.022
17	8.860180	0.021
18	9.881355	0.020

7.4 Discussion of the Results

The distribution of mean-square velocity along the length of the shell for five cases is displayed in Figs. 7.2, 7.3, 7.4, 7.5 and 7.6. In each case, the dimensionless mean-square velocity d_v^2 and its two components s_1 and s_2 are plotted as functions of $\xi = x/L$. In every case the autocorrelation sum $s_1(\xi)$ is nonnegative and symmetric about the midpoint; i.e. $s_1(\xi) = s_1(1 - \xi)$, as required by (6.27). The cross-correlation sum $s_2(\xi)$ is sometimes positive and sometimes negative, since its average is zero, but it has no simple symmetry properties.

Figure 7.2 shows the result for the beamlike limit $R/L \rightarrow \infty$ with excitation cutoff at $\Omega_c = 2\pi$. The number of modes in the excitation band is $N_c = 14$ and the local overlap ratios range from $r_2 = 0.221$ to $r_{14} = 0.025$. The distribution of mean-square velocity here is like that described in (Crandall and Wittig, 1972; and Crandall, 1974). There are small fluctuations about a uniform response level of $d_v^2 \simeq \pi N_c/B = 2200$ with the exception of two zones of intensified response, at the drive point $\xi = \eta$ and its image point $\xi = 1 - \eta$, where the peak response is about 50% greater. Here the contribution of s_2 is everywhere under 3% and the distribution of d_v^2 is essentially that of s_1. The small contribution of s_2 does however make the distribution of d_v^2 slightly unsymmetric.

Figure 7.3 shows the result for the beamlike limit $R/L \rightarrow \infty$ with the wider excitation band: $\Omega_c = 20\pi$. The number of modes in the excitation band is $N_c = 47$

Table 7.2 Values of $I_{jk}(j \leq k)$ for shell with $\frac{R}{L} = 0.5$, $\frac{h}{L} = 0.01$, and $\Omega_c = 2\pi(N_c = 14)$

k	1	2	3	4	5	6	7	8	9	10	11	12	13	14	15	16	17	18
j																		
1	156.738	152.471	87.485	17.388	3.052	0.512	0.072	−0.245	−0.312	−0.35	−382	−0.424	−0.5	−0.833	−0.422	−0.27	−0.192	−0.144
2		156.738	103.022	19.414	3.223	0.533	−0.068	−0.244	−0.312	−0.35	−0.382	−0.424	−0.5	−0.833	−0.42	−0.27	−0.192	−0.144
3			156.737	33.295	4.13	0.635	−0.05	−0.24	0.311	−0.349	−0.382	−0.425	−0.501	−0.836	−0.42	−0.269	−0.191	−0.143
4				156.736	9.225	1.012	0.005	−0.228	−0.308	−0.349	−0.382	−0.425	0.501	−0.836	−0.42	−0.269	−0.191	−0.143
5					156.734	2.987	0.183	−0.197	−0.301	−0.347	−0.383	−0.426	−0.503	−0.841	−0.421	−0.269	−0.191	−0.143
6						156.73	1.095	−0.101	−0.282	−0.343	−0.353	4418	0.507	0.549	0.422	0.265	0.19	0.141
7							156.723	0.392	−0.225	−0.333	−0.383	−0.431	−0.512	0.864	−0.423	0.266	−0.187	−0.139
8								156.711	0.076	−0.298	−0.379	−0.436	−0.522	0.888	−0.426	0.264	−0.183	−0.135
9									156.663	−0.098	−0.36	−0.441	−0.536	−0.927	−0.428	−0.259	−0.177	−0.129
10										156.662	−0.223	−0.438	−0.577	−0.989	−0.43	−0.25	−0.167	−0.119
11											156.609	−0.352	−0.583	−1.092	−0.428	−0.232	−0.148	−0.102
12												156.502	−0.565	−1.28	−0.407	−0.192	−0.111	−0.072
13													156.21	−1.686	−0.3	−0.084	−0.027	−0.008
14														151.515	0.929	0.542	0.36	0.258
15															0.442	0.242	0.159	0.114
16																0.136	0.091	0.065
17																	0.061	0.044
18																		0.032

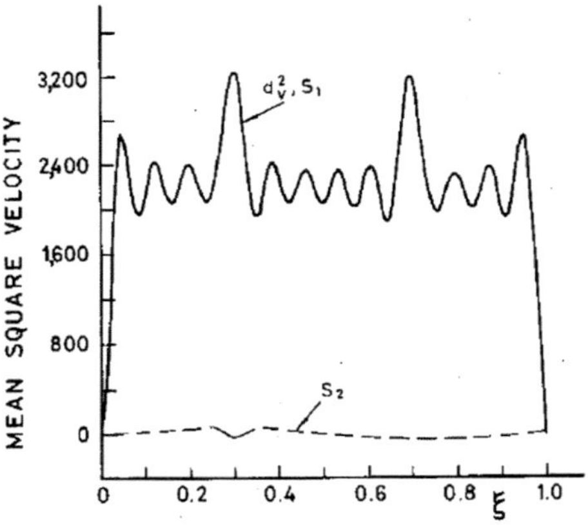

Fig. 7.2 Dimensionless mean-square velocity d_v^2 and component sums s_1 and s_2 for beamlike limit $R/L \to \infty$ with $h/L = 0.01$ and $\Omega_c = 2\pi\,(N_c = 14)$

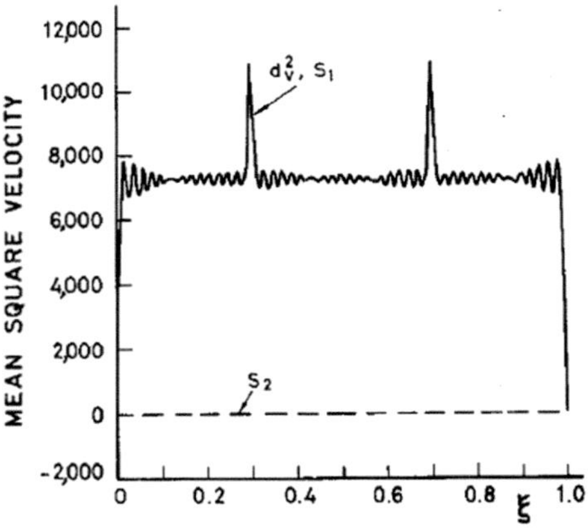

Fig. 7.3 Dimensionless mean-square velocity d_v^2 and component sums for beamlike limit $R/L \to \infty$ with $h/L = 0.01$ and $\Omega_c = 20\pi\,(N_c = 47)$

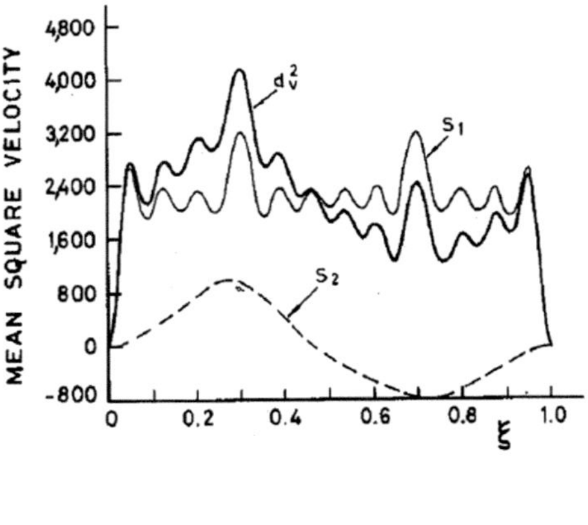

Fig. 7.4 Dimensionless mean-square velocity d_v^2 and component sums, for shell with $R/L = 0.5$, $h/L = 0.01$, and $\Omega_c = 2\pi\,(N_c = 14)$

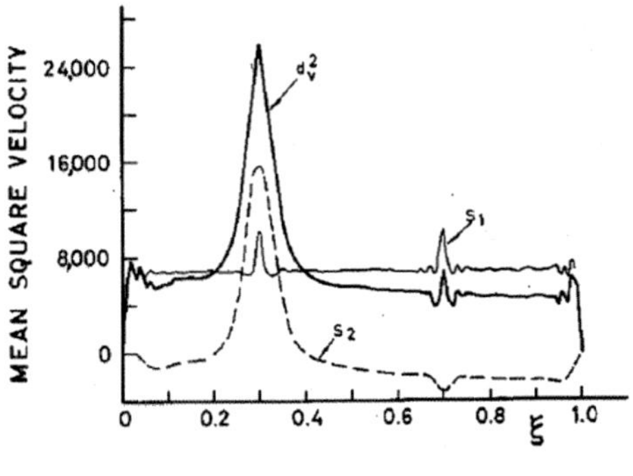

Fig. 7.5 Dimensionless mean-square velocity d_v^2 and component sums for shell with $\frac{R}{L} = 0.5$, $\frac{h}{L} = 0.001$, and $\Omega_c = 2\pi\,(N_c = 44)$

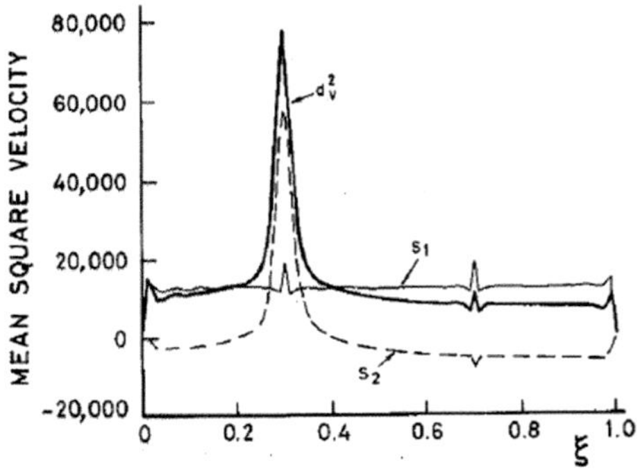

Fig. 7.6 Dimensionless mean-square velocity d_v^2 and component sums for shell with $\frac{R}{L} = 0.5$, $\frac{h}{L} = 0.0003$, and $\Omega_c = 2\pi\,(N_c = 81)$

and the local overlap ratios, for the 33 additional modes beyond those included in Fig. 7.2, range from $r_{15} = 0.023$ to $r_{47} = 0.007$. These additional modes contribute importantly to s_1 but hardly at all to s_2 because of their small overlaps. As a result, the relative contribution of the cross-correlation terms is smaller in Fig. 7.3 than in Fig. 7.2.

Figure 7.4 shows the result for the shell with $R/L = 0.5$ and $h/L = 0.01$. The number of modes within the excitation band, $\Omega_c = 2\pi$, is $N_c = 14$ and the local modal overlap ratio ranges from $r_2 = 5.984$ to $r_{14} = 0.026$ as can be seen from Table 7.1. In this case, the cross-correlation sum s_2 introduces a substantial amount of asymmetry into the distribution of d_v^2 although the zones of intensified response at $\xi = \eta$ and $\xi = 1 - \eta$ are still recognizable.

Figure 7.5 shows the result for the shell with $R/L = 0.5$ and $h/L = 0.001$. Here $N_c = 44$ for $\Omega_c = 2\pi$ and the local modal overlap ratio varies from $r_2 = 597.889$ to $r_{44} = 0.0817$, and remains greater than unity up through the 13th mode. The cross-correlation sum s_2 now dominates the response at the drive point $\xi = \eta$. A small remnant of intensified response is still noticeable at the image point $\xi = 1 - \eta$.

Finally, in Fig. 7.6, the results for the thinnest shell are shown. The shell parameters are $R/L = 0.5$, $h/L = 0.0003$, and $N_c = 81$ for $\Omega_c = 2\pi$. The local modal overlap ratio varies from $r_2 = 6470.398$ to $r_{81} = 0.147$ and remains greater than unity up through the 30th mode. The contribution from s_2 is now over three times that of s_1

Table 7.3 Components of driving-point response for shells with $R/L = 0.5$ and $\Omega_c = 2\pi$

h/L	N_c	s_1	s_2	d_v^2	d_v^2/s_1
0.01	14	3,209.4	948.5	4,157.90	1.30
0.01	44	10,279.9	15,743.0	26,022.9	2.53
0.0003	81	19,075.0	58,699.0	77,774.0	4.08

at the drive point and the intensified response at the image point has almost been erased.

The contribution to the drive-point mean-square response in the three shells with $R/L = 0.5$ are summarized in Table 7.3. The values of d_v^2 at $\xi = \eta$ are tabulated along with their component sums s_1 and s_2. As the small thickness parameter decreases from 0.01 to 0.0003, the ratio of the actual response to the prediction based on the autocorrelation sum s_1 alone increases from 1.30 to 4.08. Further details can be found in the paper by Elishakoff, Van Zanten, and Crandall (1979).

References

Bolotin, V. V. (1961). An asymptotic method for the study of the problem of eigenvalues for rectangular regions. In J. R. M. Dodds (Ed.), *Problems in continuum mechanics*, Volume dedicated to N.I. Muskhelishvili (pp. 56–58). Philadelpha, PA: Society of Industrial and Applied Mathematics.

Chyu, W. J., & Au-Yang, M. K. (1972). *Random response of rectangular panels to the pressure field beneath a turbulent boundary layer in subsonic flows*. NASA TN-D-6970.

Crandall, S. H. (1974). Wide-band random vibration of structures. In *Proceedings of the 7th U.S. national congress of applied mechanics* (pp. 131–138). New York: ASME Press.

Crandall, S. H. (1977). Structured response patterns due to wide-band random excitation. In B. L. Clarkson (Ed.), *Stochastic problems in dynamics* (pp. 366–389). London: Pitman.

Crandall, S. H., & Mark, W. D. (1963). *Random vibration in mechanical systems*. New York: Academic Press.

Crandall, S. H., & Wittig, L. E. (1972). Chladni's patterns for random vibration of a plate. In G. Herrmann, N. Perrone (Eds.), *Dynamic response of structures* (pp. 35–71). New York: Pergamon Press.

Davis, R. E. (1965) Random pressure excitation of shells and statistical dependence effect of normal mode response. *NASA CR-311*.

Dimentberg, M. F. (1961). Forced vibrations of plates under load representing spatial-temporary random process. *Inzhenernyi Zhurnal, 1*(2), 97–105. (in Russian).

Elishakoff, I. (1976). Bolotin's dynamic edge-effect method. *Shock and Vibration Digest, 8*, 95–104.

Elishakoff, I. (1977a). Random vibrations of orthotropic plates clamped or simply supported all round. *Acta Mechanica, 28*(1–4), 165–176.

Elishakoff, I. (1977b). On the role of cross-correlations in the random vibrations of shells. *Journal of Sound and Vibration, 50*(2), 239–252.

Elishakoff, I. (1977c). Random vibrations of orthotropic plates with all edges clamped or simply supported. *Acta Mechanica, 28*, 165–176.

Elishakoff, I. (1977d). Flutter and random vibration in plates. In B. L. Clarkson (Ed.), *Stochastic problems in dynamics* (pp. 390–410). London: Pitman.

Elishakoff, I., Van Zanten, A. Th., & Crandall, S. H. (1979) Wide-band random axisymmetric vibration of cylindrical shells. *Journal of Applied Mechanics, 46*(2), 417–423.

Itao, K., & Crandall H. (1978). Wide-band random vibration of circular plates. *Journal of Engineering Design, 100*(4), 690–695.
Mercer, C. A. (1965). Response of a multi-supported beam to a random pressure field. *Journal of Sound and Vibration, 2,* 293–306.
Powell, A. (1958a). On the approximation to the "Infinite" solution by the method of normal modes for random vibrations. *Journal of the Aeronautical Society of America, 30,* 1136–1139.
Powell, A. (1958b). On the response of structures to random pressures and to jet noise in particular. In S. H. Crandall (Ed.), *Random vibration* (Chap. 8, pp. 187–229). New York: Wiley.
Schweiker, J. W., & Davis, R. E. (1966). Response of complex shell structures to aerodynamic noise. *NASA CR-450.*

Chapter 8
Turbulent Flow-Excited Vibrations of a Shallow Cylindrical Shell

8.1 Introductory Comments

This chapter deals with random turbulent excitation of a cylindrical shell supported at both ends by shear diaphragms. In describing the probabilistic behavior of the external pressure, the cross-spectral density of the latter is assumed as a product of longitudinal and circumferential correlation coefficients, in turn, taken as functions of the frequency and of the appropriate separation distance only. The representation used for the longitudinal coefficient is chosen so as to not conflict with measurements on the rigid wall in the flow direction. For the circumferential coefficient, two models are used: a fully correlated pressure field, and a modification of the coefficient for the rigid wall in the direction normal to the flow. The formalism of the normal-mode method (Bolotin, 1969; Lin, 1967; Powell, 1958a, 1958b) is used in deriving the probabilistic response characteristics of the shell.

Related works on random excitation of cylindrical shells include Fedorov (1963), Nemat-Nasser (1968), and Kana (1971). The effect of air within the cylinder was taken into account by Bolotin (1968), Hwang and Pi (1969), Bolotin and Elishakoff (1971).

8.2 Analytical Formulation

Consider a cylindrical shell of uniform thickness, supported at both ends by shear diaphragms and subjected to random pressure fluctuations in a turbulent layer with intensity $q(x, \theta, t)$, x and θ denoting the space coordinates, and $t =$ time. Assuming that the correlation scale of the pressure is much larger, and the mean-square deviation of the shell displacements much smaller than the thickness of the shell—the

I. Elishakoff, *Dramatic Effect of Cross-Correlations in Random Vibrations of Discrete Systems, Beams, Plates, and Shells*,
https://doi.org/10.1007/978-3-030-40394-2_8

problem may be considered in a statistically linear setting. In these circumstances, the normal deflection $w(x, \theta, t)$ must satisfy the Donnell–Vlasov equations (Bolotin & Elishakoff, 1971)

$$D\left(\frac{\partial}{\partial t}\right)\Delta_2\Delta_2 w - \frac{1}{R}\frac{\partial^2\chi}{\partial x^2} + \rho h\frac{\partial^2 w}{\partial t^2} = q(x, \theta, t) \tag{8.1}$$

$$\Delta_2\Delta_2\chi + \frac{h}{R}E\left(\frac{\partial}{\partial t}\right)\frac{\partial^2 w}{\partial x^2} = 0,$$

$$\Delta_2 = \frac{\partial^2}{\partial x^2} + \frac{1}{R^2}\frac{\partial^2}{\partial \theta^2}$$

where ρ, h, and R are, respectively, the mass density, thickness, and radius of the shell, Δ_2 denotes the Laplace operator on middle surface, $D(\partial/\partial t)$ are viscoelastic operators of the time-differentiation parameter $\partial/\partial t$, corresponding to the stiffness modulus and to Young's modulus, respectively; N_{xx}, $N_{\theta\theta}$, and $N_{x\theta}$ are stress resultants $N_{xx} = R^2\partial^2\chi/\partial\theta^2$, $N_{\theta\theta} = \partial^2\chi/\partial x^2$, $N_{x\theta} = -\partial^2\chi/\partial x\partial\theta$, and χ is the stress resultant function.

Equation (8.1) are known in the literature as those for shallow cylindrical shells or those for stress–strain fields varying rapidly on the middle surface. They are valid for primarily flexural modes of vibration (including the momentless axisymmetric mode); at large wave numbers, they have to be corrected for the effects of rotatory inertia and shear deformations.

The supplementary boundary conditions are

$$w = \frac{\partial^2 w}{\partial x^2} = \frac{\partial^2\chi}{\partial x^2} = \frac{\partial^2\chi}{\partial\theta^2} = 0 \tag{8.2}$$

at $x = 0$ and $x = l$, where l is the shell length.

The problem is formulated as follows: *Given the probabilistic behavior of the ergodic random space-time function $q(x, \theta, t)$, find that of the displacement field of the shell, dispensing with initial conditions.*

For the stochastic boundary value problem Eqs. (8.1) and (8.2), we resort to the normal-mode approach (Bolotin, 1969; Lin, 1967; Powell, 1958a, 1958b) which consists in expanding the (probabilistically) given and sought functions in stochastic Fourier integrals and in series in terms of the modes of the undamped shell in vacuo.

The turbulent pressure may be resolved in a stochastic Fourier integral:

$$q(x, \theta, t) = \langle q\rangle + \int_{-\infty}^{\infty} Q(x, \theta, \omega)e^{i\omega t}d\omega \tag{8.3}$$

where $Q(x, \theta, \omega)$ is the complex random spectrum. The displacement- and force-resultant functions may be represented in the same manner:

$$w(x,\theta,t) = \langle w\rangle + \int\limits_{-\infty}^{\infty} Q(x,\theta,\omega)e^{i\omega t}d\omega$$
$$\chi(x,\theta,t) = \langle \chi\rangle + \int\limits_{-\infty}^{\infty} X(x,\theta,\omega)e^{i\omega t}d\omega \qquad (8.4)$$

Deviation of the mathematical expectations of the displacements, stresses, etc., reduces to the solution of the corresponding deterministic problem. Therefore, in the following, we confine ourselves to centered functions:

$$\langle q\rangle = \langle w\rangle = \langle \chi\rangle = 0. \qquad (8.5)$$

We formulate the random spectra Q, W, and X for each fixed frequency as expansions in terms of the modes of undamped system in vacuo (using complex notation):

$$Q(x,\theta,\omega) = \sum_{m=1}^{\infty}\sum_{n=-\infty}^{\infty} Q_{mm}(\omega)\psi_m(x)e^{in\theta}$$
$$W(x,\theta,\omega) = \sum_{m=1}^{\infty}\sum_{n=-\infty}^{\infty} W_{mm}(\omega)\psi_m(x)e^{in\theta}$$
$$X(x,\theta,\omega) = \sum_{m=1}^{\infty}\sum_{n=-\infty}^{\infty} X_{mn}(\omega)\psi_m(x)e^{in\theta} \qquad (8.6)$$

where $Q_{mn}(\omega)$, $W_{mn}(\omega)$, and $X_{mn}(\omega)$ are random complex functions of ω, and

$$\psi_m(x) = \sin(m\pi x/l),\ m = 1, 2, 3, \ldots \qquad (8.7)$$

Substitution of Eqs. (8.3)—(8.7) in Eq. (8.1) yields

$$W_{mn}(\omega) = \frac{Q_{mn}(\omega)}{L_{mn}(\omega)},\quad X_{mn}(\omega) = \frac{Q_{mn}(\omega)}{L_{mn}(\omega)}H_{mn}(\omega) \qquad (8.8)$$

where

$$L_{mn}(\omega) = D(i\omega)\left[\frac{m^2\pi^2}{l^2} + \frac{n^2}{R^2}\right]^2 + E(i\omega)\frac{h}{R}\frac{m^4\pi^4}{l^4}\left[\frac{m^2\pi^2}{l^2} + \frac{n^2}{R^2}\right]^{-2} - \rho h\omega^2. \qquad (8.9)$$
$$H_{mn}(\omega) = -\frac{h}{R}E(i\omega)\left[\frac{m^2\pi^2}{l^2} + \frac{n^2}{R^2}\right]^{-2}$$

For the pressure space-time correlation functions of the external pressure

$$K_q(x,\theta,t;x',\theta',t') = \langle q^*(x,\theta,t)q(x',\theta,t')\rangle (\equiv K_q) \tag{8.10}$$

(the asterisk denoting the complex conjugate, and the triangular brackets—the expected value), we find

$$\begin{aligned} K_q = \sum_{m=1}^{\infty}\sum_{n=-\infty}^{\infty}\sum_{p=1}^{\infty}\sum_{q=-\infty}^{\infty}\int_{-\infty}^{\infty}\int_{-\infty}^{\infty} &\langle Q_{mn}^*(\omega)Q_{pq}(\omega')\rangle \\ &\times \psi_m(x)\psi_p(x')e^{-in\theta}e^{lq\theta}e^{i\omega(t'-t)}d\omega d\omega' \end{aligned}. \tag{8.11}$$

By the assumption of stationarity of $q(x,\theta,t)$, we have the following condition for statistical orthogonality of $Q_{mn}(\omega)$:

$$\langle Q_{mn}^*(\omega)Q_{pq}(\omega')\rangle = S_{mn,pq}(\omega)\delta(\omega-\omega') \tag{8.12}$$

[$\delta(\ldots)$ denoting Dirac's delta function, and $S_{mn,pq}$ the cross-spectral densities of the generalized forces], whereby Eq. (8.11) takes the form:

$$\begin{aligned} K_q = \sum_{m=1}^{\infty}\sum_{n=-\infty}^{\infty}\sum_{p=1}^{\infty}\sum_{q=-\infty}^{\infty}\int_{-\infty}^{\infty} &S_{mn,pq}(\omega)\psi_m(x) \\ &\times \psi_p(x')e^{i(q\theta'-n\theta)}e^{i\omega\tau}d\omega \end{aligned} \tag{8.13}$$

where $\tau = t' - t$ is the time delay.

The Wiener–Khintchine theorem (Lin, 1967) in Eq. (8.13) permits determination of the cross-spectral density of the external pressure:

$$\begin{aligned} S_q(x,\theta;x',\theta',\omega) &= \frac{1}{2\pi}\int_{-\infty}^{\infty} K_q e^{-i\omega\tau}d\tau (\equiv S_q) \\ S_q &= \sum_{m=1}^{\infty}\sum_{n=-\infty}^{\infty}\sum_{p=1}^{\infty}\sum_{q=-\infty}^{\infty} S_{mn,pq}(\omega)\psi_m(x) \\ &\quad \times \psi_p(x')e^{i(q\theta'-n\theta)} \end{aligned}. \tag{8.14}$$

Equation (8.14) and the orthogonality property of the eigenfunctions yield the cross-spectral densities of the generalized forces:

$$S_{mn,pq} = \frac{I}{4\pi^2\|\psi_m\|^2\|\psi_p\|^2}\int_0^l\int_{-\pi}^{\pi}\int_0^l\int_{-\pi}^{\pi} S_q\psi_m(x)e^{in\theta}$$

$$\times \psi_p(x')e^{-iq\theta'}dxd\theta dx'[d\theta' \tag{8.15}$$

where

$$\|\psi_m\|^2 = \int_0^l \psi_m^2(x)dx, \quad \|\psi_p\|^2 = \int_0^l \psi_p^2(x)dx. \tag{8.16}$$

For the correlation function of the displacements

$$K_w(x, \theta, t; x', \theta', t') = \langle w^*(x, \theta, t)w(x', \theta', t')\rangle(\equiv K_w) \tag{8.17}$$

we find

$$K_w = \int_{-\infty}^{\infty} S_w(x, \theta; x', \theta'; \omega)e^{i\omega\tau}d\omega \left(\equiv \int_{-\infty}^{\infty} S_w e^{i\omega\tau}d\omega\right) \tag{8.18}$$

the cross-spectral density of the displacements being

$$S_w = \sum_{m=1}^{\infty}\sum_{n=-\infty}^{\infty}\sum_{p=1}^{\infty}\sum_{q=-\infty}^{\infty} \frac{S_{mn,pq}}{L_{mn}(\omega)L_{pq}(\omega)}. \times \psi_m(x)\psi_p(x')e^{i(q\theta'-n\theta)} \tag{8.19}$$

8.3 Excitation Cross-Spectral Densities

As established experimentally (Corcos, 1963; Hay, 1964), pressure fluctuations in a turbulent boundary layer may be approximated by a random function stationary in time

$$S_q(x, \theta; x', \theta'; \omega) = S_q(\xi, \eta, \omega) \tag{8.20}$$

where $\xi = x' - x$ and $\eta = \theta' - \theta$ are the separation distances in the longitudinal and circumferential directions, respectively. S_q is most commonly represented in product form

$$S_q(\xi, \eta, \omega) = \langle q^2\rangle\psi(\omega)A(\xi, \omega)B(\eta, \omega) \tag{8.21}$$

where $\langle q^2\rangle = S_q(0, 0, 0)$ is the mean-square value for the external pressure, and $\psi(\omega)$ the spectral density normalized so that $\int_{-\infty}^{\infty}\psi(\omega)d\omega = 1$. The correlation coefficients in the longitudinal and circumferential directions are $A(\xi, \omega)$ and $B(\eta, \omega)$, respectively; generally, they are complex functions of the frequency and of the appropriate separation distance, respectively.

By virtue of the separability assumption, Eq. (8.21), the cross-spectral densities of the generalized forces Eq. (8.15) may also be represented as the product:

$$S_{mn,pq}(\omega) = \langle q^2\rangle \psi(\omega) S_{mp} S_{nq} \tag{8.22}$$

where

$$S_{mp} = \frac{1}{\|\psi_m\|^2 \|\psi_p\|^2} \int_0^l \int_0^l A(\xi,\omega)\psi_m(x)\psi_p(x')dxdx', \tag{8.23}$$

$$S_{nq} = \frac{1}{4\pi^2} \int_{-\pi}^{\pi} \int_{-\pi}^{\pi} B(\eta,\omega) e^{i(n\theta - q\theta')} d\theta d\theta'. \tag{8.24}$$

The functions S_{mp} and S_{nq} may be interpreted as the cross-spectral densities of a corresponding beam strip of length l and a ring of radius R, respectively.

The function $A(\xi,\omega)$ may be taken in the following form: (Crocker, 1969; Elishakoff, 1974a, 1974b)

$$A(\xi,\omega) = e^{-0.265\frac{|\xi|}{\delta_0}} e^{-0.1\left|\frac{\omega\xi}{u_c}\right|} e^{-i\frac{\omega\xi}{u_c}} \tag{8.25}$$

where u_c is the convection velocity and δ_0 the thickness of the boundary layer.

For the function $B(\eta,\omega)$, we use two models: (a) a fully correlated pressure field in the circumferential direction with

$$B_1(\eta,\omega) \equiv 1 \tag{8.26}$$

and (b) the correlation coefficient for rigid-wall measurements in the direction normal to the flow, originally reading (Crocker, 1969):

$$B_{11}(\eta,\omega) = e^{-b_\eta(\omega)|\eta|}, \quad b_\eta(\omega) = -0.715\frac{\omega}{u_c} - \frac{2.0}{\delta_0} \tag{8.27}$$

and modified here, for our purpose, so as to have the following periodicity property:

$$B_{11}(\eta,\omega) = B_{11}(\eta \pm 2\pi k,\omega), \quad k = 1, 2, 3, \ldots \tag{8.28}$$

The expression for $B_{11}(\eta,\omega)$ is (Bolotin and Elishakoff, 1971)

$$B_{11}(\eta,\omega) = e^{-b_\eta(\omega)|\eta|} e^{2\pi b_\eta(k-1)}$$
$$\text{for } \pi(2k-2) \le |\eta| \le (2k-1)$$

$$B_{11}(\eta, \omega) = e^{b_\eta(\omega)|\eta|} e^{2\pi b_\eta(k-1)}$$
$$\text{for } \pi(2k-1) \leq |\eta| \leq 2\pi k \tag{8.29}$$

Now the formula for the cross-spectral density coefficients S_{nq}, Eq. (8.24) may be specified, taking into account the homogeneity property in the circumferential direction

$$S_{nq} = S_{nn}\delta_{nq}, \quad n, q = 0, 1, 2, \ldots \tag{8.30}$$

where S_{nq} is the Kronecker delta. The formula for the cross-spectral density of the external forces, Eq. (8.14), thus becomes

$$S_q = \sum_{m=1}^{\infty} \sum_{n=-\infty}^{\infty} \sum_{p=1}^{\infty} S_{mm,pn}(\omega)\psi_m(x)\psi_p(x')e^{in(\theta'-\theta)} \tag{8.31}$$

and

$$S_{mn,pn} = S_{mp}S_{nn}. \tag{8.32}$$

The function S_{mp} is given by Eqs. (8.23) and (8.24) becomes

$$S_{nn} = \frac{1}{4\pi^2} \int_{-\pi}^{\pi} \int_{-\pi}^{\pi} B(\eta, \omega)e^{in(\theta-\theta')} d\theta d\theta'. \tag{8.33}$$

Equation (8.33) is valid for both models of the correlation coefficients in the circumferential direction. For the first model, however, further simplification is possible since we have

$$S_{nn} = \delta_{n0} \tag{8.34}$$

or, instead of Eq. (8.32)

$$S_{mn,pn} = S_{mp}\delta_{no}. \tag{8.35}$$

8.4 Shell Response Characteristics

We first consider the case

$$S_q(\xi, \eta, \omega) = \langle q^2 \rangle \psi(\omega) A(\xi, \omega) B_1(\eta, \omega). \tag{8.36}$$

Taking into account the statistical orthogonality conditions, Eqs. (8.12) and (8.35), we arrive at the following expression for the cross-spectral density of the displacements:

$$S_w = \sum_{n=1}^{\infty}\sum_{p=1}^{\infty} \frac{S_{mp}}{L^*_{mo}(\omega)L_{po}(\omega)}\psi_m(x)\psi_p(x') \tag{8.37}$$

whereby only axisymmetric modes are excited. The spectral density of the displacements at point (x, θ)

$$S_w(x, \theta, \omega) = S_w(x, \theta; x', \theta', \omega)\Big|_{\substack{\theta' = \theta \\ x' = x}} \tag{8.38}$$

is given by

$$S_w(x, \theta, \omega) = \sum_{m=1}^{\infty}\sum_{p=1}^{\infty} \frac{S_{mp}}{L^*_{mo}(\omega)L_{po}(\omega)}\psi_m(x)\psi_p(x). \tag{8.39}$$

Concentrating on the area-averaged spectral density, we have

$$\begin{aligned} s_w(\omega) &= \frac{1}{2\pi l}\int_{-\pi}^{\pi}\int_{0}^{l} S_w(x, \theta, \omega)dxd\theta \\ &= \sum_{m=1}^{\infty} \frac{\|\psi_m\|^2}{l}\frac{S_{mm}}{|L_{mo}(\omega)|^2} \end{aligned} \quad . \tag{8.40}$$

Note that the following simple expression (from Powell, 1965) is obtained for the cross-spectral density coefficient S_{mm} from the general formula, Eq. (8.23), taking into account the homogeneity property of the external pressure in the longitudinal direction

$$S_{mm} = 4\int_0^1 \mathrm{Re}A(\zeta l\omega)\left[(1-\zeta)\cos m\pi\zeta - \frac{1}{m\pi}\sin m\pi\zeta\right]d\zeta. \tag{8.41}$$

We now attempt to specify the form of the operators $E(\partial/\partial t)$ and $D(\partial/\partial t)$, assuming them to have, for the set of solutions containing the time multiplier $e^{i\omega t}$, the property

$$\begin{aligned} E(d/dt)e^{i\omega t} &= e^{i\omega t}E(i\omega), \quad E(i\omega) = E_R + iE_I \\ D(\partial/\partial t)e^{i\omega t} &= e^{i\omega t}D(i\omega), \quad D(i\omega) = D_R + iD_I \end{aligned} \tag{8.42}$$

the dissipation measure [the ratio of the imaginary and real parts of $E(i\omega)$] being generally a function of the frequency $\psi(i\omega) = E_I/E_R$.

If the system is slightly damped, $\psi(i\omega) \ll 1$ and $D_I/D_R \ll 1$, then the area-averaged spectral density has maxima near the roots of the equation

$$\mathrm{Re}L_{mo}(\omega) = D_R \frac{m^4\pi^4}{l^4} + E_R \frac{h}{R} - \rho h\omega^2 = 0. \tag{8.43}$$

There are the eigenfrequencies of the corresponding undamped shell

$$\omega_{mo} = \left(\frac{D_R}{\rho h} \frac{m^4\pi^4}{l^4} + \frac{E_R}{\rho R} \right)^{1/2}. \tag{8.44}$$

As for the function S_{mm} for fixed $m \neq 1$, it has an absolute maximum at the frequencies close to the "convection" of the external pressure $\Omega_m = m\pi/l$. When this velocity coincides with the phase velocity of the free flexural waves in the shell, the response of the latter is maximal. This phenomenon is analogous to the familiar coincidence effect in acoustics (Mead, 1968). For the case under consideration, the coincidence frequency ω_c is determined as the root of the equation

$$\mathrm{Re}L_{mo}(\omega_c) = 0. \tag{8.45}$$

Noting that $m\pi u_c/l = \omega_c$ and making the appropriate substitution in Eq. (8.44), we have

$$D_R \left(\frac{\omega_c^4}{u_c^4} \right) + E_R \left(\frac{h}{R} \right) - \rho h\omega_c^2 = 0. \tag{8.46}$$

This equation has two roots:

$$\omega_c^2 = 1/2 \left[\omega_{oo}^2 \pm \left(\omega_{00}^4 - 4\omega_o^2\omega_{oo}^2 \right)^{1/2} \right] \tag{8.47}$$

where

$$\omega_o = R^{-1}(E_R/\rho)^{1/2}, \quad \omega\omega_{oo} = u_c^2(\rho h/D_R)^{1/2} \tag{8.48}$$

denote the ring- and coincidence frequencies for the corresponding beam strip, respectively.

For $h/R \ll 1$ and/or $u_c^2 \gg E/\rho$, these two roots approximate to

$$\omega_{c,1} \simeq \omega_o, \quad \omega_{c,2} \simeq \omega_{oo}. \tag{8.49}$$

An axisymmetrical flow-excited cylindrical shell thus has two coincidence frequencies, one very close numerically to the ring frequency and the other to the coincidence frequency of the corresponding beam strip. Note that the beam formally also has two coincidence frequencies, 0 and ω_{oo}. The zero is the counterpart of the ring frequency and obtainable from the latter as the limit for $R \to \infty$; the other may be interpreted as a nontrivial coincidence frequency of the beam strip.

In the case

$$S_q(\xi, \eta, \omega) = \langle q^2 \rangle \psi(\omega) A(\xi, \omega) B_{11}(\eta, \omega) \tag{8.50}$$

the area-averaged spectral density has the form

$$\begin{aligned} s_w(\omega) = &\sum_{m=1}^{\infty} \frac{\|\psi_m\|^2}{l} \frac{S_{mm} S_{oo}}{|L_{mo}(\omega)|^2} \\ &+ \sum_{m=1}^{\infty} \sum_{\substack{n=-\infty \\ n \neq 0}}^{\infty} \frac{\|\psi_m\|^2}{l} \frac{S_{mm} S_{nn}}{|L_{mm}(\omega)|^2}, \end{aligned} \tag{8.51}$$

the first sum representing the part of the response due to axisymmetric modes of vibration, and the second part due to nonaxisymmetric ones. These conclusions regarding two nonzero coincidence frequencies are also valid for the first sum. In the general case, the coincidence effect occurs if

$$\mathrm{Re} L_{mn}(\omega_c) = 0, \tag{8.52}$$

the counterpart of Eq. (8.46) here being

$$\begin{aligned} &\omega_c^2 - \frac{D_R}{\rho h} \left(\frac{\omega_c^2}{u_c^2} - \frac{h^2}{R^2} \right)^2 \\ &- \frac{E_R}{\rho R^2} \frac{\omega_c^4}{u_c^4} \left(\frac{\omega_c^2}{u_c^2} + \frac{h^2}{R^2} \right)^{-2} = 0 \end{aligned} \tag{8.53}$$

with the roots

$$\begin{aligned} \omega_c^2 = 1/2 \Bigg[&- 2\frac{h^2}{R^2} u_c^2 - 1/2\omega_{oo}^2 \\ &\mp 1/2\omega_{oo}^2 \left[1 - 4\frac{\omega_o^2}{\omega_{oo}^2} \right]^{1/2} \\ &\mp \Bigg(2\frac{h^2}{R^2} u_c^2 - 1/2\omega_{oo}^2 \mp \omega_{oo}^2 \\ &\times \left[1 - 4\frac{\omega_o^2}{\omega_{oo}^2} \right]^{1/2} - 4\frac{n^4}{R^4} u_c^4 \Bigg)^{1/2} \Bigg]. \end{aligned} \tag{8.54}$$

As in the axisymmetric case, there are again two coincidence frequencies. The higher one is close to ω_{oo} and the lower one not exceeding the ring frequency and for the low circumferential mode numbers n practically equals ω_o. It should also be noted that near this frequency, the modal density increases substantially (Elishakoff, 1973; Skudrzyk, 1968).

8.5 Random Vibrations in Cylindrical Shells Containing an Acoustic Medium

In this section, we deal with the problems of random vibrations associated with supersonic aircraft. In 1947 Charles E. Yeager flew his Bell X-1 experimental plane, named *Glamorous Glennis,* faster than sound and into history. In the interview by F. Sietzen, Jr. (2003) Yeager said: "In my opinion, there are going to be only three airplanes in aviation to be remembered. The first one, the Wright brothers' – the first airplane to fly. The X-1, the first airplane to go supersonic. And X-15, the first airplane to get into space. That covers the spectrum: Everything in between just had to happen—" (see also Chuck Yeager et al. 1997).

The Nobel-Prize winning Polish poet and essayist Wislawa Szymborska (1923–2012) wrote in her poem (Szymborska, 1998) *Greeting the Supersonics* about the human quest to fly supersonic:

Faster than sound today,
faster than light tomorrow,
we'll turn sound into the Tortoise
and light into the Hare.
Two venerable creatures
From the ancient parable,
a noble team, since ages past
competing fair and square.

The human dream was put into practice by the Concorde British-French turbojet supersonic passenger airliner that was operated between the years 1976 and 2003 (Trubshaw, 2000). It had a maximum speed over twice the speed of sound at Mach 2.04 (1,354 mph or 2,180 km/h at cruise altitude), with seating for 92–128 passengers. First flown in 1969, Concorde entered service in 1976 and continued flying for the next 27 years. The other supersonic transport was the Soviet built Tupolev Tu-144, which operated in the late 1970s. Unfortunately, neither is flying nowadays (Azzerad, 2018; Melik-Karamov, 2000; Orlebar, 2011). However, this situation will not continue indefinitely, taking into account human ingenuity and audacity.

The problem of defining the acoustic pressure field within shells oscillating at random was examined by Bolotin (1968) and many other authors. This section mainly follows the papers by Bolotin, Elishakoff, Efimtsov, Moskalenko and Sharii (1969), and Bolotin and Elishakoff (1971). In these studies, a method was proposed for calculating correlation functions and spectral densities for pressure in an acoustic medium and the noise level within the shell. The following substantial limitation was introduced: the surface functions corresponding to the form of natural oscillations in the acoustic medium were assumed to coincide with the form of shell natural oscillations. This limitation is met by, e.g. closed spherical shells and circular cylindrical shells under specially chosen boundary conditions. The method is extended below to a wider class of shells. Two systems of basic functions, one for the shell and one for the acoustic medium, are used simultaneously. Smooth and reinforced circular cylindrical shells with sound insulation are used as examples. The calculation was made in conformity with oscillations in a pulsating pressure field from a turbulent boundary layer.

Let us examine an elastic shell loaded by normal external forces whose intensity depends on coordinates and time. Let the internal surface of this shell, which will subsequently be called the carrying surface, be in contact with what is broadly speaking a multilayer shell. The purpose of the latter is to reduce transmission of vibrations to the inner region, which is filled with a compressible fluid. The inner shell is of reduced rigidity and is subsequently called the insulating shell. We retain the assumptions made by Bolotin (1968) as to the properties of the carrying and insulating shells: we assume that the carrying shell is subject to classical thin shell theory equations and that normal movements on the insulating shell surfaces are linear functions of the pressures on these surfaces. The fluid within the shell will be treated as an acoustic medium. Sound radiation into the inner medium is disregarded or, more accurately, the losses due to radiation are assumed either to be small (which will be the case with, e.g. a low-density medium) or included in the total losses when the carrying shell oscillates. If a high-speed gas stream flows round the shell, aeroelastic effects must be taken into account. Some results relating to the allowance for these effects may be found by Elishakoff and Khromatoff (1971).

In contrast to the work by Bolotin, it is not assumed here that the form of natural oscillations in the acoustic medium can be obtained from the corresponding form for the shell by multiplying the latter form by certain normal coordinate functions. However, the assumption is retained that the normal coordinate is isolated in expressions for the form of natural oscillations in the medium. Thus the boundary conditions examined are more general than those in the study by Bolotin (1968), but the assumption is introduced that the effect of the medium on carrying shell behavior is negligible. This makes it possible to bring the analytical and numerical calculations to a conclusion without narrowing the class of boundary conditions. Thus it becomes possible to calculate the pressure fields within a cylindrical cavity of finite length bounded by a reinforced radiative shell and acoustically stiff sidewalls.

Let Ω be the middle surface of the carrying shell, Ω_H the inner surface of the insulating shell, and V the volume occupied by the acoustic medium. Let us designate the curved coordinates on surface Ω in terms of $\rho = \{x_1, x_2\}$. Let us take the curved coordinates in V by supplementing x_1 and x_2 by the third coordinate x_3, taken along the outer normal to Ω. Thus $r = \{x_1, x_2, x_3\}$. Oscillations in the carrying shell are described by the resolvent

$$lw = q - m_{11}w - m_{12}w_H \tag{8.55}$$

where $w(\rho, t)$ is the normal movement of the carrying shell middle surface, $w_H(\rho, t)$ the normal movement of the insulating shell inner surface, $q(\rho, t)$ the intensity of the normal external forces, l a linear operator taking account of inertial and viscous forces for the carrying shell, and m_{11} and m_{12} are linear operators taking account of the properties of the sound-insulation layer (Bolotin, 1968). The medium within V is subject to the wave equation

$$\Delta\varphi - \frac{1}{c^2}\frac{\partial^2\varphi}{\partial t^2} = 0 \tag{8.56}$$

where $\varphi(r, t)$ is the velocity potential and c the speed of sound in the acoustic medium. The condition of impermeability is imposed on Ω_H, and also the condition linking pressure on the sound-insulating layer inner surface to movements w and w_H on its outer and inner surfaces

$$\left.\frac{\partial\varphi}{\partial x_3}\right|_{\Omega_H} = \frac{\partial w_H}{\partial t}, \quad -\rho\left.\frac{\partial\varphi}{\partial t}\right|_{\Omega_H} = m_{21}w + m_{22}w_H, \tag{8.57}$$

ρ is the density of the acoustic medium and m_{21} and m_{22} are the linear operators introduced by Bolotin (1968). Equations (8.55) and (8.56) and boundary conditions (8.57) are supplemented by boundary conditions for the carrying shell and, if required, by conditions for potential $\varphi(r, t)$. Thus if part Ω_0 of the surface bounding V is acoustically stiff or acoustically soft, one of the following two conditions is imposed on this part, as appropriate

$$\left.\frac{\partial\varphi}{\partial n}\right|_{\Omega_0} = 0 \text{ and } \left.\frac{\partial\varphi}{\partial t}\right|_{\Omega_0} = 0. \tag{8.58}$$

In general, the conditions (8.58) are replaced by conditions of impedance type.

Let us examine steady oscillations in the system under load $q(\rho, t)$, which forms a random space-time field static in time and in general inhomogeneous on Ω. Since the problem is linear and the mathematical expectations of fields $q(\rho, t)$, $w(\rho, t)$, $w_H(\rho, t)$, and $\varphi(r, t)$ do not depend on time, we may without loss of generality dwell on the case in which

$$\langle q(\rho, t)\rangle = \langle w(\rho, t)\rangle = \langle w_H(\rho, t)\rangle = \langle\varphi(r, t)\rangle = 0$$

(the angled brackets indicate averaging over the set of realizations). Let us solve the stochastic boundary value problem in a correlation approximation, assuming the external pressure correlation function to be given. A Fourier transform in time for this function will be needed subsequently

$$S_q\left(\rho, \rho'; \omega\right) = \frac{1}{2\pi} \int_{-\infty}^{\infty} \langle q^*(\rho, t) q\left(\rho', t + \tau\right)\rangle e^{-i\omega\tau} d\tau. \tag{8.59}$$

The asterisk indicates the transition to a complex conjugate magnitude.

The problem is to find a similar function $S_p\left(r, r'; \omega\right)$ for the pulsating component of pressure $p = -\rho\partial\varphi/\partial t$ within the shell

$$S_p\left(r, r'; \omega\right) = \frac{1}{2\pi} \int_{-\infty}^{\infty} \langle p^*(r, t) p\left(r', t + \tau\right)\rangle e^{-i\omega\tau} d\tau. \tag{8.60}$$

The contribution of pulsations with frequencies belonging to segment $[\omega_*, \omega_{**}]$ to the pulsation pressure RMS value is

$$\langle p^2(r, \omega_*, \omega_{**})\rangle = \int_{\omega_*}^{\omega_{**}} S_p(r, r, \omega) d\omega. \tag{8.61}$$

The noise level in decibels is the final result of the calculation

$$L(r, \omega_*, \omega_{**}) = 10lg \frac{\langle p^2(r, \omega_*, \omega_{**})\rangle}{p_0^2} \tag{8.62}$$

where p_0 is a threshold pressure (normally $p_0 = 0.2 \times 10^{-5} nm^{-2}$).

Let us use Pugachev's method (1960) to solve the stochastic boundary value problem depicted by Eqs. (8.55)–(8.56) with supplementary conditions (8.57) and (8.58). We seek the movement functions $w(\rho, t)$ and $w_H(\rho, t)$ as Fourier series for the forms of shell natural oscillations and a stochastic Fourier integral for time

$$\omega(\rho, t) = \sum_\alpha \int_{-\infty}^{\infty} W_\alpha(\omega) \psi_\alpha(\rho) e^{i\omega t} d\omega, \tag{8.63}$$

$$\omega_H(\rho, t) = \sum_\alpha \int_{-\infty}^{\infty} W_{H\alpha}(\omega) \psi_\alpha(\rho) e^{i\omega t} d\omega.$$

Here $W_a(\omega)$, $W_{Ha}(\omega)$ are generalized random functions (spectra of the appropriate movements). The system of functions $\psi_a(\rho)$ is complete and orthonormal on Ω and satisfies all necessary boundary conditions. Let us represent the external load function in similar form

$$q(\rho, t) = \sum_\alpha \int_{-\infty}^{\infty} Q_\alpha(\omega) \psi_\alpha(\rho) e^{i\omega t} d\omega. \tag{8.64}$$

Let us expand the velocity potential $\varphi(r, t)$ into a Fourier series for solutions $\chi(r, \omega)$ of the Helmholtz equation $\Delta\chi - \left(\omega^2/c^2\right)\chi = 0$, which satisfies conditions of (8.58) type and conditions of limitedness within V. The curved coordinates within V are selected so that the normal coordinate x_3 is isolated, i.e. so that $\chi_\alpha(r; \omega) = \varphi_\alpha(\rho)Z_\alpha(x_3, \omega)$. As before, we use expansion into a stochastic Fourier series for a time. Thus

$$\varphi(r, t) = \sum_\alpha \int_{-\infty}^{\infty} \Phi_\alpha(\omega)\varphi_\alpha(\rho)Z_\alpha(x_3, \omega)e^{i\omega t}d\omega. \tag{8.65}$$

To calculate the functions $S_p\left(r, r'; \omega\right)$, it is essential to establish the relationships between the velocity potential spectrum $\Phi_a(\omega)$ and the load spectrum $Q_a(\omega)$

$$\Phi_\alpha(\omega) = \sum_\beta F_{\alpha\beta}(i\omega)Q_\beta(\omega). \tag{8.66}$$

This includes the system transfer matrix $F_{\alpha\beta}(1\omega)$. We find Relationships (8.66) by substituting Formulas (8.63)—(8.65) in Eq. (8.55) and boundary conditions (8.57). Substitution into (8.55) gives

$$[L_\alpha(\omega) + M_{11}(\omega)]W_\alpha(\omega) + M_{12}(\omega)W_{H\alpha}(\omega) = Q_\alpha(\omega). \tag{8.67}$$

Here $L_a(\omega)$, $M_{11}(\omega)$, $M_{12}(\omega)$ are representations of operators l, m_{11}, and m_{12} in the corresponding Fourier space (impedances). We obtain two other relationships from (8.57) by projecting the results of substituting Formulas (8.63)—(8.65) in (8.57) onto the axes of the functional basis $\varphi_a(\rho)$ or $\psi_a(\rho)$. If functions $\varphi_a(\rho)$ are orthogonal on Ω_H, we obtain the following instead of (8.57)

$$\zeta'_\alpha(\omega)\Phi_\alpha(\omega) = i\omega\sum_\beta a_{\alpha\beta}(\omega) \tag{8.68}$$

$$-i\omega\rho\zeta_\alpha(\omega)\Phi_\alpha(\omega) = \sum_\beta a_{\alpha\beta}\left[M_{21}(\omega)W_\beta(\omega) + M_{22}(\omega)W_{H\beta}(\omega)\right].$$

Here $M_{21}(\omega)'$, $M_{22}(\omega)$ are Fourier representations of operators m_{21} and m_{22}; in addition, the functions of frequency ω are introduced

$$\zeta_\alpha = Z_\alpha(x_3, \omega)\big|_{x_3=-H}\, \zeta_{\alpha'} = \left.\frac{\partial Z_\alpha(x_3, \omega)}{\partial x_3}\right|_{x_3=-H}$$

and coefficients of reexpansion $a_{\alpha\beta}$, i.e.

$$a_{\alpha\beta} = \left(\psi_\beta, \varphi_\alpha\right)/(\varphi_\alpha, \varphi_\alpha). \tag{8.69}$$

If functions $\varphi_a(\rho)$ do not satisfy the conditions of orthogonality on Ω_H, a reexpansion on basis $\psi_a(\rho)$ should be carried out. Henceforth we will discuss the former case.

To calculate the transfer matrix $F_{\alpha\beta}(i\omega)$, it is essential to exclude spectra $W_\alpha(\omega)$, $W_{H\alpha}(\omega)$ from Eqs. (8.67) and (8.68) and to solve the equations in relation to $\Phi_\alpha(\omega)$. Generally speaking, the system of equations in infinite, and the elements of the transfer matrix $F_{\alpha\beta}(i_\omega)$ are functions of frequency ω with an infinite number of poles lying close to the true axis. Each pole for $\mathrm{Re}\omega > 0$ corresponds to one of the free oscillation frequencies in a linked elastic-acoustic system. If the load spectrum $q(\rho, t)$ is broadband, an extremely large number of poles may lie within the frequency range which interests us. This makes it technically difficult to calculate $F_{\alpha\beta}(i\omega)$.

A class of problems is indicated by Bolotin (1968) for which the elements of the matrix $F_{\alpha\beta}(i\omega)$ are readily calculated. Let $\varphi_\alpha(\rho) \equiv \psi_\alpha(\rho)$, i.e. the forms of shell natural oscillations coincide with the surface functions of acoustic forms $\chi_\alpha(r, \omega)$; then $a_{\alpha\beta} = \delta_{\alpha\beta}$. We find from Eqs. (8.67) and (8.68)

$$F_{\alpha\beta}(i\omega) = -\frac{i\omega M_{21}(\omega)\delta_{\alpha\beta}}{M_\alpha\left[L_\alpha(\omega) + M_{11}(\omega) - M_{12}(\omega)M_{21}(\omega)\zeta_\alpha'(\omega)M_\alpha^{-1}\right]}. \tag{8.70}$$

The symbols

$$M_\alpha(\omega) = M_{22}(\omega)\zeta_\alpha'(\omega) - \rho\omega^2\zeta_\alpha(\omega) \tag{8.71}$$

are used in Formula (8.70) and subsequently.

The following approximation method can be used where $\varphi_\alpha(\rho) \not\equiv \psi_\alpha(\rho)$. Terms containing w_H movements are disregarded in Eq. (8.55), and Eq. (8.67) are then solved in succession. As a result, we arrive at Relationships (8.66) with the transfer matrix

$$F_{\alpha\beta}(i\omega) = -\frac{i\omega M_{21}(\omega)a_{\alpha\beta}}{M_\alpha(\omega)[L_\alpha(\omega) + M_{11}(\omega)]} \tag{8.72}$$

where $a_{\alpha\beta} = \delta_{\alpha\beta}$, Formula (8.72) differs from (8.70) in that it lacks the last term in the brackets. We can use this to evaluate the error in the approximation method (at least for a certain class of shells).

We obtain an expression for the pressure mean square if we use Formulas (8.61), (8.65), (8.66), (8.70), and (8.72)

$$\langle p^2(r; \omega_*, \omega_{**})\rangle = \sum_\alpha\sum_\beta\sum_\gamma\sum_\delta \rho^2\omega^* \int\limits_{\omega_*}^{\omega_{**}} F_{\alpha\beta}(-i\omega)F_{\beta\delta}(i\omega)S_{Q_\gamma Q_\delta}(\omega)\chi_\alpha^*(r, \omega)\chi_\beta(r, \omega)d\omega_\beta. \tag{8.73}$$

As was shown above, functions $F_{\alpha\beta}(i\omega)$ have a denumerable infinite set of poles. Due to the high quality of the elastic-acoustic system, functions $F_{\alpha\beta}(i\omega)$ have sharp spikes near each natural frequency. There are therefore serious difficulties involved in the numerical determination of the integrals in (8.73). An approximation method related to the method of steepest descents is proposed by Bolotin (1968). In the vicinity of each pole, or more precisely in the vicinity of each real root $\omega_{\alpha 1}, \ldots, \omega_{\alpha 2}, \ldots,$ of the equation

$$\mathrm{Re} M_\alpha(\omega) = 0 \tag{8.74}$$

the rapidly changing functions $M_\alpha(\omega)$ and $\zeta'(\omega)$ are expanded into power series for $\omega^2 - \omega_{\alpha\mu}^2$. Then the mean-value theorem is applied, after within the integrals are taken in finite form. The roots $\omega_{\alpha\beta}$ have the meaning of natural oscillation frequencies in an "acoustic medium-insulating shell" system with an absolutely stiff carrying shell. Limiting ourselves to terms not higher than $\omega^2 - \omega_{\alpha\beta}^2$ and calculating $F_{\alpha\beta}(i\omega)$ according to (8.70), we arrive at the formula

$$\begin{aligned} &\langle p^2(r;\omega_*, \omega_{**})\rangle \\ &\sim \sum_\alpha \sum_\mu \left| \frac{M_{21}(\omega_{\alpha\mu})\chi(r;\omega_{\alpha\mu})}{L_2(\omega_{\alpha\mu}) + M_{11}(\omega_{\alpha\mu})} \right|^2 \frac{\pi b_{\alpha\mu}\omega_{\alpha\mu}}{g_{\alpha\mu}} S_{Q_\alpha Q_\alpha}(\omega_{\alpha\mu}) \end{aligned} \tag{8.75}$$

in which summation is carried out for $\omega_{\alpha\mu}$ belonging to the segment $[\omega_*, \omega_{**}]$. The following symbols are adopted (all functions in the right-hand parts are calculated for $\omega = \omega_{\alpha\mu}$):

$$b_{\alpha\mu} = -\frac{1}{\mathrm{Re} M_{22}} \frac{\partial \zeta_\alpha}{\partial \zeta'_\alpha / \partial(\omega^2)}, \tag{8.76}$$

$$g_{\alpha\mu} = \frac{\mathrm{Im} M_{22}}{\mathrm{Re} M_{22}} - \mathrm{Im} \frac{M_{12} M_{21}}{M_{22}(L_\alpha + M_{11})}.$$

If $\varphi_\alpha(\rho) \not\equiv \psi_\alpha(\rho)$, we use Formula (8.72). The calculations give

$$\begin{aligned} &\langle p^2(r;\omega_*, \omega_{**})\rangle \\ &\sim \sum_\alpha \sum_\beta \sum_\mu \left| \frac{M_{21}(\omega_{\alpha\mu})\chi(r;\omega_{\alpha\mu})}{L_\beta(\omega_{\alpha\mu}) + M_{11}(\omega_{\alpha\mu})} \right|^2 \frac{\pi b_{\alpha\mu}\omega_{\alpha\mu} a_{\alpha\beta}^2}{g_{\alpha\mu}} S_{Q_\beta Q_\beta}(\omega_{\alpha\mu}) \end{aligned} \tag{8.77}$$

where $b_{\alpha\mu}$ and $\omega_{\alpha\mu}$ are calculated by Formulas (8.76), in which $M_{12} \equiv 0$ should be assumed.

Let us apply Eqs. (8.75) and (8.77) to the calculation of the noise level within a round cylindrical shell loaded by random pressure pulses. Let us first discuss the case of a smooth isotropic shell. Let us introduce the cylindrical system of coordinates x, r, and θ, and take equations for oscillations in the carrying shell in the form

$$D\Delta_2\Delta_2 w - \frac{1}{R}\frac{\partial^2 f}{\partial x^2} + \rho_0 h\frac{\partial^2 w}{\partial t^2} = q - m_{11}w - m_{12}w_H \tag{8.78}$$

$$\Delta_2\Delta_2 f + E\frac{h}{R}\frac{\partial^2 w}{\partial x^2} = 0, \quad \Delta_2 = \frac{\partial^2}{\partial x^2} + \frac{1}{R^2}\frac{\partial^2}{\partial \theta^2}.$$

Here $f(x, \theta, t)$ is the function of forces in the middle surface, h is the thickness of the shell, R is the radius of its middle surface, ρ_0 is the density of the shell material, and D and E are linear viscoelastic operators corresponding to cylindrical stiffness and modulus of elasticity. Let us impose Vlasov's (1949) conditions on the ends $x = 0$ and $x = l$

$$w = \frac{\partial^2 w}{\partial x^2} = \frac{\partial^2 f}{\partial x^2} = \frac{\partial^2 f}{\partial \theta^2} = 0. \tag{8.79}$$

Basic functions $\psi_\alpha(\rho)$ take the form

$$\psi_{mn}(\rho) = \sin\frac{m\pi x}{l}e^{\mathrm{i}n\theta}\begin{pmatrix} m = 1, 2, \ldots \\ n = 0, \pm 1, \pm 2, \ldots \end{pmatrix}. \tag{8.80}$$

A pair of indices $\alpha = (m, n)$ is introduced, one of which characterizes the wavelength along the shell generatrix, the other the wavelength in the peripheral direction. We obtain coincidence of basis functions $\varphi_{mn}(\rho) \equiv \psi_{mn}(\rho)$ when the end wall $x = 0$ and $x = l$ are acoustically soft, i.e. the second of Conditions (8.58) is satisfied. Here

$$\chi_{mn}(r; \omega) = \sin\frac{m\pi x}{l}e^{\mathrm{i}n\theta}J_n(x_m r), \quad (m = 1, 2, \ldots; n = 0 \pm 1, \pm 2, \ldots) \tag{8.81}$$

where

$$x_m^2 = \frac{\omega^2}{c^2} - \frac{m^2\pi^2}{l^2} \tag{8.82}$$

is introduced.

In this case, the calculation is made with Formulas (8.75). If, however, the end walls $x = 0$ and $x = l$ are acoustically stiff, i.e. the first of Conditions (8.58) is imposed,

$$\chi_{mn}(r; \omega) = \cos\frac{m\pi x}{l}e^{\mathrm{i}n\theta}J_n\left(x_m^r\right) \quad (m = 1, 2, \ldots; n = 0, \pm 1, \pm 2, \ldots). \tag{8.83}$$

The calculation requires the application of Eq. (8.77), the coefficients of (8.69) for reexpansion of functions $\psi_{\mu\nu}(\rho)$ on basis $\varphi_{\mu\nu}(\rho)$ being determined by the formula

$$a_{(mn)(\mu\nu)} = \frac{\varepsilon_\mu m\left[1 - (-1)^{m+\mu}\right]\delta_{\mu\nu}}{\left(m^2 - \mu^2\right)\pi}, \tag{8.84}$$

$$\varepsilon_0 = 1, \quad \varepsilon_\mu = 2 \quad (\mu = 1, 2, \ldots).$$

Let us examine a round cylindrical shell reinforced with a system of stringers and formers. Let the stiffness of the reinforcement be sufficient for stringer and former deformations to be disregarded in calculating oscillations. Let the characteristic space scales of external load correlation be small compared with the panel dimensions; we will treat the shell as a combination of independent panels resting freely on the framework. Let the framework form a rectangular network with side lengths a_1 and a_2 on Ω.

The end walls $x = 0$ and $x = l$ are assumed acoustically stiff. Let us designate the middle surface of the jth panel Ω_j, and the local coordinates for the panel x_1 and x_2. Index α in formulas of the (8.77) type is replaced by the three indices m_1, m_2, and j, so that

$$\psi_{n_1 m_2 j}(\rho) = \sin\frac{m_1 \pi x_1}{a_1} \sin\frac{m_2 \pi x_2}{a_2} \quad \left(\rho \in \Omega_j\right), \tag{8.85}$$

$$\psi_{m_1 m_2 j}(\rho) = 0 \quad (\rho \notin \Omega_j).$$

We designate the position of the panel on the shell using the two number N_1 and N_2 and assume that $x_1 = x_1 - N_1 a_1, x_2 = R\theta - N_2 a_2 (N_1, N_2 = 0, 1, 2, \ldots)$. We calculate coefficients (8.69)

$$a_{(mn)(m_1 m_2 j)} = \frac{2}{\pi R l} \int\limits_0^{a_1} \int\limits_0^{a_2} \sin\frac{m_1 \pi x_1}{a_1} \sin\frac{m_2 \pi x_2}{a_2} \cos\frac{m\pi (N_1 a_1 + x_1)}{l} \times \exp\left[\frac{in}{R}(N_2 a_2 + x_2)\right] dx_1 dx_2 \tag{8.86}$$

The obvious expression for these coefficients is not written out. If the characteristic scales of external load correlation are comparable with the plan dimensions of the panel, allowance must be made in a (8.72) type formula for the mutual correlation between identical forms of oscillation in all the panels. In Eq. (8.77), summation should be carried out over the indices $\alpha = (m, n)$, $\beta = (m_1, m_2)$, $f = (N_1, N_2)$, $f' = \left(N_1', N_2'\right)$, and μ

$$\langle p^2(r; \omega_*, \omega_{**}) \rangle \sim \sum_\alpha \sum_\beta \sum_j \sum_{j'} \sum_\mu \times \frac{\left|M_{21}\left(\omega_{\alpha\mu}\right) \chi\left(r, \omega_{\alpha\mu}\right)\right|^2}{\left[L_{j\beta}\left(\omega_{\alpha\mu}\right) + M_{11}\left(\omega_{\alpha\mu}\right)\right]^*} \frac{\pi b_{\alpha\mu} \omega_{\alpha\mu} a^*_{\alpha j\beta} a_{\alpha j'\beta}}{\left[L_{j'\beta}\left(\omega_{\alpha\mu}\right) + M_{11}\left(\omega_{\alpha\mu}\right)\right] g_{\alpha\mu}} S_{Q_{j\beta} Q_{j'\beta}}\left(\omega_{\alpha\mu}\right) \tag{8.87}$$

If the external force field is homogeneous, the spectral densities of the generalized forces corresponding to identical forms of oscillation in each separate panel $S_{Q_{j\beta}Q_{j\beta}}$ do not depend on index j. Impedance $L_{j\beta}$ does not depend on j, since all the panels are assumed to be identical. The preceding formula is replaced by the following:

$$\langle p^2(r;\omega_*,\omega_{**})\rangle \sim \sum_\alpha\sum_\beta\sum_\mu \left|\frac{M_{21}(\omega_{\alpha\mu})\chi(r,\omega_{\alpha\mu})}{L_\beta(\omega_{\alpha\mu})+M_{11}(\omega_{\alpha\mu})}\right|^2 \frac{\pi b_{\alpha\mu}\omega_{\alpha\mu}}{g_{\alpha\mu}} \times \left[c_{\alpha\beta}S_{Q_\beta Q_\beta}(\omega_{\alpha\mu})+d_{\alpha\beta}(\omega_{\alpha\mu})\right]. \quad (8.88)$$

Here

$$c_{(mn)(m_1m_2)} = \sum_j \left|a_{(mn)(m_1m_2j)}\right|^2,\quad d_{(mn)(m_1m_2)} = \sum_{j\neq j'}\sum a^*_{(mn)(m_1m_2)}a_{(mn)(m_1m_2j')}S_{Q_{j\beta}Q_{j'\beta}}. \quad (8.89)$$

The term $d_{\alpha\beta}$ makes allowance for mutual correlations between identical forms of oscillation in various panels. Having Eq. (8.86), the first of the (8.89) series can be summed

$$C_{(mn)(m_1n_2)} = \frac{2\varepsilon_{m^2}\varepsilon_{n^2}\left[1-(-1)^{m_1}\cos(m\pi a_1/l)\right]}{\pi Rla_1^3a_2^3\left[(m_1/a_1)^2-(m/l)^2\right]} \times \frac{\left[1-(-1)^{m_1}\cos(na_2/R)\right]}{\left[(m_2\pi/a_2)-(n/R)^2\right]}.$$

Particular cases in which $m_1/a_1 = m/l$ or $m_2\pi/a_2 = n/R$ can be obtained by passage to the limit.

Let us define the fluctuation component of pressure on the shell by the spectrum of space correlations $S_q(\rho,\rho';\omega)$. For this function we take the multiplicative presentation (Crocker, 1969), modified to take account of the condition of periodicity in the peripheral direction

$$S_q(\rho,\rho';\omega) = \langle q^2\rangle\varphi(\omega)f_1(\xi,\omega)f_2(\eta,\omega)$$

$$f_1(\xi,\omega) = e^{-a|\xi|}e^{-ib\xi},\ \varphi_k(\eta) = e^{-c|\eta|}e^{2\pi c(k-1)}, \quad (8.90)$$

$$f_2(\eta,\omega) = \begin{cases}\varphi_k(\eta)\pi pH(2k-2)\pi \le |\eta| \le \pi(2k-1)\\ \varphi_k^{-1}(\eta)\pi pH(2k-1)\pi \le |\eta| \le 2\pi k(k=1,2,\ldots)\end{cases},$$

Here $\langle q^2\rangle$ is the RMS value of the external load, $\varphi(\omega)$ is the spectral density, and $a(\omega)$, $b(\omega)$, $c(\omega)$ are empirical functions of frequency. The mutual spectral densities of the generalized forces $S_{Q_\alpha Q_\beta}(\omega)$ in (8.75) are calculated by the formula ($\alpha =$

$m, n; \beta = m_1, n_1)$

$$S_{Q_\alpha Q_\beta}(\omega) = \langle q^2 \rangle \varphi(\omega) I_{mm_1}(\omega) I_n(\omega) \delta_{nn_1}. \tag{8.91}$$

Here the following symbols are used

$$d_m = \frac{al}{m\pi},\ e_m = \frac{bl}{m\pi},\ E_m = 1 + d_m^2 + e_m^2,\ D_m = E_m^2 - 4e_m^2$$

$$\begin{aligned} I_{mm}(\omega) &= \frac{4d_m}{\pi m} \frac{E_m}{D_m} + \frac{8\left[E_m^2 - 4e_m^2\left(1 + 2d_m^2\right)\right]}{\pi^2 m^2 D_m^2}\left[1 - (-1)^m e^{-al} \cos bl\right] \\ &+ \frac{32 d_m e_m}{\pi^2 m^2 D_m^2}\left(E_m^2 - 2e_m^2\right)(-1)^m e^{-al} \sin bl \end{aligned} \tag{8.92}$$

$$I_n(\omega) = \frac{\varepsilon_n c R}{\pi\left(c^2 R^2 + n^2\right)}\left[1 - (-1)^n e^{-\pi c R}\right].$$

The expressions for $I_{mm_1} (m \neq m_1)$ are not written out to save space.

The mutual spectral densities of generalized forces for similar forms of oscillation in an individual panel in Eq. (8.77) are ascertained as in (8.91) ($\beta = m, n$)

$$S_{Q_\beta Q_\beta}(\omega) = \langle q^2 \rangle \varphi(\omega) I_{mm}(\omega) I_{nn}(\omega).$$

The function I_{mm} is obtained from (8.92) by the formal replacement of l by a_1. The relationships for $S_{Q_{j\beta} Q_{j'\beta}}(\omega)$ ($j \neq j'$; $j = p, q$; $j' = k, r$) take the form

$$I_{(pk)(m)} = e^{-aa_1|p-h|}[\cos ba_1(p-k) A(d_m, e_m) - \sin ba_1|p-k| B(d_m, e_m)], \tag{8.93}$$

$$I_{(qr)(n)} = C(d_n, e_n) e^{ca_2|q-r|} \text{ for } |q - r| \leq \pi R/a_2 - 1.$$

Here

$$k_1 = \left(d_m^2 - e_m^2 + 1\right)^2 - 4d_m^2 e_m^2,\ k_2 = 1 - (-1)^m \operatorname{ch} aa_1 \cos ba_1$$

$$k_3 = 4d_m e_m\left(d_m^2 - e_m^2 + 1\right),\ k_4 = (-1)^m \operatorname{sh} aa_1 \sin ba_1$$

$$A(d_m, e_m) = 8m^{-2}\pi^{-2} D_m^{-2}(k_1 k_2 - k_3 k_4), \tag{8.94}$$

$$B(d_n, e_n) = 8m^{-2}\pi^{-2} D_m^{-2}(k_1 k_4 + k_2 k_3),$$

$$C(d_n, e_n) = 8m^2\pi^2\left(c^2 a_2^2 \pi^2\right)^{-2}\left[1 - (-1)^n \operatorname{ch} ca_2\right].$$

An expression can be obtained for the evaluation of the pressure mean square with allowance for mutual correlations of various forms of oscillation in a linked elastic-acoustic system, within the framework of the approximation approach set out in Sect. 8.3. As a supplement to the term written out in Eq. (8.75), we obtain

$$2\mathrm{Re}\sum_{\alpha<\beta}\sum\sum_{\mu}\sum_{\delta}\left[H_\beta\left(r,\omega_{\alpha\mu}\right)+H_\alpha\left(r,\omega_{\beta\delta}\right)\right] \tag{8.95}$$

where

$$D=(1+\eta)\left(\Omega^2+\eta\omega^2\right)+2\delta\omega^2(1+\eta)\left(\Omega^2+\eta\omega^2\right)+\delta^4\omega^4(1+\delta)$$

$$B,C=2^{-1/2}\left[D\pm(1+\eta)\left(\Omega^2+\eta\omega^2\right)+\delta^2\omega^2\right]^{1/2},\ \omega^2\approx\Omega^2(1-b)$$

$$H_\beta\left(r,\omega_{\alpha\mu}\right)=\frac{C_{\alpha\mu}\mathrm{Re}G_\beta\left(\omega_{\alpha\mu}\right)+B_{\alpha\mu}\mathrm{Im}G_\beta\left(\omega_{\alpha\mu}\right)}{\left(B_{\alpha\mu^2}+C_{\alpha\mu^2}\right)\left|G_\beta\left(\omega_{\alpha\mu}\right)\right|^2}, \tag{8.96}$$
$$\times\pi S_{Q_\alpha Q_\beta}\left(\omega_{\alpha\mu}\right)\chi_{\alpha^*}\left(r,\omega_{\alpha\mu}\right)\chi_\beta\left(r,\omega_{\alpha\mu}\right)$$
$$G_\beta=\left[\left(L_\beta+M_{11}\right)M_\beta-M_{12}M_{21}\zeta_\beta'\right]\left(\rho\omega^2\zeta_\beta\right)^{-1}.$$

The indices have been omitted in Expressions (8.96) defining $B_{\alpha\mu}$, $C_{\alpha\mu}$, and $D_{\alpha\mu}$.

The principal contribution to series (8.56) and (8.95) is made by terms corresponding to "triple coincidence," i.e. to the proximity of the shell natural oscillation frequencies to the acoustic oscillation frequencies and the frequency of shell radial oscillations. The latter corresponds on the one hand to the coincidence of the load wave numbers and the form of oscillations at identical frequencies and on the other, according to the asymptotic evaluations by Bolotin (1963), to bunching of shell natural oscillation frequencies.

8.6 Conclusion

In this chapter, new analytical results on random vibrations of shallow cylindrical shells in a turbulent boundary layer are presented. It was shown that there exist two coincidence frequencies, one closer to the ring frequency and the other to the coincidence frequency in a beam strip of the same length and elastic properties. This find may account for the minima in the noise-reducing behavior of the cylindrical shell, referred to by Rattayya and Junger (1969), Manning and Maidank (1964), White (1966) and Manning (1969). Additionally, the theory was presented to evaluate noise levels inside cylindrical shells containing acoustic medium. Further works include those by Crocker (1969), Crocker and White (1969), Elishakoff and Khromatoff (1970, 1971), Elishakoff and Efimtsov (1972), Efimtsov and Moskalenko (1973),

Elishakoff (1974a), Efimtsov (1982, 1984, 1987), Efimtsov, Kazlov, Kravchenko and Anderson (1999), Cunningham and White (2003), Cunningham, Langley and White (2003), Efimtsov and Lizarev (2011), and possibly others. Space shuttle-oriented research was conducted by Elishakoff, Lin and Zhu (1994), Elishakoff, Baruch, Zhu and Caimi (1995) and Margasahayam and Caimi (1997).

References

Azzerad, L. (2018). *Supersonic: The design and lifestyle of concorde*. NewYork: Prestel Publishing.

Bolotin, V. V., & Elishakoff, I. (1971). Random vibrations of elastic shells containing an acoustic medium. *Mechanics of Solids, 6,* 99–107.

Bolotin, V. V. (1963). On the density of the distribution of natural frequencies of thin elastic shells. *PMM: Journal of Applied Mathematics and Mechanics, 27*(2), 362–364. (in Russian).

Bolotin, V. V. (1968). Pressure fields within shells performing random vibration. *Mechanics of Solids, 3,* 50–58.

Bolotin, V. V. (1969). *Statistical methods in structural mechanics* (English translation by S. Aroni, of the 2nd ed., Chap. 5, pp. 102–153). San Francisco: Holden-Day Inc.

Bolotin, V. V., Elishakoff, I., Efimtsov, B. M., Moskalenko, V. N., & Sharii, N. V. (1969). Method of analysis of acoustic fields inside stiffened shells. In V. V. Bolotin (Ed.), *Proceedings of the Scientific Technical Conference Based on Results of Scientific Investigations in Years 1968–1969, Section of Energetic Machinery, Subsection of Dynamics and Strength of Machines* (pp. 5–26). Moscow: Power Engineering Institute, Press. (in Russian).

Corcos, G. M. (1963). Resolution of pressure in turbulence. *Journal of the Acoustical Society of America, 35,* 192–198.

Crocker, M. J., & White, R. W. (1969). Response of uniform beams to homogeneous random pressure fields. *Journal of the Acoustical Society of America, 45,* 1097–1103.

Crocker, M. J. (1969). The response of a supersonic transport fuselage to boundary layer and to reverberant noise. *Journal of Sound and Vibration, 9,* 6–20.

Cunningham, P. R., & White, R. G. (2003). Dynamic response of doubly curved honeycomb sandwich panels to random acoustic excitation. Part 1: Experimental study. *Journal of Sound and Vibration, 264*, 579–603.

Cunningham, P. R., Langley, R. S., & White, R. G. (2003). Dynamic response of doubly curved honeycomb sandwich panels to random acoustic excitation Part 2: Theoretical study. *Journal of Sound and Vibration, 264,* 605–637.

Efimtsov, B. M., & Lizarev, L. A. (2011). Sound pressure in cylindrical shells with regular orthogonal system of stiffener, excited by a random fields of forces. *Journal of Sound and Vibration, 330*(15), 3684–3697.

Efimtsov, B. M., & Moskalenko, V. N. (1973). *Excitation of multispan plates in a random acoustic field, FTD-MT-23-231-73*. Wright-Patterson AFB, Ohio: Foreign Technology Division.

Efimtsov, B. M. (1982). Characteristics of the field of turbulent wall pressure fluctuations at large Reynolds-numbers. *Soviets Physics Acoustics, 28*(4), 289–292.

Efimtsov, B. M., Kazlov, N. M., Kravchenko, S. V., & Anderson A. O. (1999). Wall pressure-fluctuation spectra at small forward-facing steps. *AIAA Paper 1964*. Washington, D.C.: AIAA Press.

Efimtsov, B. M. (1984). Similarity criteria for the spectra of wall pressure fluctuations in a turbulent boundary layer. *Akusticheskii Zhurnal, 30,* 58–61. (in Russian).

Efimtsov, B. M. (1987). Vibration of fuselage wing of airplane, due to turbulent layer. *Scientific Notes of TSAGI, 18*(1), 90–99. (in Russian).

Elishakoff, I., & Efimtsov, B. (1972). On the vibrations of unbounded plates and the field of random forces. *Soviet Applied Mechanics, 8,* 903–906.

Elishakoff, I., & Khromatoff, V. E. (1970). Effect of the supersonic flow on the behavior of panels in the field of random forces. In *Proceedings of the Moscow Power Engineering Institute, Dynamics and Strength of Machines* (Vol. 74, pp. 54–60). Moscow: Power Engineering Institute Press. (in Russian).

Elishakoff, I., & Khromatoff, V. V. (1971). Statistical analysis of the vibrations of a panel in a supersonic flow. *Mechanics of Solids, 6*(1), 41–44.

Elishakoff, I., Baruch, M., Zhu, L. P., & Caimi, R. E. (1995). Random vibration of space shuttle weather protection systems. *Journal of Shock and Vibration, 2,* 111–118.

Elishakoff, I., Lin, Y. K., & Zhu, L. P. (1994). *Probabilistic and convex modelling of acoustically excited structures*. Amsterdam: Elsevier.

Elishakoff, I. (1973). Random vibrations of a two-span beams. *Israel Journal of Technology, 11*(5), 317–320.

Elishakoff, I. (1974a). Mean-square stability of elastic bodies in supersonic flow. *Journal of Sound and Vibration, 33,* 67–78.

Elishakoff, I. (1974b). Vibration analysis of clamped square orthotropic plates. *AIAA Journal, 12,* 921–924.

Fedorov, Yu A. (1963). Vibration of closed circular cylindrical shell in a field of random acoustic pressures. *Inzhenernyi Zhurnal (Engineering Journal), 3,* 498–503. (in Russian).

Hay, J. A. (1964). Problems of cabin noise estimation for supersonic transports. *Journal of Sound and Vibration, 1,* 113–126.

Hwang, C., & Pi, W. S. (1969). Random acoustic response of a cylindrical shell. *AIAA Journal, 7,* 2204–2210.

Kana, D. D. (1971). Response of a cylindrical shell to random acoustic excitation. *AIAA Journal, 9,* 425–431.

Lin, Y. K. (1967). *Probabilistic theory of structural dynamics* (Chap. 7, pp. 203–252). New York: McGraw-Hill.

Manning, J. E., & Maidanik, G. (1964). Radiation properties of cylindrical shells. *The Journal of the Acoustical Society of America, 36,* 1691–1698.

Manning, J. E. (1969). Comment on paper by White, P.M., sound transmission through a finite, closed, cylindrical shell. *The Journal of the Acoustical Society of America, 43*, 886–887.

Margasahayam, R. N., & Caimi, R. E. (1997). Random vibration response of a cantilever beam to acoustic forcing by supersonic rocket exhausts during a space shuttle launch. In *Proceeding of the 5th International Congress on Sound and Vibration, Adelaide, South Australia* (Vol. 1518).

Mead, D. J. (1968). Elements of sound transmission. In E. J. Richards, & D. J. Mead (Eds.), *Noise and acoustic fatigue in aeronautics* (pp. 87–88). New York: Wiley.

Melik-Karamov, V. (2000). Life and death of TU-144. *Flame,* 3.

Nemat-Nasser, S. (1968). On the response of shallow thin shells to random excitation. *AIAA Journal, 6,* 1327–1330.

Orlebar, C. (2011). *The concorde story* (7th ed.). Osprey Publishing.

Powell, A. (1965). An introduction to acoustic fatigue. In W. J. Trapp, & D. M. Jr. Forney (Eds.), *Acoustical fatigue in aerospace structures* (pp. 1–15). Syracuse University Press.

Powell, A. (1958a). On the approximation to the "Infinite" solution by the method of normal modes for random vibrations. *Journal of the Aeronautical Society of America, 30,* 1136–1139.

Powell, A. (1958b). On the response of structures to random pressures and to jet noise in particular. In S. H. Crandall (Ed.), *Random vibration* (Chap. 8, pp. 187–229). New York: Wiley.

Pugachev, V. S. (1960). *Theory of random functions and its application to automatic control problems*. Moscow: "Fizmatgiz" Publishing House, Moscow. (in Russian).

Rattayya, J. V., & Junger, M. C. (1969). Flow excitation of cylindrical shells and associated coincidence effects. *The Journal of the Acoustical Society of America, 36*, 878–884.

Sietzen, F., Jr (2003). Conversation with Chuck Yeager, *Aerospace America*. pp. 12–15, January issue.

Skudrzyk, E. (1968). *Simple and complex vibratory systems* (Chap. 9, pp. 309–316). University Park, PA: Pennsylvania State University Press.

Szymborska, W. (1998). *Poems, new and collected 1957–1997*. New York: Harcourt Brace and Co.
Trubshaw, B. (2000). *Concorde: The Inside story*, Sutton Publishing Limited.
Vlasov, V. Z. (1949). *General theory of shells and its technological applications*. Moscow-Leningrad: Gostekhizdat. (in Russian).
White, R. H. (1966). Transduction of boundary layer noise by a rectangular panel. *The Journal of the Acoustical Society of America, 40,* 1354–1362.
Yeager, Ch., Cardenas, B., Hoover, B., Russell, J., & Young, J. (1997). *The Quest for Mach One*, New York: Penguin Studio.

Chapter 9
On the Role of Cross-Correlations in the Random Vibrations of Spherical Shells

9.1 Introductory Comments

The formalism of the normal mode approach (Bolotin, 1961; Crandall and Mark, 1963; Lin, 1967; Powell, 1965; Parkus, 1969) yields the following formula for the cross-spectral density of the displacement at any point $\vec{\rho}$ on the surface of the structure:

$$G(\vec{\rho},\omega) = \sum_{r=1}^{\infty}\sum_{s=1}^{\infty} E_{rs}(\omega) f_r(\vec{\rho}) f_s(\vec{\rho}), \tag{9.1}$$

where $\vec{\rho}$ is a vector defining the position of points on the surface, $f_r(\vec{\rho})$ and $f_s(\vec{\rho})$ are mode shapes, and $E_{rs}(\omega)$ are complex valued functions of frequency characterizing both the structure and loading; each of the indices r and s may replace a pair of indices if the structure is a plate or a shell. The explicit form of $E_{rs}(\omega)$ will be derived later.

Formula (9.1) maybe rewritten as follows:

$$G(\vec{\rho},\omega) = \sum_{r=1}^{\infty} E_{rr}(\omega) f_r^2(\vec{\rho}) + \sum_{r=1}^{\infty} \sum_{\substack{s=1 \\ r \neq s}}^{\infty} E_{rs}(\omega) f_r(\vec{\rho}) f_s(\vec{\rho}). \tag{9.2}$$

The first sum represents the interaction of similar modes in formulating the response cross-spectral density, whereas the second sum represents the interaction between separate modes of free vibration. The mean square level of displacement at $\vec{\rho}$, which is obtained by integrating Eqs. (9.1) or (9.2) overall frequencies, may also be derived in a similar manner

I. Elishakoff, *Dramatic Effect of Cross-Correlations in Random Vibrations of Discrete Systems, Beams, Plates, and Shells*,
https://doi.org/10.1007/978-3-030-40394-2_9

$$\langle w^2(\vec{\rho}, t)\rangle = \sum_{r=1}^{\infty} F_{rr} f_r^2(\vec{\rho}) + \sum_{r=1}^{\infty} \sum_{\substack{s=1 \\ r \neq s}}^{\infty} F_{rs} f_r(\vec{\rho}) f_s(\vec{\rho}),$$

$$F_{rr} = \int_{-\infty}^{\infty} E_{rr}(\omega) d\omega, \quad F_{rs} = \int_{-\infty}^{\infty} E_{rs}(\omega) d\omega, \tag{9.3}$$

Very often cross-terms in response cross-spectral densities are neglected and formulae (9.2) and (9.3) are approximated in the following way (which becomes exact for the external forces, delta-correlated in space)

$$G(\vec{\rho}, \omega) \cong \sum_{r=1}^{\infty} E_{rr}(\omega) f_r^2(\vec{\rho}), \quad \langle w^2(\vec{\rho}, t)\rangle \cong \sum_{r=1}^{\infty} F_{rr} f_r^2(\vec{\rho}). \tag{9.4}$$

Generally, such an approximation is incorrect. However, one special type of random vibration problem maybe noted where cross-correlations do not occur. Let the acceptability of the structure for actual use be determined by the surface-averaged mean square level of displacements, as defined by

$$\langle w_2(t)\rangle = \int_{\vec{R}} \langle w^2(\vec{\rho}, t)\rangle d\vec{R}, \tag{9.5}$$

$\vec{R}$ being the surface of the structure. Taking into account the orthogonality property of the mode shapes,

$$\int_{\vec{R}} f_r(\vec{\rho}) f_s(\vec{\rho}) d\vec{R} = \delta_{rs} \int_{\vec{R}} f_r^2(\vec{\rho}) d\vec{R},$$

(δ_{rs} is Kronecker's delta), one obtains

$$\langle w^2(t)\rangle = \sum_{r=1}^{\infty} F_{rr} \int_{\vec{R}} f_r^2(\vec{\rho}) d\vec{R}, \tag{9.6}$$

i.e., the surface-averaged mean square level of displacements is determined only by correlations between similar modes and not by cross-correlations between separate modes. If the external forces have a broad frequency bandwidth, a large number of terms will be significant in Eq. (7.6). Bolotin (1963, 1966) Ribner (1966), White (1966), Stearn (1970) an important simplification with regard to the large number of excited modes is indicated. Summation of the series is approximated by integration over the wave numbers. Integral approximation of the surface-averaged mean-value is equivalent to the use of an infinite model of the structure. Where an integral probabilistic characterization (9.5) is inadequate, a detailed analysis according to Eqs. (9.2) or (9.3) is called for by Elishakoff (1975a, 1975b). Thus, when the integral characterization is not enough, a finite-system model in problems of structural integrity. Consequently, the general formulae of type (9.2) or (9.3) have to be used instead of formula (9.6). In 1961 Bolotin (1, p. 130) showed that the terms due to cross-correlations maybe neglected, and consequently Eqs. (9.2) and (9.3) maybe replaced by Eq. (9.4), if the following strong inequality holds

$$\left|\omega_\alpha^2 - \omega_\beta^2\right| \gg \max\left(\omega_\alpha \varepsilon_\alpha, \omega_\beta \varepsilon_\beta\right), \tag{9.7}$$

where ω_α is the eigenfrequency and ε_α is the coefficient of viscous damping corresponding to the α th mode.

Inequality (9.7) is not fulfilled when the eigenfrequencies occur close to each other. This may take place, for example, when multiple eigenfrequencies exist (for example, a simply supported square isotropic plate has a double eigenfrequency of type $\omega_{jk} = \omega_{kj}$, $k \neq j$, a triple eigenfrequency $\omega_{55} = \omega_{17} = \omega_{71}$, etc.); as a result, cross-terms corresponding to different modes with equal eigenfrequencies cannot be neglected.

In 1966, apparently without the knowledge of Bolotin's work, Davis (1965) obtained a formula for the percentage error incurred by neglecting the cross-terms in a random vibration response

$$R(f) \equiv 1 - C_w^{aa}(I, J, f)/C_w^{aa}(I, J, IJ, f) = 200\xi/(1+\xi), \tag{9.8}$$

where $C_w^{aa}(I, J, IJ, f)$ denotes the response power spectrum at point a, including the cross-term for modes I and J, at frequency f, $C_w^{aa}(I, J, f)$ denotes the power spectrum at point a with the cross-term omitted and ξ is defined as follows:

$$\xi = \left\{\left(1/4\xi_J^2\right)[1 - (f_I/f_J)^2 + (f_I/f_J)^2]\right\}^{-1/2}. \tag{9.9}$$

For the eigenfrequencies $f_I \approx f_J$, when Bolotin's inequality is not fulfilled, the error incurred by neglecting the cross-term reached 50% (see also Fig. 3 by Davis (1965)).

In this book, *inter alia*, a criterion is presented for the omission of the cross-terms for a structure possessing hysteretic damping.

An investigation of the effect of cross-terms was made by Mercer (1965), who considered random vibrations of a multispan beam excited by acoustic pressure. As a numerical example, a nine-span beam whose first nine eigenfrequencies range in close intervals from 500 to 612 Hz was analyzed. The cross-terms contributed only 20% to the overall r.m.s. response level in the case with maximal coupling. In more typical cases the contribution was of the order of 5–10%. However, the contribution of the cross-terms can be expected to increase with the number of spans. Note that as the latter increases the normal mode approach becomes cumbersome, and integral canonical expansion has to be used instead (cf. the papers by Moskalenko (1968) and by Vaicaitis and Lin (1972)).

Schweiker and Davis (1966) made two response calculations, one with the cross-terms neglected, i.e., when $I \neq J$, and the other with the cross-terms included. Nineteen modes were used in calculations, many of them very closed spaced (for example, 277 and 280 Hz, 336 and 337 Hz). The maximum error incurred by neglecting the cross-terms in this problem was 15% of the spectral amplitude at any frequency, and 3% based on r.m.s. levels.

Dimentberg (1961) considered the random vibrations of a clamped plate, using Bolotin's (1961) asymptotic method. The r.m.s. value of the maximum normal stresses, including the cross-terms due to multiple frequencies, was 670kg cm^{-2}, whereas its counterpart when neglecting these cross-terms was 564kg cm^{-2}. Mkhitarov (1972) showed that interactions between coincident or closely spaced frequencies have to be taken into account for plates in a liquid.

The pertinent question arises of the possible generalization of Bolotin's conditions that are both sufficient and necessary in order to be able to neglect cross-correlations. The answer turns out to be, as was clearly demonstrated in definitive studies by Dahlberg (1982, 1999) that satisfaction of Bolotin's conditions is not sufficient to justify the neglect of the cross-correlations.

Dahlberg (1982, 1998, 1999) extended Bolotin's (1961) criteria for neglecting cross-correlations. He notes (Dahlberg, 1999): "The first one to point out conditions under which the modal cross-spectral densities maybe neglected seems to have been Bolotin, who already in 1961 stated that the conditions for possible neglection of the modal cross-terms are violated in two cases, namely when the damping is very large or when the elastic system has multiple or closely spaced eigenfrequencies. The conditions to be satisfied read (Bolotin and Elishakoff, 1971) $\max(\omega_j\zeta_j, \omega_k\zeta_k \ll \left|\omega_j^2 - \omega_k^2\right|$; where ζ_i is relative damping in mode i, and ω_i is the ith eigenfrequency of the structure. This statement seems to have been interpreted so that the cross-terms may always be neglected if the conditions are satisfied. As will be shown below this is not the case. One indication of this problem is given in reference [by] Wirsching, Paez and Ortiz (1995), where the modal closeness is commented upon: "No general operational definition of closeness exists."

Moreover, Dahlberg (1999) stresses: "In reference [by] Dahlberg (1982), examples were given where modal cross-spectral densities were shown to play an important role in the response spectral densities of a structure, although the Bolotin conditions were satisfied. These examples concerned structural responses to narrow-band random excitation. The structural parameters were such that as anti-resonance of the structure appeared at, or in the vicinity of, the frequency range of the random excitation. In such a case, none of the modes of the structure dominates in the response, and consequently, the contributions from the cross-spectral densities maybe important. It was demonstrated that the modal cross-spectral densities contribute to the response of a structure mainly in the frequency range close to antiresonances of the structure. This has also been pointed out in, for example, reference [by] Keltie and Peng (1987), where the effects of modal coupling on the acoustic power radiation from panels were investigated." Specifically, Keltie and Peng (1987) arrived at the conclusion that under off-resonant excitation the contributions due to modal coupling maybe important.

In another publication, Dahlberg (1998) provided some additional examples where, according to Dahlberg (1999, p. 158), "modal cross-spectral densities played an important role in the structural response although the conditions of small damping and well separated eigenfrequencies were satisfied. The bandwidth of the excitation was broadband so that it covered several resonance frequencies of the structure (a taut spring). The excitation was distributed along the whole structure and not, as in Crandall's problem, concentrated to one single point." Dahlberg (1998) studied the case of the band-limited white noise excitation, covering eigenfrequencies numbered 8–12 of the string. Remarkably, neglect of cross-correlations resulted in response differing from the exact value of the mean square displacement by a factor of three (or even more)!

Dahlberg (1999, p. 171) derived the additional important result; he emphasizes: "By studying the vibration of a simply supported been subjected to a stationary random loading it has been demonstrated that the modal cross-spectral densities may play an important role in a structural response even if the system is lightly damped and the eigenfrequencies are well separated."

Dahlberg (1999, p. 174) made the following conclusion: "By a simple example, it has been demonstrated that the two conditions: (a) the system should be lightly damped and (b) the eigenfrequencies should be well separated are not sufficient to ensure that the modal cross-spectral densities maybe neglected when calculating the response spectral density of a structure even if the random excitation of the structure is broadband so that several eigenmodes of the structure are excited... Although the excitation frequency range covered several eigenfrequencies of the structure (a beam), an error of more than 60% of the standard deviation of the beam deflection has been demonstrated. Factors that influence the cross-modal contributions are not only the damping and the frequency separation, but also the loading (its frequency content and its location on the structure,), and the location where the response is calculated." It is impossible to disagree with Dahlberg (1999, p. 174) when he summarizes: "...there is no reason to recommend any rejection of the modal cross-spectral densities."

A significant case where the cross-terms cannot be omitted is a structure with very high modal density (given approximately by the number of eigenfrequencies per unit interval (Bolotin 1972; Bourgine, 1970; Elishakoff, 1971a, 1971b, 1971c; Hart and Shah, 1971; Kornev, 1972). Then the interaction of different modes with very close eigenfrequencies has to be taken into account. Here such an example is considered: a spherically curved plate (spherical shell, rectangular in plane), supported all round by shear diaphragms.

9.2 Response Cross-Spectral Densities of a Spherical Shell

Consider a spherically curved plane of uniform thickness, supported at all edges by shear diaphragms and subjected to random pressure with the intensity $q(x, y, t)$, where x and y denote the space coordinates and t the time. On the assumption that the correlation scale of the external forces is very large, and the mean square deviation of the panel displacements is very small in comparison to the thickness of the panel, the problem maybe considered in a statistically linear setting. The normal displacement $w(x, y, t)$ and the stress resultant function $\chi(x, y, t)$ must satisfy the Donnell-Vlasov equations

$$D\Delta\Delta w + (1/R)\Delta\chi + \rho h\partial^2 w/\partial t^2 + L_1 w = q(x, y, t), \qquad (9.10)$$

$$\Delta\Delta\chi - (Eh/R)\Delta w - L_2 w = 0, \qquad (9.11)$$

where ρ, h, and R are the mass density, thickness, and radius of the panel, respectively, Δ is the Laplace operator in the panel middle surface, and D and E are stiffness and Young's moduli, respectively. The operators

$$L_1 w = L_1(\partial/\partial t, \partial/\partial x, \partial/\partial y)w, \quad L_2 w = L_2(\partial/\partial t, \partial/\partial x, \partial/\partial y)w, \qquad (9.12)$$

take into account the damping in the shell material and at this stage will not be specified.

The supplementary boundary conditions are

$$w = \partial^2 w/\partial x^2 = \partial^2\chi/\partial x^2 = \partial^2\chi/\partial y^2 = 0 \quad \text{at } x = 0, \quad x = a, \qquad (9.13)$$

$$w = \partial^2 w/\partial y^2 = \partial^2\chi/\partial y^2 = \partial^2\chi/\partial x^2 = 0 \quad \text{at } y = 0, \ y = b, \qquad (9.14)$$

where a and b are the lengths of panel sides.

The cross-spectral density of the external forces is assumed to be given. The cross-spectral density of the displacements will be determined below.

For Eqs. (9.10) and (9.11) a solution of the following type is assumed:

$$w(x, y, t) = \int_{-\infty}^{\infty} W(x, y, \omega)\exp(i\omega t)d\omega, \qquad (9.15)$$

$$\chi(x, y, t) = \int_{-\infty}^{\infty} X(x, y, \omega)\exp(i\omega t)d\omega. \qquad (9.16)$$

and the external pressure is expressed in a similar way

$$q(x, y, t) = \int_{-\infty}^{\infty} Q(x, y, \omega) \exp(i\omega t) d\omega, \tag{9.17}$$

where w, W; χ, X; q, Q are Fourier transform pairs. Thus, the following frequency equation in the frequency domain is obtained

$$D\Delta\Delta\Delta W + \left(Eh/R^2\right)\Delta W + (1/R)l_2 W - \rho h\omega^2 \Delta W + \Delta l_1 W = \Delta Q(x, y, \omega), \tag{9.18}$$

where

$$l_1 W = \exp(-i\omega t) L_1\left[W(x, y, \omega) \exp(i\omega t)\right], \tag{9.19}$$

$$l_2 W = \exp(-i\omega t) L_2\left[W(x, y, \omega) \exp(i\omega t)\right]. \tag{9.20}$$

$W(x, y, \omega)$ and $P(x, y, \omega)$ maybe expressed in terms of the modes of free vibration as

$$W(x, y, \omega) = \sum_{m=1}^{\infty} \sum_{n=1}^{\infty} W_{mn}(\omega) f_{mn}(x, y), \tag{9.21}$$

$$Q(x, y, \omega) = \sum_{m=1}^{\infty} \sum_{n=1}^{\infty} Q_{mn}(\omega) f_{mn}(x, y). \tag{9.22}$$

The normal modes, $f_{mn}(x, y)$, which satisfy the boundary conditions (9.13) and (9.14) and the differential equation

$$D\Delta\Delta\Delta f_{mn} + \left(Eh/R^2\right)\Delta f_{mn} - \rho h\omega_{mn}^2 \Delta f_{mn} = 0, \tag{9.23}$$

are given by

$$f_{mn} = \sin(m\pi x/a) \sin(n\pi y/b), \tag{9.24}$$

ω_{mn} being the mnth eigenfrequency. Let the operators l_α permit the following solution of Eq. (9.18)

$$W_{mn}(x, y, \omega) = \sum_{m=1}^{\infty} \sum_{n=1}^{\infty} Q_{mn}(\omega) \phi_{mn}(\omega) f_{mn}(x, y). \tag{9.25}$$

The excitation cross-correlation function $R_q\left(x', y'; x'', y''; \tau\right)$ and cross-spectral density $G_q\left(x', y'; x'', y''; \omega\right)$ are defined by

$$R_q\left(x', y'; x'', y''; \tau\right) = \lim_{T\to\infty} \frac{1}{2T} \int_{-T}^{T} q^*\left(x', y', t'\right) q\left(x'', y'', t' + \tau\right) dt', \tag{9.26}$$

$$G_q\left(x', y', x'', y''; \omega\right) = \frac{1}{2\pi} \int_{-\infty}^{\infty} R_q\left(x', y'; x'', y''; \tau\right) \exp(-i\omega\tau) d\tau. \tag{9.27}$$

Equations (9.26) and (9.27) maybe used to derive the following expression for the displacement cross-spectral density

$$G_w\left(x', y'; x, y; \omega\right) = \sum_{m=1}^{\infty}\sum_{n=1}^{\infty}\sum_{p=1}^{\infty}\sum_{q=1}^{\infty} S_{mn,pq}(\omega)\phi_{mn}^*(\omega)\phi_{pq}(\omega) f_{mn}\left(x', y'\right) f_{pq}\left(x'', y''\right), \tag{9.28}$$

where

$$S_{mn,pq} = \frac{1}{v_{mn}^2 v_{pq}^2} \int_0^a \int_0^b \int_0^a \int_0^b G_q\left(x', y'; x'', y''; \omega\right) f_{mn}\left(x', y'\right) f_{pq}\left(x'', y''\right) dx' dy' dx'' dy'',$$

$$v_{mn}^2 = \int_0^a \int_0^b f_{mn}^2\left(x', y'\right) dx' dy', \quad v_{pq}^2 = \int_0^a \int_0^b f_{pq}\left(x'', y''\right) dx'' dy'', \tag{9.29}$$

$S_{mn,pq}$ are referred to as joint acceptances when $m = p$ and $n = q$, and as cross-acceptances when $m \neq p$ or $n \neq q$ by Mercer (1965). The formulae (9.2) and (9.3) transform, respectively, to $(r \to m, n; s \to p, q)$

$$G_w(x, y; x, y; \omega) = \sum_{m=1}^{\infty}\sum_{n=1}^{\infty} S_{mn,mn}(\omega)\left|\phi_{mn}(\omega)\right|^2 f_{mn}^2(x, y) + \sum_{m=1}^{\infty} \sum_{\substack{n=1 \\ m\neq p}}^{\infty} \sum_{\substack{p=1 \\ or}}^{\infty} \sum_{\substack{q=1 \\ n\neq q}}^{\infty} S_{mn,pq}\phi_{mn}^*(\omega)\phi_{pq} f_{mn}(x, y) f_{pq}(x, y), \tag{9.30}$$

$$\left\langle w^2(x, y, t)\right\rangle = \sum_{m=1}^{\infty}\sum_{n=1}^{\infty} F_{mn,mn} f_{mn}^2(x, y) + \sum_{\substack{m=1 \\ m\neq p}}^{\infty} \sum_{\substack{n=1 \\ or}}^{\infty} \sum_{\substack{p=1 \\ n\neq q}}^{\infty} \sum_{q=1}^{\infty} F_{mn,pq} f_{mn}(x, y) f_{pq}(x, y), \tag{9.31}$$

where

$$F_{mn,mn} = \int_{-\infty}^{\infty} S_{mn,mn}(\omega)|\phi_{mn}(\omega)|^2 d\omega, \tag{9.32}$$

$$F_{mn,pq} = \int_{-\infty}^{\infty} S_{mn,pq}(\omega)\phi_{mn}^*(\omega)\phi_{pq}(\omega)d\omega. \tag{9.33}$$

Upon taking into account the Hermitian property of the joint acceptances $S_{mn,pq} = S_{pq,mn}^*$, the formula (9.31) maybe rewritten in the following way

$$\begin{aligned}\left\langle w^2(x,y,t)\right\rangle = &\sum_{m=1}^{\infty}\sum_{n=1}^{\infty} F_{mn,mn} f_{mn}^2(x,y) + 2\sum_{m=1}^{\infty}\sum_{\substack{n=1\\ m<p}}^{\infty}\sum_{p=1}^{\infty} \mathrm{Re} F_{mn,pq} f_{mn}(x,y) f_{pn}(x,y)\\ &+2\sum_{m=1}^{\infty}\sum_{\substack{n=1\\ n<q}}^{\infty}\sum_{q=1}^{\infty} \mathrm{Re} F_{mn,mq} f_{mn}(x,y) f_{mq}(x,y)\\ &+2\sum_{\substack{m=1\\ m\neq p}}^{\infty}\sum_{n=1}^{\infty}\sum_{\substack{p=1\\ n<q}}^{\infty}\sum_{q=1}^{\infty} \mathrm{Re} F_{mn,pq} f_{mn}(x,y) f_{pq}(x,y).\end{aligned} \tag{9.34}$$

Two different models of panel material damping are now considered.
For the case of viscous damping we have

$$L_2 \equiv 0,\; L_1 = 2\rho h\varepsilon\partial/\partial t, \tag{9.35}$$

where ε is a viscous damping factor, $l_1 W$ becomes

$$l_1 W = 2\rho h\varepsilon i\omega, \tag{9.36}$$

and

$$\phi_{mn}(\omega) = 1/\rho h\left(\omega_{mn}^2 - \omega^2 + 2\varepsilon i\omega\right). \tag{9.37}$$

9.3 Hysteretic Damping

For this case

$$l_1 W = D i\mu\Delta W, \quad l_2 W = E i\mu(h/R)\Delta W, \tag{9.38}$$

where μ is the hysteretic damping factor, and $\phi_{mn}(\omega)$ takes the form

$$\phi_{mn}(\omega) = 1/\rho h\left(\omega_{mn}^2 - \omega^2 + \mathrm{i}\mu\omega_{mn}^2\right). \tag{9.39}$$

9.4 Example Problem and Discussion

Consider an excitation which is a function of time only, the correlation function being

$$R_q\left(x'y'; x''y''; \tau\right) = 2\pi c\delta(\tau), \tag{9.40}$$

where c is some positive constant and $\delta(\tau)$ is Dirac's delta function. The cross-spectral densities of the generalized forces—the joint and cross-acceptances—are

$$S_{mn,pq} = \begin{cases} 256/\pi^5 mnpq, & \text{for } m, n, p \text{ and } q \text{ odd,} \\ 0, & \text{otherwise} \end{cases} \tag{9.41}$$

With $H_{mn,pq}$ defined by

$$H_{mn,pq} = \begin{cases} (\rho h)^2 F_{mn,pq}/S_{mn,pq}, & \text{for } m, n, p \text{ and } q \text{ odd,} \\ 0, & \text{otherwise} \end{cases} \tag{9.42}$$

one obtains

$$H_{mn,mn} = \pi/2\varepsilon\omega_{mn}^2, \tag{9.43}$$

for viscous damping and

$$H_{mn,mn} = \alpha/\mu\sqrt{2\left(1 + \mu^2\right)}\omega_{mn}^3, \tag{9.44}$$

for hysteretic damping, where

$$\alpha = \sqrt{1 + \sqrt{1 + \mu^2}}. \tag{9.45}$$

In both formulae (9.43) and (9.44) the eigenfrequency is determined by

$$\omega_{mn}^2 = \left(E/\rho R^2\right) + (D/\rho h)\left\{\left(m^2\pi^2/a^2\right) + \left(n^2\pi^2/b^2\right)\right\}^2, \tag{9.46}$$

where m and n are positive integers denoting the number of half-waves in the coordinate directions. For the case of viscous damping

$$\operatorname{Re}H_{mn,pq} = \int_{-\infty}^{\infty} \frac{\left(\omega_{mn}^2 - \omega^2\right)\left(\omega_{pq}^2 - \omega^2\right) + 4\varepsilon^2\omega^2}{\left[\left(\omega_{mn}^2 - \omega^2\right)^2 + 4\varepsilon^2\omega^2\right]\left[\left(\omega_{pq}^2 - \omega^2\right)^2 + 4\varepsilon^2\omega^2\right]} d\omega$$
$$= \frac{8\pi\varepsilon}{\left(\omega_{mn}^2 - \omega_{pq}^2\right)^2 + 8\varepsilon^2\left(\omega_{mn}^2 + \omega_{pq}^2\right)}, \tag{9.47}$$

and for the case of hysteretic damping

$$\operatorname{Re}H_{mn,pq} = \int_{-\infty}^{\infty} \frac{\left(\omega_{mn}^2 - \omega^2\right)\left(\omega_{pq}^2 - \omega^2\right) + \mu^2\omega_{mn}^2\omega_{pq}^2}{\left[\left(\omega_{mn}^2 - \omega^2\right)^2 + \mu^2\omega_{mn}^4\right]\left[\left(\omega_{pq}^2 - \omega^2\right)^2 + \mu^2\omega_{pq}^4\right]} d\omega$$
$$= \frac{\pi}{\omega_{mn}\omega_{pq}} \frac{\sqrt{1+\mu^2}}{2} \frac{\left(\omega_{mn} + \omega_{pq}\right)\left[\alpha\eta\left(\omega_{mn}^2 + \omega_{pq}^2\right) - \beta\left(\omega_{mn} - \omega_{pq}\right)^2\right]}{\left(\omega_{mn}^2 - \omega_{pq}^2\right)^2 + 2\mu^2\left(\omega_{mn}^4 + \omega_{pq}^4\right) + \mu^4\left(\omega_{pq}^2 + \omega_{mn}^2\right)^2}, \tag{9.48}$$

where

$$\beta = \sqrt{\sqrt{1+\mu^2} - 1}. \tag{9.49}$$

These integrals were evaluated by means of Cauchy's theorem of residues. For $\mu \ll 1$ one obtains $\alpha \to \sqrt{2}$ and $\beta \to \mu/\sqrt{2}$.

A comparison between formulae (9.43) and (9.47) leads to criterion (9.9) for the negligibility of the cross-terms when the structure possesses viscous damping. For the structure with hysteretic damping, an analogous criterion maybe obtained by comparing the formulae (9.44) and (9.48)

$$\left|\omega_{mn}^2 - \omega_{pq}^2\right| \gg \mu \max\left(\omega_{mn}^2, \omega_{pq}^2\right). \tag{9.50}$$

For the actual calculations, a shell possessing viscous damping was chosen. The non-dimensional mean square value of displacements,

$$d(\xi,\eta) = \pi^4(\rho h)^2 w^2(x,y,t)/128c, \quad \xi = x/a, \eta = y/b, \tag{9.51}$$

was calculated by using formula (9.34), which maybe rewritten as

$$\mathrm{d}(\xi,\eta) = \sum_{i=1}^{4} B_i, \tag{9.52}$$

where

$$B_1 = \sum_{m=1}^{N}\sum_{n=1}^{N} \frac{(\sin m\pi\xi)^2(\sin n\pi\eta)^2}{m^2n^2\bar{\varepsilon}\bar{\omega}_{mn}^2}, \tag{9.53}$$

$$B_2 = 2\sum_{m}^{N} \sum_{\substack{n \\ m<p}}^{N} \sum_{p}^{N} \frac{16\bar{\varepsilon}(\sin m\pi\xi)(\sin p\pi\xi)(\sin n\pi\eta)^2}{mn^2p\left[\left(\bar{\omega}_{mn}^2 - \bar{\omega}_{pn}^2\right)^2 + 8\bar{\varepsilon}^2\left(\bar{\omega}_{mn}^2 + \bar{\omega}_{pn}^2\right)\right]}, \tag{9.54}$$

$$B_3 = 2\sum_{m}^{N} \sum_{\substack{n \\ n<q}}^{N} \sum_{q}^{N} \frac{16\bar{\varepsilon}(\sin m\pi\xi)^2(\sin n\pi\eta)(\sin q\pi\eta)}{m^2nq\left[\left(\bar{\omega}_{mn}^2 - \bar{\omega}_{mq}^2\right)^2 + 8\bar{\varepsilon}^2\left(\bar{\omega}_{mn}^2 + \bar{\omega}_{mq}^2\right)\right]}, \tag{9.55}$$

$$B_4 = 2\sum_{\substack{m \\ m\neq p}}^{N} \sum_{n}^{N} \sum_{\substack{p \\ n<q}}^{N} \sum_{q}^{N} \frac{16\bar{\varepsilon}(\sin m\pi\xi)(\sin p\pi\xi)(\sin n\pi\eta)(\sin q\pi\eta)}{mnpq\left[\left(\bar{\omega}_{mn} - \bar{\omega}_{pq}\right)^2 + 8\bar{\varepsilon}^2\left(\bar{\omega}_{mn}^2 + \bar{\omega}_{pq}^2\right)\right]}. \tag{9.56}$$

In Eqs. (9.53)–(9.56)

$$\bar{\omega}_{mn} = \omega_{mn}\sqrt{\rho R^2/E}, \quad \bar{\varepsilon} = \varepsilon\sqrt{\rho R^2/E}, \tag{9.57}$$

$$\bar{\omega}_{mn}^2 = 1 + (h/a)^2(R/a)^2\{\pi^4/12(1-\nu^2)\}\left(m^2 + n^2a^2/b^2\right)^2. \tag{9.58}$$

ν is Poisson's ratio, $\bar{\omega}_{mn}$ is a non-dimensional eigenfrequency, $\bar{\varepsilon}$ is a non-dimensional factor of viscous damping and N is the number of modes taken into account.

In Tables 9.1, 9.2, 9.3, 9.4, 9.5 and 9.6 the calculated values of B_i are presented. In Tables 9.7 and 9.8 the first hundred eigenfrequency coefficients, $\bar{\omega}_{mn}$, are listed

Table 9.1 Mean square value of displacements in a spherically curved panel ($a/b = 1, \bar{\varepsilon} = 0.01, d = 367.0$)[a]

η	ξ	B_1	B_2	B_3	B_4	$d(\xi, \eta)$
0.25	0.1	5.893	0.003	0.001	0.764	6.660
	0.2	18.159	0.007	0.003	1.947	20.116
	0.3	31.582	0.003	0.009	1.068	32.661
	0.4	43.848	−0.007	0.014	−1.947	41.909
	0.5	49.741	−0.012	0.017	−3.606	46.140
0.50	0.1	11.785	0.006	−0.001	−1.631	10.160
	0.2	36.318	0.013	−0.005	−3.834	32.492
	0.3	63.163	0.006	−0.012	−2.132	61.024
	0.4	87.696	−0.013	−0.020	3.779	91.442
	0.5	99.481	−0.024	−0.024	7.457	106.891

[a] $d(\xi, \eta) = d(1-\xi, \eta) = d(\xi, 1-\eta)$

Table 9.2 Mean square value of displacements in a spherically curved panel ($a/b = 1, \bar{\varepsilon} = 0.05, d = 73$)

η	ξ	B_1	B_2	B_3	B_4	$d(\xi, \eta)$
0.25	0.1	2.453	3.747	0.165	−0.306	6.059
	0.2	5.334	3.846	0.918	0.638	10.736
	0.3	8.211	−0.511	2.890	0.959	11.549
	0.4	11.093	−4.047	5.665	1.103	13.814
	0.5	13.546	−6.087	7.724	1.518	16.701
0.50	0.1	4.906	7.493	−0.392	−3.476	8.531
	0.2	10.669	7.692	−1.745	−3.756	12.860
	0.3	16.422	−1.021	−4.507	−2.635	8.260
	0.4	22.186	−8.093	−8.749	−2.645	2.699
	0.5	27.091	−12.173	−12.173	−1.941	0.804

Table 9.3 Mean square value of displacements in a spherically curved panel ($a/b = 1, \bar{\varepsilon} = 0.1, d = 36.0$)

η	ξ	B_1	B_2	B_3	B_4	$d(\xi, \eta)$
0.25	0.1	1.226	2.390	−0.025	−0.223	3.369
	0.2	2.667	1.674	0.101	0.116	4.558
	0.3	4.106	−0.603	0.966	0.368	4.836
	0.4	5.546	−1.972	2.213	0.623	6.411
	0.5	6.773	−3.007	3.108	0.782	7.656
0.50	0.1	2.453	4.781	−0.171	−2.025	5.037
	0.2	5.334	3.348	−0.756	−1.936	5.991
	0.3	8.211	−1.206	−1.996	−1.538	3.471
	0.4	11.093	−3.944	−4.125	−1.444	1.580
	0.5	13.546	−6.014	−6.014	−1.301	0.217

for odd m and n. The following structural data were chosen: $h/a = 0.005$, $R/a = 1$, $\nu = 0.3$. The non-dimensional surface-averaged mean square level of displacements is given by

$$d = \int_0^1 \int_0^1 d(\xi, \eta) d\xi d\eta. \tag{9.59}$$

As maybe seen from these tables, B_1 may not be used even for a rough estimation of the true mean square value of the displacements $d(\xi, \eta)$ in a spherically curved panel. The percentage difference between them

Table 9.4 Mean square value of displacements in a spherically curved panel ($a/b = 2, \bar{\varepsilon} = 0.01, d = 362.0$)

η	ξ	B_1	B_2	B_3	B_4	$d(\xi, \eta)$
0.25	0.1	3.972	0.000	0.000	−0.005	3.966
	0.2	12.240	0.000	0.001	−0.003	12.238
	0.3	21.356	0.000	0.004	−0.004	21.355
	0.4	29.624	0.000	0.006	0.005	29.635
	0.5	33.596	0.000	0.007	0.000	33.603
0.50	0.1	22.848	0.321	−1.167	−0.201	21.801
	0.2	49.995	0.696	−7.743	1.230	44.177
	0.300	77.085	0.298	−20.131	−0.146	57.105
	0.400	104.232	−0.696	−33.539	−1.326	68.671
	0.500	127.080	−1.238	−39.704	−2.287	83.850

Table 9.5 Mean square value of displacements in a spherically curved panel ($a/b = 2, \bar{\varepsilon} = 0.05, d = 72.0$)

η	ξ	B_1	B_2	B_3	B_4	$d(\xi, \eta)$
0.25	0.1	2.285	0.555	0.145	−0.196	2.789
	0.2	4.999	1.186	0.965	−0.149	7.001
	0.3	7.708	0.499	2.948	0.231	11.387
	0.4	10.423	−1.187	5.777	0.524	15.538
	0.5	12.708	−2.107	7.423	0.345	18.369
0.50	0.1	4.570	1.110	−0.306	−1.022	4.352
	0.2	9.999	2.372	−1.699	−1.734	8.938
	0.3	15.417	0.999	−4.492	−1.426	10.497
	0.4	20.846	−2.374	−8.755	−0.468	9.250
	0.5	25.416	−4.214	−11.488	0.036	9.751

$$\kappa = [1 - B_1/d(\xi, \eta)] \times 100\% \quad (\text{for}\, d(\xi, \eta) \neq 0), \tag{9.60}$$

may sometimes exceed 250% (e.g., see Table 9.3: $\xi = 0.5, \eta = 0.5$; or Table 9.6: $\xi = 0.5, \eta = 0.5$). Furthermore, B_1 represents neither an upper bound nor a lower bound of the true mean square value since it maybe either positive or negative (e.g., see Table 9.6: $\xi = 0.3, \eta = 0.25$ and $\xi = 0.4, \eta = 0.25$). This indicates that for shells with high modal density *at the low end of the frequency domain* the cross-terms must be considered.

Table 9.6 Mean square value of displacements in a spherically curved panels ($a/b = 2, \bar{\varepsilon} = 0.1, d = 36.0$)

η	ξ	B_1	B_2	B_3	B_4	$d(\xi, \eta)$
0.25	0.10	1.142	0.582	0.004	−0.094	1.636
	0.20	2.500	1.197	0.177	0.007	3.880
	0.30	3.854	0.476	1.001	0.197	5.528
	0.40	5.212	−1.199	2.283	0.288	6.584
	0.50	6.354	−2.113	3.052	0.223	7.516
0.50	0.10	2.285	1.165	−0.141	−0.788	2.520
	0.20	4.999	2.394	−0.728	−1.285	5.380
	0.30	7.708	0.952	−1.930	−0.954	5.777
	0.40	10.423	−2.398	−4.061	−0.402	3.563
	0.50	12.708	−4.226	−5.652	−0.161	2.668

Table 9.7 First hundred eigen frequency coefficients ($a/b = 1; m, nodd$)

m	n	$\bar{\omega}_{mn}$	m	n	$\bar{\omega}_{mn}$
1	1	1.04365	9	13	37.34687
3	1	1.79724	5	15	37.34687
1	3	1.79724	15	7	40.92971
3	3	2.86800	7	15	40.92971
5	1	4.00939	17	1	43.31838
1	5	4.00939	13	11	43.31838
5	3	5.17489	11	13	43.31838
3	5	5.17489	1	17	43.31838
7	1	7.53336	17	3	44.51274
5	5	7.53336	3	17	44.51274
1	7	7.53336	15	9	45.70712
7	3	8.71890	9	15	45.70712
3	7	8.71890	17	5	46.90151
7	5	11.09586	5	17	46.90151
5	7	11.09586	17	7	50.48477
9	1	12.28614	13	13	50.48477
1	9	12.28614	7	17	50.48477
9	3	13.47720	15	11	51.67921

(continued)

Table 9.7 (continued)

m	n	$\bar{\omega}_{mn}$	m	n	$\bar{\omega}_{mn}$
3	9	13.47720	11	15	51.67921
7	7	14.66885	19	1	54.06812
9	5	15.86095	1	19	54.06812
5	9	15.86095	19	3	55.26259
11	1	18.24616	17	9	55.26259
1	11	18.24616	9	17	55.26259
11	3	19.43915	3	19	55.26259
9	7	19.43915	19	5	57.65156
7	9	19.43915	5	19	57.65156
3	11	19.43915	15	13	58.84606
11	5	21.82567	13	15	58.84606
5	11	21.82567	19	7	61.23507
9	9	24.21275	17	11	61.23507
13	1	25.40645	11	17	61.23507
11	7	25.40645	7	19	61.23507
7	11	25.40645	21	1	66.01316
1	13	25.40645	19	9	66.01316
13	3	26.60024	9	19	66.01316
3	13	26.60024	1	21	66.01316
13	5	28.98803	21	3	67.20770
5	13	28.98803	15	15	67.20770
11	9	30.18202	3	21	67.20770
9	11	30.18202	17	13	68.40224
13	7	32.57015	13	17	68.40224
7	13	32.57015	21	5	69.59679
15	1	33.76428	5	21	69.59679
1	15	33.76428	19	11	71.98589
15	3	34.95844	11	19	71.98589
3	15	34.95844	21	7	73.18045
11	11	36.15264	7	21	73.18045
15	5	37.34687	17	15	76.76414
13	9	37.34687	15	17	76.76414

Table 9.8 First hundred eigenfrequency coefficients ($a/b = 2$; $m, nodd$)

m	n	$\bar{\omega}_{mn}$	m	n	$\bar{\omega}_{mn}$
1	1	1.248005	5	11	76.01754
3	1	2.183759	23	1	79.60125
5	1	4.444639	7	11	80.79583
1	3	5.615117	21	5	81.99041
3	3	6.794023	15	9	83.185
7	1	7.977621	19	7	84.37958
5	3	9.164093	23	3	90.35255
9	1	12.73271	9	11	91.54715
7	3	12.73271	11	11	95.13096
1	5	15.11584	17	9	97.52017
3	5	16.30808	25	1	101.104
9	3	17.50066	23	5	102.2986
11	1	18.6935	21	7	102.2986
5	5	18.6935	13	11	104.6878
7	5	22.27321	25	3	105.8825
11	3	23.46674	1	13	108.2717
13	1	25.85411	19	9	113.0502
9	5	27.04793	3	13	114.2448
1	7	29.43577	5	13	115.4394
13	3	30.62978	15	11	119.0233
3	7	30.62978	25	5	126.1911
11	5	33.01794	23	7	126.1911
5	7	33.01794	7	13	134.5536
15	1	34.21208	27	1	134.5536
7	7	36.60047	9	13	135.7482
15	3	38.98898	27	3	138.1375
13	5	40.18327	21	9	138.1375
9	7	41.37758	17	11	141.7214
17	1	43.76626	11	13	144.1107
11	7	47.34941	25	7	146.5
17	3	48.54382	27	5	152.4732
15	5	48.54382	29	1	154.8625
1	9	48.54382	19	11	159.6411
3	9	49.73825	13	13	166.809
5	9	52.12713	23	9	168.0036
19	1	54.51605	29	3	172.7822

(continued)

Table 9.8 (continued)

m	n	$\bar{\omega}_{mn}$	m	n	$\bar{\omega}_{mn}$
13	7	54.51605	15	13	173.9769
7	9	55.71052	1	15	176.3662
17	5	58.0995	3	15	177.5608
19	3	59.29399	27	7	179.9501
9	9	60.4885	21	11	184.7288
15	7	62.87753	5	15	188.3127
21	1	66.46111	29	5	190.702
11	9	66.46111	25	9	197.87
19	5	68.85019	7	15	200.2593
21	3	71.23929	31	1	206.2326
17	7	72.43385	17	13	215.7898
1	11	72.43385	9	15	215.7898
13	9	73.62841	31	3	216.9845
3	11	73.62841	23	11	219.3738

9.5 Conclusion

The normal mode approach was applied in this chapter to the analysis of the random vibrations of thin-elastic shells. The cross-spectral densities of responses were formulated in terms of cross-spectral densities of external forces. The response cross-spectral density was written as a sum of joint and cross-correlations, the first of which represents the autocorrelations between similar modes of free vibration, and the second, the cross-correlations between the separate modes. The effect of cross-terms, which are usually neglected in the literature, on the response cross-spectral density was studied for shells with a *high modal density of eigenfrequencies*. Omission of cross-terms in such cases turned out to be unsatisfactory. A numerical example for a spherically curved panel, supported at all edges, was evaluated

References

Bolotin, V. V. (1961). An asymptotic method for the study of the problem of eigenvalues for rectangular regions. In J. R. M. Dodds (Ed.), *Problems in continuum mechanics*, Volume dedicated to N. I. Muskhelishvili (pp. 56–58). Philadelpha, PA: Society of Industrial and Applied Mathematics.

Bolotin, V. V. (1963). On the density of the distribution of natural frequencies of thin elastic shells. *PMM: Journal of Applied Mathematics and Mechanics, 27*(2), 362–364. (in Russian).

Bolotin, V. V. (1966). Density of eigenvalues in problems of shell and plates' vibration. In *Proceedings of the 6th All-Union Conference on Theory of Shells and Plates*, Baku, Azerbaijan (pp. 161–167). (in Russian).

Bolotin, V. V. (1972). Theory of distribution of the characteristic frequencies of elastic bodies and its application to problems of random vibrations. *Soviet Applied Mechanics, 8*(4), 343–364.
Bolotin, V. V., & Elishakaff, I. (1971). Random oscillation in elastic shells containing an acoustic medium. *Mechanics of Solids*, 6 (5), 99–106.
Bourgine, A. (1970). Méthodes de Calcul de la Response d'une Structure Soumise à un Environnement Aléatoire. *La Recherche Aerospatiale,* 6, 301–314. (in French).
Crandall, S. H., & Mark, W. D. (1963). *Random vibration in mechanical systems*. New York: Academic Press.
Dahlberg, T. (1982). Modal cross-spectral terms may be important and an alternative method of analysis be preferable. *Journal of Sound Vibration, 84,* 503–508.
Dahlberg, T. (1998). On the importance of modal cross-spectral densities in random vibration analysis. In *Proceedings of the ESA International Workshop on Advanced Mathematical Methods in the Dynamics of Flexible Bodies* (pp. 131–142). Noordwijk, The Netherlands: European Space Agency.
Dahlberg, T. (1999). The effect of modal coupling in random vibration analysis. *Journal of Sound and Vibration, 228*(1), 157–176.
Davis, R. E. (1965). Random pressure excitation of shells and statistical dependence effect of normal mode response. *NASA CR-311.*
Dimentberg, M. F. (1961). Forced vibrations of plates under load representing spatial-temporary random process. *Inzhenernyi Zhurnal, 1*(2), 97–105. (in Russian).
Elishakoff, I. (1971a). On evaluation of spectral densities of external forces for plates and shells. In V. V. Bolotin & A. Chyras (Eds.), *Problems of Reliability in Structural Mechanics* (pp. 37–44). Vilnius, Luthuania. (in Russian).
Elishakoff, I. (1971b). On the determination of spectral densities of external forces of plates and shells. In V. V. Bolotin & A. A. Čyras (Eds.), *Reliability problems in structural mechanics* (pp. 36–44). Vilnius.
Elishakoff, I. (1971c). *Vibration fields in cylindrical shells, excited by random forces*. Ph.D. Thesis, Department of Dynamics and Strength of Machines, Moscow Power Engineering Institute and State University. (in Russian).
Elishakoff, I. (1975a). Distribution of natural frequencies in certain structural elements. *The Journal of the Acoustical Society of America, 57*(2), 361–369.
Elishakoff, I. (1975b). Turbulent flow excited vibration of a shallow cylindrical shell. *AIAA Journal, 13*(11), 09–1118.
Hart, F. D., & Shah, K. C. (1971). Compendium of modal densities for structures. *NASA CR*-1773.
Keltie, R. F., & Peng, H. (1987). The effects of modal coupling on the acoustic power radiation from panels. *Journal of Vibration, Acoustics, Stress and Reliability in Design, 109,* 48–54.
Kornev, V. M. (1972). Approximate solution of problems on the stability and vibration of elastic shells under condensation of eigenvalues. *Mechanics of Solids, 7,* 119–129.
Lin, Y. K. (1967). *Probabilistic theory of structural dynamics* (Chap. 7, pp. 203–252). New York: McGraw-Hill.
Mercer, C. A. (1965). Response of a multi-supported beam to a random pressure field. *Journal of Sound and Vibration, 2,* 293–306.
Mkhitarov, R. A. (1972). Interaction of the vibrational modes of a thin plate in a liquid. *Soviet Physics-Acoustics, 18,* 123–126.
Moskalenko, V. N. (1968). Random vibrations of multi-span plates. *Mechanics of Solids, 3,* 79–84.
Parkus, H. (1969) *Random Process in Mechanical Sciences*. International Centre for Mechanical Sciences. Udine 1969.
Powell, A. (1965). An introduction to acoustic fatigue. In W. J. Trapp & D. M. Jr. Forney (Eds.), *Acoustical fatigue in aerospace structures* (pp. 1–15). Syracuse University Press.
Ribner, H. S. (1966). Response of flexible panel to turbulent flow: Running wave versus modal density analysis. *The Journal of the Acoustical Society of America, 40*(3), 721–726.
Schweiker, J. W., & Davis, R. E. (1966). Response of complex shell structures to aerodynamic noise. *NASA CR-450.*

Stearn, S. M. (1970). Spatial variation of stress, strain and acceleration in structures subject to broad frequency band excitation. *Journal of Sound and Vibration, 12,* 85–97.
Vaicaitis, R., & Lin, Y. K. (1972). Response of finite periodic beam to turbulent boundary-layer pressure excitations. *AIAA Journal, 16,* 1020–1024.
White, R. H. (1966). Transduction of boundary layer noise by a rectangular panel. *The Journal of the Acoustical Society of America, 40,* 1354–1362.
Wirsching, P. H., Paez, T. L., & Ortiz, H. (1995). *Random vibration: Theory and practice*. New York: Wiley.

Chapter 10
Dramatic Effect of Cross-Correlations in Random Vibration of Point-Driven Flat Plate and Spherically Curved Panel

10.1 Introductory Comments

The first study on random vibration of point-driven structures was undertaken by Crandall and Wittig (1972). Subsequently. several investigations were devoted to this subject by Crandall (1977; 1979a, 1979b), Elishakoff, van Zanten and Crandall (1979), Elishakoff (1999, 1977a, 1977b, 1977c, 1977d) and others. Briefly, results can be summarized as follows: the distribution of the mean-square velocity turned out to be unexpectedly uniform except for small zones or narrow bands of enhanced response. For a string or a beam, for example, the enhanced response zones include the vicinities of the driving point and of its image, and symmetrical about the midpoint of the string or the beam. As the bandwidth of the excitation increases without limit, the relative enhancement approaches 1.5 asymptotically. In a rectangular, non-square plate, the bands of enhanced response form a tic-tac-toe pattern with the driving point at one of the intersections. Crandall (1978) showed that the relative enhancement in the bands again approaches 1.5, while the relative enhancement at the driving point and its three images approaches 2.25. The picture turns out to be different for a square plate. If the driving point is off the diagonals, the maximum enhancement reaches 3.25 (Fig. 10.1 shows the intensification factors at various characteristic points on the plate, the driving point being denoted by the symbol D and its images points by I); if it lies on a diagonal the intensification there becomes 4.5 (see Fig. 10.3); finally, when the load is at the center of the square, the resulting pattern resembles the Union Jack, the intensification factor being 2 along the bands and 8 at the center (Fig. 10.2).

These results were obtained by normal-mode analysis Crandall and Wittig (1972), as well as by the image-sum technique Crandall (1977, 1978), Langley and Taylor (1979). For string and beams, the natural frequencies are widely spaced, so that under light damping the terms involving modal cross-correlations can be neglected relative to the autocorrelations. This fact was the origin of the symmetrical distribution of the mean-square velocity response.

I. Elishakoff, *Dramatic Effect of Cross-Correlations in Random Vibrations of Discrete Systems, Beams, Plates, and Shells*,
https://doi.org/10.1007/978-3-030-40394-2_10

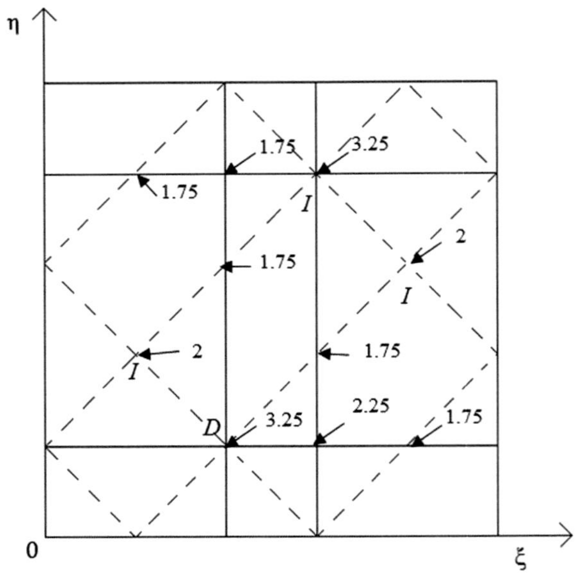

Fig. 10.1 Asymptotic values reported by Crandall (1977, 1978), for the intensification factor $r(\xi, \eta|\xi_0, \eta_0)$ at the driving point off the diagonals (*D*-driving point, *I*-image points in the plate)

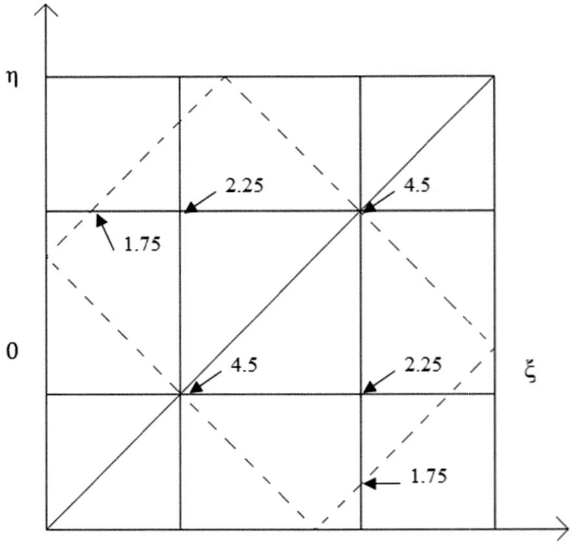

Fig. 10.2 Asymptotic values of $r(\xi, \eta|\xi_0, \eta_0)$ when the driving point is at the diagonal derived by Crandall (1977)

Analysis of the rectangular plate was likewise based on the *omission* of the modal cross-correlations. As is known, such omission is inappropriate when the eigenfrequencies lie close together, especially when their spacing is small compared with the modal bandwidth, but such occurrences are infrequent for most aspect ratios, provided the average spacing $(4 \cdot \pi / L_X L_Y) \cdot (D/\rho \cdot \eta)^{1/2}$ exceeds the bandwidth. In the case of the square plate$(L_X = L_Y)$, however, there exist series of coincidences, for

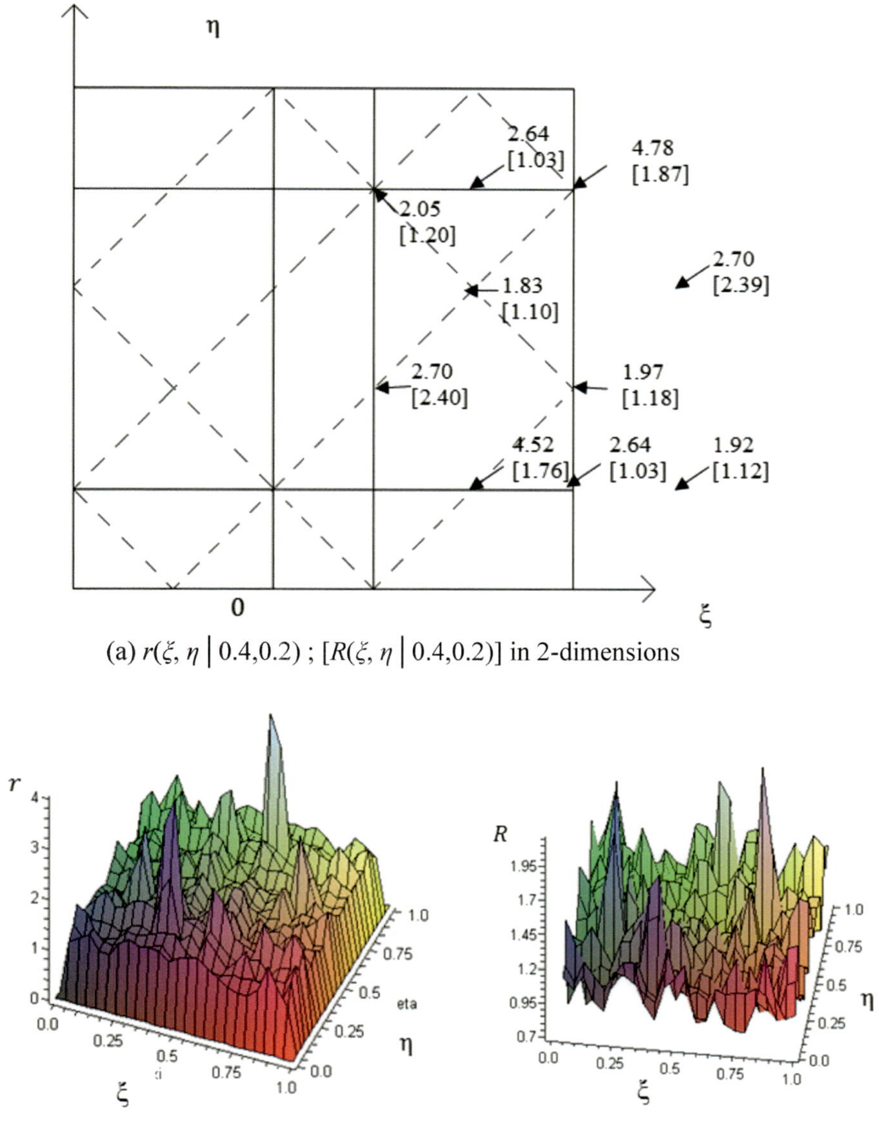

(a) $r(\xi, \eta \mid 0.4, 0.2)$; $[R(\xi, \eta \mid 0.4, 0.2)]$ in 2-dimensions

(b) $r(\xi, \eta \mid 0.4, 0.2)$ (c) $R(\xi, \eta \mid 0.4, 0.2)$

Fig. 10.3 **a** The intensification factor $r(\xi, \eta|0.4, 0.2)$, and the factor $[R(\xi, \eta|0.4, 0.2)]$ of error, caused by the neglect of the cross-correlation terms, for the square plate $(h/L_X = 0.01)$, **b** variation of factor $r(\xi, \eta|0.4, 0.2)$ with ξ, η, **c** variation of factor $\mathrm{R}(\xi, \eta|0.4, 0.2)$ with ξ, η

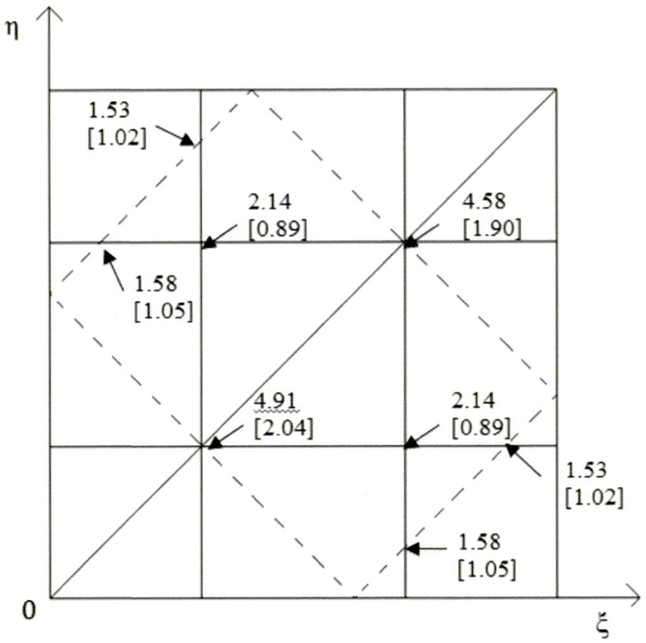

(a) $r(\xi, \eta \mid 0.3{,}0.3)$; $[R(\xi, \eta \mid 0.3{,}0.3)]$ in 2-dimensions

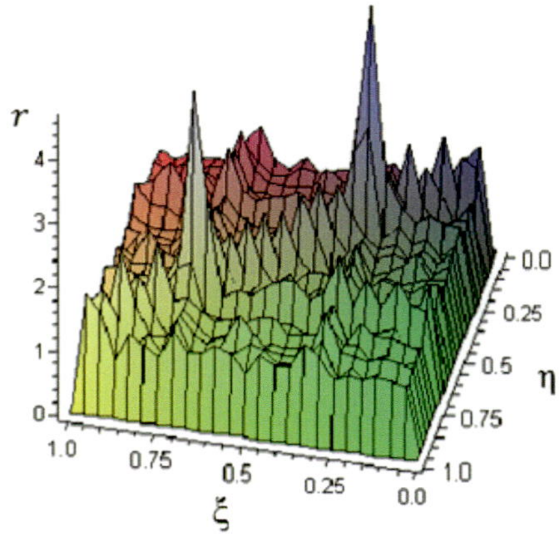

(b) $r(\xi, \eta \mid 0.3{,}0.3)$

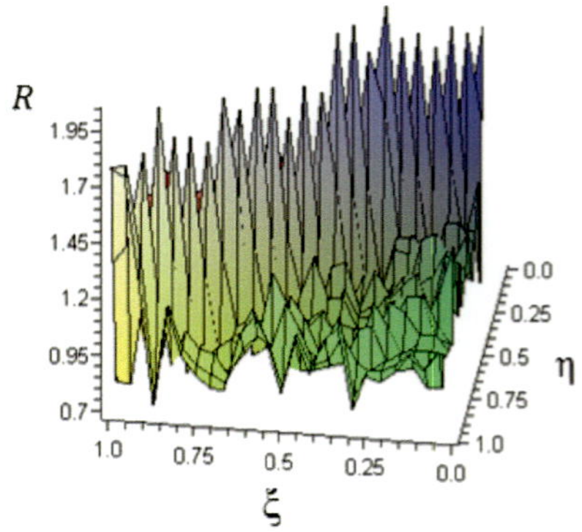

(c) $R(\xi, \eta \mid 0.3{,}0.3)$

Fig. 10.4 **a** The intensification factor $r(\xi, \eta|0.3, 0.3)$, and the factor $[R(\xi, \eta|0.3, 0.3)]$ of error, caused by the neglect of the cross-correlation terms, for the square plate $(h/L_X = 0.01)$; **b** variation of factor $r(\xi, \eta|0.3, 0.3)$ with ξ, η; **c** variation of factor $R(\xi, \eta|0.3, 0.3)$ with ξ, η

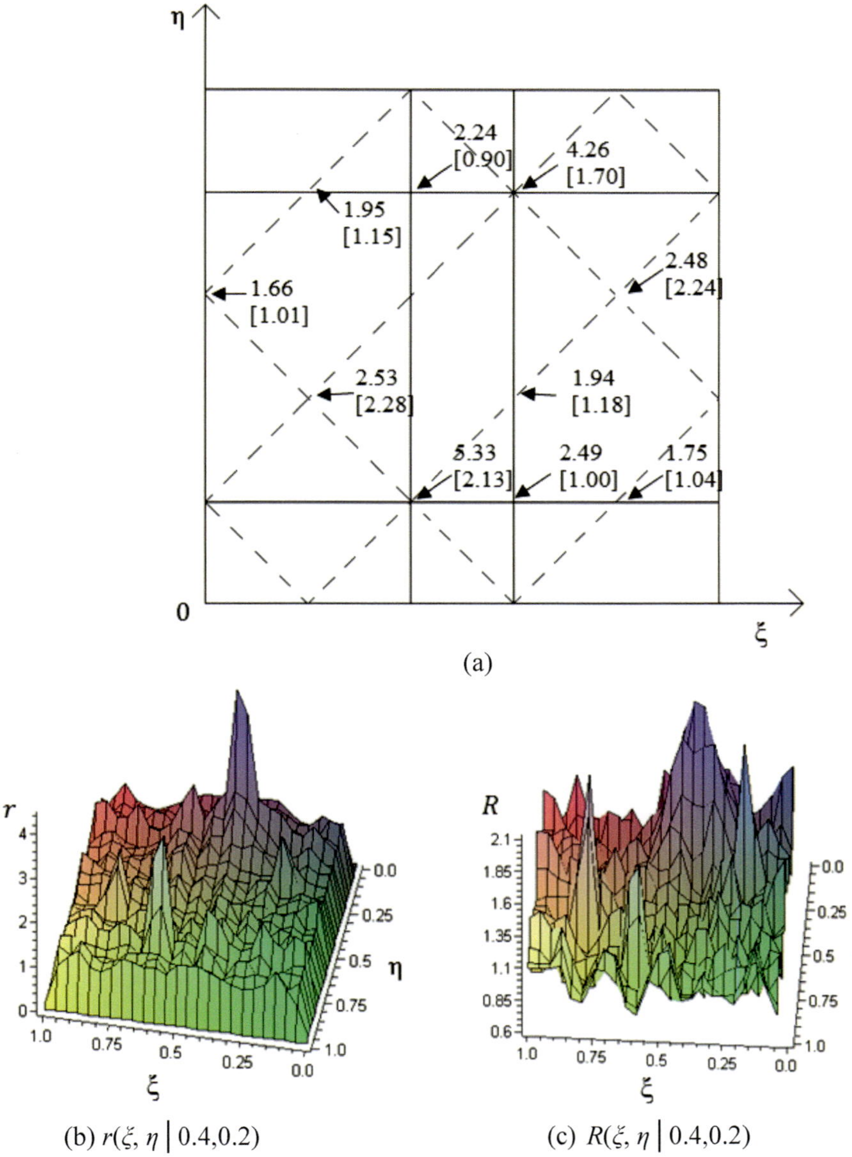

Fig. 10.5 **a** The intensification factor $r(\xi, \eta|0.4, 0.2)$, and the factor $[R(\xi, \eta|0.4, 0.2)]$ of error, caused by the neglect of the cross-correlation terms, for the square plate ($h/L_X = 0.01$, $L_X/R = 2$); **b** variation of factor $r(\xi, \eta|0.4, 0.2)$ with ξ, η; **c** variation of factor $R(\xi, \eta|0.4, 0.2)$ with ξ, η

example, double eigenfrequencies of the type $\omega_{jp} = \omega_{pj}$, $j \neq p$, so that the modes with $j = p$ have unrepeated eigenfrequencies. Square plate results included the modal cross-correlation terms associated with double eigenfrequencies, but multiple resonances (three or more coincident eigenfrequencies) have been neglected.

Wide-band random vibration of a uniform circular cylindrical shell excited by a ring load, uniform along the circumference and random in time, was considered by Elishakoff, Crandall and van Zanten (1979). In this problem, the contributions of the modal cross-correlations introduce *asymmetry* into the distribution of the mean-square velocity response and alter the magnitude of the local response considerably. For example, in this shell with a radius-to-length ratio of 0.5, excited by a band-limited "white noise" with 81 modes, this contribution at a driven section may exceed that of the modal autocorrelations by a factor of 3 or more.

The structure considered by Elishakoff, van Zanten, and Cranten (1979) was actually one-dimensional, although considerably different from its counterparts for point-excited uniform strings and beams. The object of the present investigation, dealing with a point-driven spherically curved panel, is twofold: extension of the results of our earlier investigation (just referred to) to the two-dimensional structure, and investigation of the effect of omission of different types of modal cross-correlations for the shell, plus the verification of the plate results (which have to be obtained from those for the shell with the panel radius tending to infinity).

Related works on random excitation of spherical shells include Elishakoff's (1977a, 1977b, 1977c, 1977d), Jullien's (1966), and Palmov's (1965)and those on the role of cross-correlations include Powell's (1968), Bolotin's (1969), Khabbaz's (1964), Davis's (1966), Mercer's (1965), and Garrelick and Chayes's (1978).

10.2 Random Vibration of a Curved Spherical Panel

Consider a uniform spherically curved panel supported at all its edges by shear diaphragms. The shell is subject to a viscous damping force c per unit area per unit velocity, and excited by a transverse force $f(x, y, t)$ per unit area. The normal displacement $w(x, y, t)$ and the stress resultant function $\chi(x, y, t)$ must satisfy the Donnell–Vlasov equations

$$
\begin{aligned}
&D\nabla^2\nabla^2 w + \frac{1}{R}\nabla^2\chi + \rho h\frac{\partial^2 w}{\partial t^2} + c\frac{\partial w}{\partial t} = f(x, y, t)\\
&\nabla^2\nabla^2\chi - \frac{Eh}{R}\nabla^2 w = 0,
\end{aligned}
\tag{10.1}
$$

where ρ, h and R are the mass density, thickness, and radius of the panel respectively, ∇^2 is the Laplace operator at the middle surface, D and E are the stiffness and Young moduli, respectively.

Equation (10.1) are supplemented by the boundary conditions

$$w = \frac{\partial^2 w}{\partial x^2} = \frac{\partial^2 \chi}{\partial x^2} = \frac{\partial^2 \chi}{\partial y^2} = 0 \quad \text{at} \quad x = 0,\ x = L_X,$$
$$w = \frac{\partial^2 w}{\partial y^2} = \frac{\partial^2 \chi}{\partial y^2} = \frac{\partial^2 \chi}{\partial x^2} = 0 \quad \text{at} \quad y = 0,\ y = L_Y, \tag{10.2}$$

where L_X and L_Y are the panel sides.

The time history of loadings is taken to be a stationary wide-band random process. The space-time correlation function of the excitation is

$$E[f(x_1, y_1, t_1) f(x_2, y_2, t_2)] = \delta(x_1 - x_0)\delta(x_2 - x_0)\delta(y_1 - y_0)\delta(y_2 - y_0) R_f(t_2 - t_1). \tag{10.3}$$

x_0 and y_0 being the coordinates of the driven point.

The problem is formulated as follows: Given the autocorrelation function $R_f(t_2 - t_1)$, find the probabilistic behavior of the panel displacements and velocity, dispensing with initial conditions.

For the stochastic boundary value problem defined by Eqs. (10.1)–(10.3), we resort to the normal-mode approach by Elishakoff (1999) and Lin (1967) which consists of expanding the probabilistically given and sought functions in stochastic Fourier integrals and in series in terms of the modes of the undamped panel *in vacuo*.

The result, for the mean-square of the transverse velocity $v(t) = \partial w/\partial t$, reads

$$E\left[v^2(x, y|x_0, y_0)\right] = \sum_{m=1}^{\infty}\sum_{n=1}^{\infty}\sum_{p=1}^{\infty}\sum_{q=1}^{\infty} \psi_m(x_0)\psi_p(x_0)\psi_n(y_0)\psi_q(y_0)\psi_m(x)\psi_p(x)\psi_n(y)\psi_q(y)\Phi'_{mn,pq}, \tag{10.4}$$

where

$$\psi_m(x) = \sqrt{2}\sin\left(\frac{m\pi x}{L_X}\right),\ \psi_n(y) = \sqrt{2}\sin\left(\frac{n\pi y}{L_Y}\right), \tag{10.5}$$

the mode shape being $\psi_m(x)\psi_n(y)$, m denoting the number of half-waves in the x-direction, n that in the y-direction, and

$$\Phi'_{mn,pq} = \int_{-\infty}^{\infty} \omega^2 S_f(\omega) H^*_{mn}(\omega) H_{pq}(\omega), \mathrm{d}\omega \tag{10.6}$$

where $H_{mn}(\omega)$ is the frequency response function for the (m, n)th mode

$$H_{mn}(\omega) = \frac{1}{M\left(\omega_{mn}^2 + \mathrm{i}\beta\omega - \omega^2\right)},\ \beta = \frac{c}{\rho h},\ M = \rho h L_X L_Y, \tag{10.7}$$

β is the modal bandwidth, M—the mass of the panel, ω_{mn}—the (m, n) the eigenfrequency

$$\omega_{mn}^2 = \frac{D}{\rho h}\left(\frac{m^2\pi^2}{L_X^2} + \frac{n^2\pi^2}{L_Y^2}\right)^2 + \frac{E}{\rho R^2}, \tag{10.8}$$

$S_f(\omega)$ the spectral density of the point-loading assumed to be a band-limited white noise spectrum

$$S_f(\omega) = \begin{cases} S_0, \ |\omega| < \omega_c \\ 0, \ \omega_c < |\omega| \end{cases} \tag{10.9}$$

ω_c. being the cutoff eigenfrequency, ω the angular frequency.

A closed-form expression is obtainable by Elishakoff (1977a, 1977b, 1977c, 1977d) for $\Phi'_{mn,pq}$

$$\Phi'_{mn,pq}(\omega_c) = \frac{S_0}{M^2}\frac{F'(\omega_{mn}, \omega_{pq}; \omega_c) + F'(\omega_{pq}, \omega_{mn}; \omega_c)}{\left(\omega_{mn}^2 - \omega_{pq}^2\right)^2 + 2\beta^2(\beta^2 + \omega_{pq}^2)}, \tag{10.10}$$

where

$$\begin{aligned} F'(\omega_{mn}, \omega_{pq}; \omega_c) = & \frac{\omega_{mn}^2\left(\omega_{mn}^2 - \omega_{pq}^2\right) + \frac{\beta^2\left(\omega_{mn}^2 + \omega_{pq}^2\right)}{2}}{2\left(\omega_{mn}^2 - \frac{\beta^2}{4}\right)^{\frac{1}{2}}} \\ & + \ln \frac{\omega_c^2 + \omega_{mn}^2 - 2\omega_c\left(\omega_{mn}^2 - \frac{\beta^2}{4}\right)^{\frac{1}{2}}}{\omega_c^2 + \omega_{mn}^2 + 2\omega_c\left(\omega_{mn}^2 - \frac{\beta^2}{4}\right)^{\frac{1}{2}}} \\ & \times \beta\left(\omega_{mn}^2 + \omega_{pq}^2\right)\left\{\tan^{-1}\left[\frac{\omega_c - \left(\omega_{mn}^2 - \frac{\beta^2}{4}\right)^{\frac{1}{2}}}{\frac{\beta}{2}}\right]\right. \\ & \left. + \tan^{-1}\left[\frac{\omega_c + \left(\omega_{mn}^2 - \frac{\beta^2}{4}\right)^{\frac{1}{2}}}{\frac{\beta}{2}}\right]\right\}. \end{aligned} \tag{10.11}$$

Equation (10.4) can be put in the following no dimensional form

$d_v^2(\xi, \eta|\xi_0, \eta_0)$

$$= \sum_{m=1}^{\infty}\sum_{n=1}^{\infty}\sum_{p=1}^{\infty}\sum_{q=1}^{\infty} \psi_m(\xi_0)\psi_p(\xi_0)\psi_n(\eta_0)\psi_q(\eta_0)\psi_m(\xi)\psi_p(\xi)\psi_n(\eta)\psi_q(\eta)I'_{mn,pq}(\Omega_c), \tag{10.12}$$

where

$$d_v^2(\xi,\eta|\xi_0,\eta_0) = \frac{E\left[v^2(x,y|x_0,y_0)\right]m^2\left(\frac{E}{\rho}\right)^{\frac{1}{2}}}{S_0 L_X}$$

$$\xi = \frac{x}{L_x},\ \xi_0 = \frac{x_0}{L_x},\ \eta = \frac{y}{L_y},\ \eta_0 = \frac{y_0}{L_y},\ \Omega = \frac{\omega \mathrm{L_x}}{(E/\rho)^{\frac{1}{2}}},\ \Omega_c = \frac{\omega_c L_X}{(E/\rho)^{\frac{1}{2}}},$$

$$\Omega_{mn} = \frac{\omega_{mn}\mathrm{L_x}}{(E/\rho)^{\frac{1}{2}}} = \left[\frac{1}{12\left(1-\upsilon^2\right)}\left(\frac{h}{L_x}\right)^2\left(m^2\pi^2 + n^2\pi^2\frac{L_x^2}{L_y^2}\right)^2 + \frac{L_x^2}{R^2}\right]^{\frac{1}{2}}, \tag{10.13}$$

$$B = \frac{\beta L_x}{\sqrt{E/\rho}},\ \mathrm{I}'_{mn,pq} = \int_{-\Omega_c}^{\Omega_c} \frac{\Omega^2 \mathrm{d}\Omega}{\left(\Omega_{mn}^2 - \mathrm{i}\Omega B - \Omega^2\right)\left(\Omega_{pq}^2 + \mathrm{i}\Omega B - \Omega^2\right)}.$$

The dimensionless mean-square velocity d_v^2 is resolved into four sum-type components,

$$d_v^2(\xi,\eta|\xi_0,\eta_0) = s_1 + s_2 + s_3 + s_4, \tag{10.14}$$

$$s_1 = 16\sum_{m=1}^{\infty}\sum_{n=1}^{\infty} \sin^2(m\pi\xi_0)\sin^2(n\pi\eta_0)\sin^2(m\pi\xi)\sin^2(n\pi\eta)\mathrm{I}'_{mn,mn}(\Omega_c), \tag{10.15}$$

$$s_2 = 32\sum_{\substack{m=1\\ m<p}}^{\infty}\sum_{n=1}^{\infty}\sum_{p=1}^{\infty} \sin(m\pi\xi_0)\sin(p\pi\xi_0) \times \sin^2(n\pi\eta_0)\sin(m\pi\xi)\sin(p\pi\xi)\sin^2(n\pi\eta)\mathrm{I}'_{mn,pn}(\Omega_c), \tag{10.16}$$

$$s_3 = 32\sum_{\substack{m=1\\ n<q}}^{\infty}\sum_{n=1}^{\infty}\sum_{q=1}^{\infty} \sin^2(m\pi\xi_0) \times \sin(n\pi\eta_0)\sin(q\pi\eta_0)\sin^2(m\pi\xi)\sin(n\pi\eta)\sin(q\pi\eta)\mathrm{I}'_{mn,mq}(\Omega_c), \tag{10.17}$$

$$s_4 = 32 \sum_{\substack{m=1 \\ m \neq p, n < q}}^{\infty} \sum_{n=1}^{\infty}\sum_{p=1}^{\infty}\sum_{q=1}^{\infty} \sin(m\pi\xi_0)\sin(p\pi\xi_0)\sin(n\pi\eta_0) \times \sin(q\pi\eta_0)\sin(m\pi\xi)\sin(p\pi\xi)\sin(n\pi\eta)\sin(q\pi\eta)I'_{mn,mq}(\Omega_c). \quad (10.18)$$

The first sum covers the modal autocorrelations, and the other three the modal cross-correlations, s_2 referring to the modes with an equal number of half-waves in the y-direction ($n = q$), s_3—to those with equal number of half-waves in the x-direction ($m = p$), and s_4—those with $m \neq p$ and $n \neq q$.

Regarding the distribution of the mean-square velocity over the panel surface, we first define its spatial mean (surface-averaged mean-square level of velocity)

$$\tilde{E}\left[v^2(x_0, y_0)\right] = \frac{1}{L_x L_y}\int_0^{L_x}\int_0^{L_y} E\left[v^2(x, y|x_0, y_0)\right]dxdy, \quad (10.19)$$

The plateau level of the mean-square velocity (denoted by P) is in turn defined as the average $\tilde{E}\left[v^2(x_0, y_0)\right]$ with respect to the variation surface of the location of the driving point

$$P = \frac{1}{L_x L_y}\int_0^{L_x}\int_0^{L_y} \tilde{E}\left[v^2(x_0, y_0)\right]dx_0 dy_0, \quad (10.20)$$

or, in the non-dimensional form

$$P = \int_0^1\int_0^1 \tilde{d}_v^2(\xi_0, \eta_0)d\xi_0 d\eta_0. \quad (10.21)$$

One maybe interested in the "overshooting" ratio, or the ratio of the response to the surface average

$$r(x, y|x_0, y_0) = \frac{1}{P}E\left[v^2(x, y|x_0, y_0)\right], \quad (10.22)$$

or, in the non-dimensional form, we get

$$r(\xi, \eta|\xi_0, \eta_0) = \frac{1}{P}d_v^2(\xi, \eta|\xi_0, \eta_0), \quad (10.23)$$

The expression for P becomes

$$P=\int_0^1\int_0^1 \tilde{d}_v^2(\xi_0,\eta_0)d\xi_0 d\eta_0=\sum_{m=1}^{\infty}\sum_{n=1}^{\infty} I'_{mn,mn}(\Omega_c), \tag{10.24}$$

Note that it is straightforward to evaluate the limiting value of the quantity $I'_{mn,mn}(\Omega_c)$ when the cutoff frequency Ω_c tends to infinity

$$\lim_{\Omega_c\to\infty}\left(\mathrm{I}'_{mn,mn}(\Omega_c)\right)=\lim_{\Omega_c\to\infty}\int_{-\Omega_c}^{\Omega_c}\frac{\Omega^2 d\Omega}{\left(\left(\Omega_m^2-\Omega^2\right)^2+(\Omega B)^2\right)}=\frac{\pi}{B}, \tag{10.25}$$

Note that a simple estimate can be obtained for p. In fact, for $I_{mn,mq}(\Omega_c)$ one can use the ideal white noise approximation with sufficient accuracy, if $\Omega_{mn}\le\Omega_c$ by Crandall and Wittig (1972); and Elishakoff (1999). Thus, one obtains

$$p\simeq\frac{\pi}{B}\sum\sum_{m,n:\ \Omega_{mn}\le\Omega_c} 1. \tag{10.26}$$

The result of the double summation can be immediately recognized as the frequency distribution function at the level Ω_c, representing the number of eigenfrequencies not exceeding a fixed cutoff frequency Ω_c, or number of integer solutions of inequality $\Omega_c\le\Omega_c$ by Elishakoff (1975a, 1975b)

$$F(\Omega_c)=\sum\sum_{m,n:\ \Omega_{mn}\le\Omega_c} 1, \tag{10.27}$$

As a result, the following expression for P can be obtained

$$P\simeq\frac{\pi}{B}N_c. \tag{10.28}$$

10.3 Numerical Examples and Discussion

In order to get some insight into the mean-square velocity response of the panel, several numerical examples were evaluated. The square, spherically curved panel was considered with the driving point at different locations (ξ_0,η_0), and the intensification factors $r(\xi,\eta|\xi_0,\eta_0)$ were calculated at different points with coordinates (ξ,η). The formula (10.12) was approximated as follows:

$$d_v^2(\xi,\eta|\xi_0,\eta_0)\simeq 16\sum_m{}'\sum_n{}'\sum_p{}'\sum_q{}'\sin(m\pi\xi_0)\sin(p\pi\xi_0)\sin(n\pi\eta_0)$$

$$\times \sin(q\pi\eta_0)\sin(m\pi\xi)\sin(p\pi\xi)\sin(n\pi\eta)\sin(q\pi\eta)I'_{mn,mq}(\Omega_c), \tag{10.29}$$

The prime in summations denotes the fact that only those terms are taken into account that satisfies the inequalities $\Omega_{mn} \leq \Omega_c$ and $\Omega_{pq} \leq \Omega_c$. We also calculate the error factor $R(\xi, \eta|\xi_0, \eta_0)$ defined as

$$R(\xi, \eta|\xi_0, \eta_0) = \frac{d_v^2(\xi, \eta|\xi_0, \eta_0)}{s_1}. \tag{10.30}$$

representing the number of times by which the mean-square velocity due to both the modal autocorrelations and cross-correlations is more than due to modal autocorrelations. In all instances, the non-dimensional cutoff frequency Ω_c was fixed at 2π.

First, in order to compare with the asymptotic values of the intensification factor $r(\xi, \eta|\xi_0, \eta_0)$, reported by Crandall (1977, 1978), the flat plate was considered, with results obtained as a limit case when $L_X/R \to 0$. Two values of the thickness-to-side-length ration h/L_X were considered, 0.01 (with $N_c = 152$) and 0.005 (with $N_c = 308$). The results are displayed in Figs. 10.3 and 10.6. Figure 10.4 is associated with the driving point off the diagonals, whereas Fig. 10.4—with the driving point on the diagonal. Figures 10.3 and 10.4 show the results for $h/L_X = 0.01$. In figures, the values of r (upper line) and R (bottom line) at various points are given. The comparison with the asymptotic values, given in Fig. 10.2, shows that the cross-correlations cause, in some points, to the increase of the intensification factor r. For example, for the flat plate, driven on the diagonal, the asymptotic value $r(\xi, \eta|\xi_0, \eta_0) = 4.5$ (Fig. 10.2), whereas the actual value is 4.91 (Fig. 10.4).

Note that the asymptotic intensification factor has some symmetry properties, e.g. $r(\xi_0, \eta_0|\xi_0, \eta_0) = r(1-\xi_0, 1-\eta_0|\xi_0, \eta_0)$ (Fig. 9.3), whereas the actual intensification factor lacks this property.

With the decrease of the thickness-to-side-length ratio the number of excited modes increases, so that then the larger number of close eigenfrequencies occurs, and correspondingly, the intensification factor increases.

For the curved panel with L_X/R fixed at 2, results are given in Figs. 10.5, 10.6, 10.7, 10.8. Figures 10.5 and 10.6 are associated with $h/L_X = 0.01$ and Fig. 10.11, 8.12 with $h/L_X = 0.005$. The N_c values are, for these cases, 141 and 292, respectively. Now, however, these eigenfrequencies happen to be closer to each other, especially at the lower end of the frequency spectrum. This concentration results in increased weight of the cross-correlations, and consequently, of the intensification factors: for a driving point off the diagonal ($\xi_0 = 0.4$, $\eta_0 = 0.2$) the factor r becomes 5.33 at the driving point (Fig. 10.5), versus 4.52 for the flat plate (Fig. 10.3), whereas for a driving point on the diagonal, $r = 5.84$ for the curved plate versus $r = 4.91$ for the flat one (Fig. 10.4). As the h/L_X ratio decreases, the contribution of the cross-correlations acquires greater significance, the maximum value of r becoming 7.19 (for a driving point off the diagonal, Fig. 10.7), and 7.87 (for a driving point on the diagonal, Fig. 10.8).

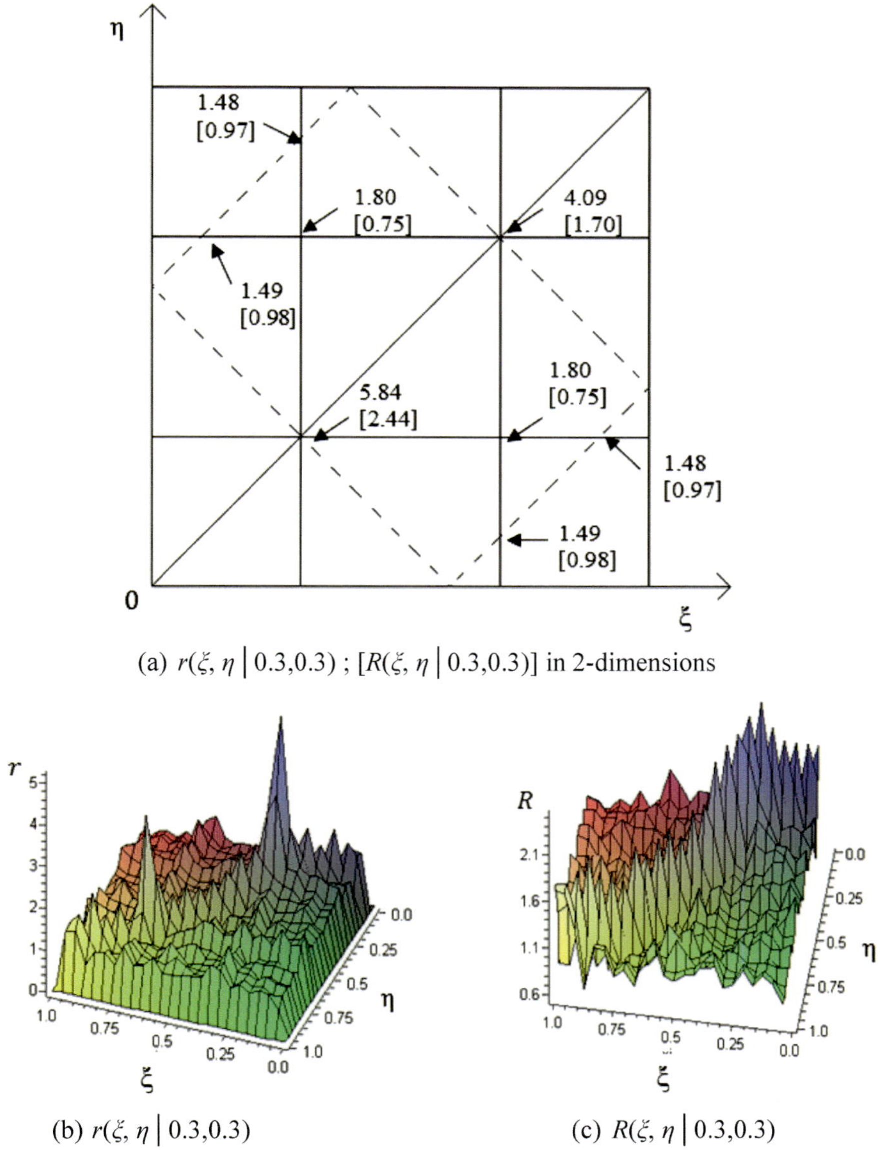

Fig. 10.6 **a** The intensification factor $r(\xi, \eta|0.3, 0.3)$, and the factor $[R(\xi, \eta|0.3, 0.3)]$ of error, caused by the neglect of the cross-correlation terms, for the square plate; **b** variation of factor $r(\xi, \eta|0.3, 0.3)$ with ξ, η; **c** variation of factor $R(\xi, \eta|0.3, 0.3)$ with ξ, η

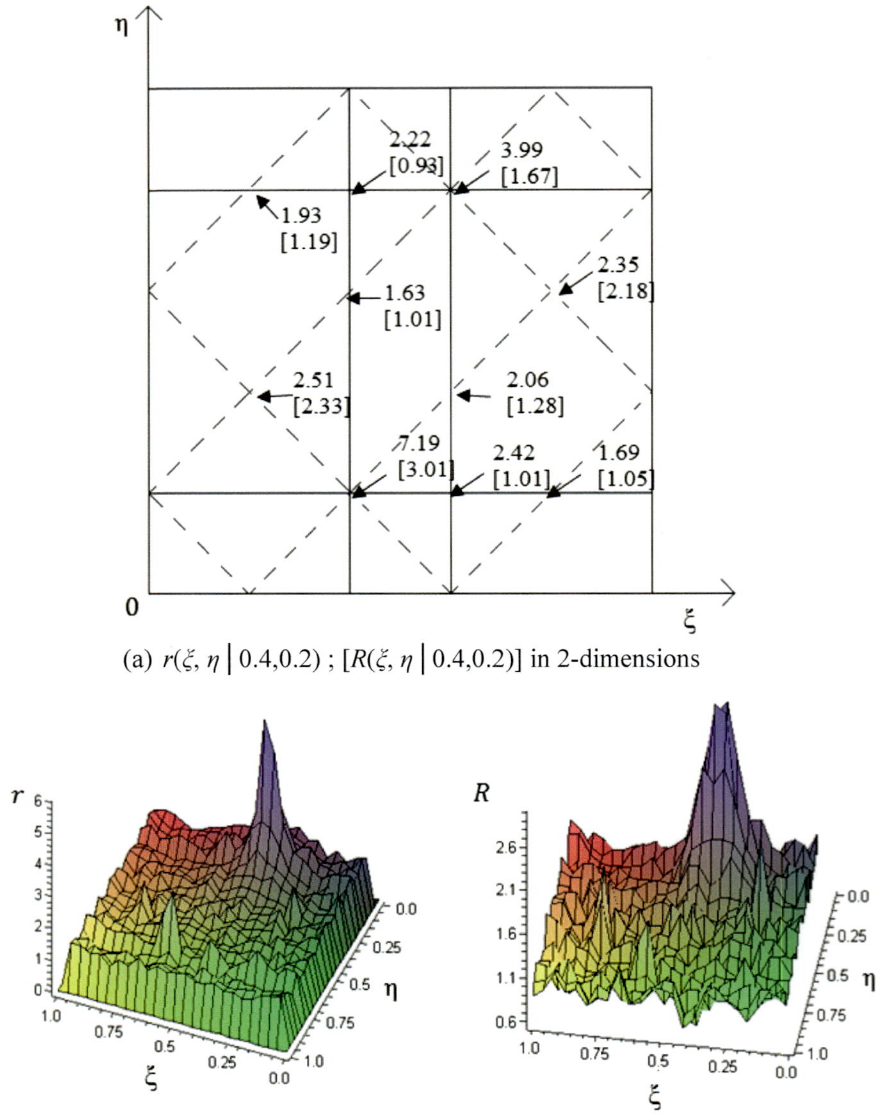

(a) $r(\xi, \eta \mid 0.4, 0.2)$; $[R(\xi, \eta \mid 0.4, 0.2)]$ in 2-dimensions

(b) $r(\xi, \eta \mid 0.4, 0.2)$

(c) $R(\xi, \eta \mid 0.4, 0.2)$

Fig. 10.7 **a** The intensification factor $r(\xi, \eta|0.4, 0.2)$, and the factor $[R(\xi, \eta|0.4, 0.2)]$ of error, caused by the neglect of the cross-correlation terms, for the square plate $(h/L_X = 0.005, L_X/R = 2)$; **b** variation of factor $r(\xi, \eta|0.4, 0.2)$ with ξ, η; **c** variation of factor $R(\xi, \eta|0.4, 0.2)$ with ξ, η

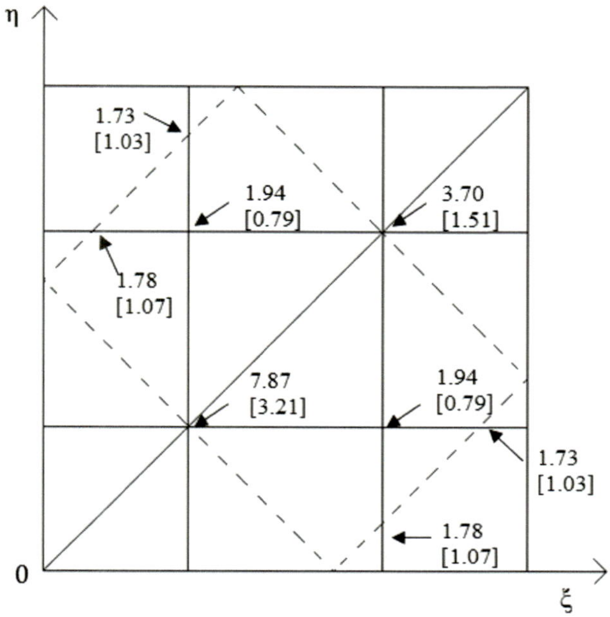

(a) $r(\xi, \eta \mid 0.3, 0.3)$; $[R(\xi, \eta \mid 0.3, 0.3)]$ in 2-dimensions

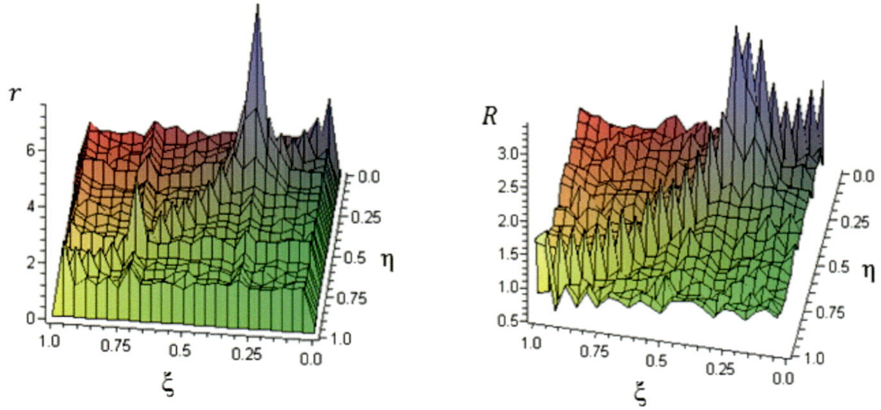

(b) $r(\xi, \eta \mid 0.3, 0.3)$ (c) $R(\xi, \eta \mid 0.3, 0.3)$

Fig. 10.8 **a** The intensification factor $r(\xi, \eta|0.3, 0.3)$, and the factor $[R(\xi, \eta|0.3, 0.3)]$ of error, caused by the neglect of the cross-correlation terms, for the square plate ($h/L_X = 0.005$, $L_X/R = 2$); **b** variation of factor $r(\xi, \eta|0.3, 0.3)$ with ξ, η; **c** variation of factor $R(\xi, \eta|0.3, 0.3)$ with ξ, η

Results for a plate excited at the center are reported in the Table 10.1. Here again the difference between the asymptotic value of r and the actual intensification is large: for the flat plate of $h/L_X = 0.005$ we have $r = 10.284$ (entry 7) instead of its asymptotic counterpart 8, whereas for the curved plate the actual intensification factor becomes 11.359 (entry 9). This discrepancy can be explained by the presence of multiple eigenfrequencies occurring with asymptotic regularity.

All eigenfrequencies Ω_{mn} (with m and n both odd, as required for a central driving point) were evaluated for $\Omega_{mn} \leq \Omega_c = 2\pi$. In the case of $h/L_X = 0.01$, there are 41 contributing modes. The corresponding eigenfrequencies could be categorized as follows:

$n_1 = 4$ single eigenfrequencies
$n_2 = 13$ double eigenfrequencies
$n_3 = 1$ triple eigenfrequencies
$n_4 = 2$ quadruple eigenfrequencies.

The natural frequencies are listed in Table 10.2. Specifically, the natural frequency Ω_{mn} is listed as numbers m and n in square brackets $[m, n]$: $\Omega_{mn} = [m, n]$.

The total number of nodes, with m, n odd or even, with $\Omega_{mn} \leq 2\pi$ is 152. Now, if we count only the contribution of the multiple eigenfrequencies

$$d_v^2 \simeq \left[16 \sum (i)^2 n_i\right] \frac{\pi}{B}, \tag{10.31}$$

then for the driving point intensification, we get an extended asymptotic value, bearing in mind Eq (10.28):

$$r = \frac{16 \sum (i)^2 n_i}{N_c} = \frac{16\left[4 + (2)^2 13 + (3)^2 + (4)^2 2\right]}{152} = 10.211. \tag{10.32}$$

whereas the actual intensification factor is 9.831, the extended asymptotic prediction representing an overestimate with 3.87% error.

In the case of $h/L_X = 0.005$, there are 83 contributing modes (m, n odd, $\Omega_{mn} \leq 2\pi$), whose eigenfrequencies could be categorized as follows (see Table 10.3):

$n_1 = 5$ single eigenfrequencies
$n_2 = 24$ double eigenfrequencies
$n_3 = 2$ triple eigenfrequencies
$n_4 = 6$ quadruple eigenfrequencies

whereas $N_c = 308$. The extended asymptotic value of r is

$$r = \frac{16\left[5 + (2)^2 24 + (3)^2 2 + (4)^2 6\right]}{308} = 11.169. \tag{10.33}$$

the exact intensification factor being 11.035, i.e. the extended asymptotic prediction is an underestimate of 1.21%.

Table 10.1 Results for the square plate and panel driven at the center

Entry	B	h/Lx	Lx/R	Number of eigenfrequencies (m, n odd) with multiplicity				Number of contributing modes (m, n odd)	Nc	Asymptotic	r extended asymptotic	r exact	R	Error in %
				1	2	3	4							
1	0.02	0.01	0	4	13	1	2	41	152	8	10.21052632	9.659680702	2.24478635	5.702524036
2	0.02	0.01	1	4	13	1	2	41	150		10.34666667	9.90667093	2.273369224	4.44140862
3	0.02	0.01	2	4	12	1	2	39	141		10.55319149	10.3593 843	2.345184862	1.870836951
4	0.002	0.01	0	4	13	1	2	41	152	8	10.21052632	10.14389577	2.351102785	0.656853612
5	0.002	0.01	1	4	13	1	2	41	150		10.34666667	10.27894705	2.351209508	0.658813616
6	0.002	0.01	2	4	12	1	2	39	141		10.55319149	10.48286783	2.369162872	0.670843709
7	0.02	0.005	0	5	24	2	6	83	308	8	11.16883117	10.28367904	2.393718805	8.607348843
8	0.02	0.005	1	5	24	2	6	83	306		11.24183007	10.83706755	2.509304209	3.734981936
9	0.02	0.005	2	5	24	2	5	79	292		10.90410959	11.3590624	2.63423528	4.17230593
10	0.002	0.005	0	5	24	2	6	83	308	8	11.16883117	11.03522441	2.56029974	1.210729876
11	0.002	0.005	1	5	24	2	6	83	306		11.24183007	11.11227327	2.561757855	1.165889213
12	0.002	0.005	2	5	24	2	5	79	292		10.90410959	10.80308799	2.496613802	0.935117803

Table 10.2 Natural frequencies of multiplicity i for the case $B = 0.02$, $h/L_X = 0.01$, $L_X/R = 0$

Multiplicity	Natural frequencies
1	[1, 1], [3, 3], [7, 7], [9, 9]
2	[1, 3] = [3, 1]; [1, 5] = [5, 1]; [1, 9] = [9, 1]; [1, 11] = [11, 1]; [3, 5] = [5, 3]; [3, 7] = [7, 3]; [3, 9] = [9, 3]; [3, 13] = [13, 3]; [5, 7] = [7, 5]; [5, 9]= [9, 5]; [5, 11] = [11, 5]; [5, 13] = [13, 5]; [9, 11] = [11, 9]
3	[1, 7] = [5, 5] = [7, 1]
4	[1, 13] = [7, 11] = [11, 7] = [13, 1]; [3, 11] = [7, 9] = [9, 7] = [11, 3]

Table 10.3 Natural frequencies of multiplicity i for the case $B = 0.02$, $h/L_X = 0.005$, $L_X/R = 0$

Multiplicity	Natural frequencies
1	[1, 1]; [3, 3]; [7, 7]; [9, 9]; [11, 11]
2	[1, 3] = [3, 1]; [1, 5] = [5, 1]; [1, 9] = [9, 1]; [1, 11] = [11, 1]; [1, 15] = [15, 1]; [1, 19] = [19, 1]; [3, 5] = [5, 3]; [3, 7] = [7, 3]; [3, 9] = [9, [3, 13] = [13, 3]; [3, 15] = [15, 3]; [3, 17] = [17, 3]; [5, 7] = [7, 5]; [5, 9] = [9, 5]; [5, 11] = [11, 5]; [5, 13] = [13, 5]; [5, 17] = [17, 5]; [5, 19] = [19, 5]; [7, 13] = [13, 7]; [7, 15] = [15, 7]; [9, 11] = [11, 9]; [9, 15] = [15, 9]; [11, 15] = [15, 11]; [13, 15] = [15, 13]
3	[1, 7] = [5, 5] = [7, 1]; [7, 17] = [13, 13] = [17, 7]
4	[1, 13] = [7, 11] = [13, 1] = [11, 7]; [1, 17] = [17, 1] = [11, 13] = [13, 11]; [3, 11] = [11, 3] = [9, 7] = [7, 9]; [3, 19] = [19, 3] = [9, 17] [17, 9]; [5, 15] = [15, 5] = [9, 13] = [13, 9]; [7, 19] = [19, 7] = [11, 17] = [17, 11]

For thin panels, however, the extended asymptotic prediction is less reliable: for example, for $B = 0.02$ and $h/L_X = 0.005$ the extended asymptotic prediction is 10.904, versus the exact value of 11.359, making a 4.2% difference. This is explained by the contribution of the neighboring eigenfrequencies, which is not taken into account by the asymptotic prediction. For example, with this data, $\Omega_{11} = 2.000036$, $\Omega_{33} = 2.002888$, and $I'_{11,33}(2\pi) = 153.608$ (close to the asymptotic value $I'_{11,11}(\infty) = \pi/B = 157.08$). The extended asymptotic prediction does not take into account the contribution of this term, although it is not negligible. It was of interest to compare the extended asymptotic intensification factors with the associated exact values for $B = 0.002$ (i.e. with the modal overlap ratios reduced by a factor of ten). The corresponding results are listed in a Table. Now the extended asymptotic predictions come much closer to the exact values, the maximum difference being 1.21%. For example, the same neighboring eigenfrequencies Ω_{11} and Ω_{33} result in a $I'_{11,33}(2\pi)$ value equal to 517.444–only about a third of the asymptotic value, $I'_{11,11}(\infty) = 157.08$.

10.4 Conclusion

In this chapter, it was demonstrated that the modal cross-correlations have a dramatic effect on the evaluation of the random response of the curved panels. Hence, the modal cross-correlations ought not to be neglected in the analysis. The evaluation of intensification factors is significantly affected by modal responses associated with coinciding natural frequencies of various multiplicities.

It is interesting to note the modern renewed interest in random vibration theory and practice from the point of view of energy harvesting. Interested readers can consult with papers by Adhikari, Friswell, and Inman (2009), Litak, Friswell, and Adhikari (2010), Lin and Alphenaar (2010), Cottone, Gammaitoni, and Vocca (2012), Cottone, Gammaitoni, Vocca, Ferrari, and Ferrari (2012), Hobeck and Inman (2012), Tang and Zuo (2012), Zhao and Erturk (2013), Yoon and Youn (2014), Li, Gao and Cai (2015), Aridogan, Basdogan and Erturk (2016), Luan, Wang, Jin and Huang (2018).

References

Adhikari, S., Friswell, M. I., Lonkar, K., & Sarkar, A. (2009). Experimental base studies for uncertainty quantification in structural dynamics. *Probabilistic Engineering Dynamics, 24*(4), 473–492.

Aridogan, U., Basdogan, I., & Erturk, A. (2016). Random vibration energy harvesting on thin plates using multiple piezopatches. *Journal of Intelligent Material Systems and Structures, 27*(20), 2744–2756.

Bolotin, V. V. (1969). *Statistical methods in structural mechanics* (English translation by S. Aroni, of the 2nd ed.), (Chap. 5, pp. 102–153). San Francisco: Holden-Day Inc.

Cottone, F., Gammaitoni, L., & Vocca, H. (2012). Piezoelectric buckled beams for random vibration energy harvesting. *Smart Materials and Structures, 21*(3), Article 035021.

Cottone, F., Gammaitoni, L., Vocca, H., Ferrari, M., & Ferrari, V. (2012). Piezoelectric buckled beams for random vibration energy harvesting. *Smart Materials and Structures, 21*(3), Article 035021.

Crandall, S. H. (1977). Structured response patterns due to wide-band random excitation. In B. L. Clarkson (Ed.), *Stochastic problems in dynamics* (pp. 366–389). London: Pitman.

Crandall, S. H. (1978). Introduction to random processes. In I. Holand, D. Kavlie, G. Moe & R. Sigbjornsson (Eds.), *Safety of structures under dynamic loading* (pp. 19–34).

Crandall, S. H. (1979a). Random vibration of one- and two-dimensional structures. In P. R. Krishnaiah (Ed.), *Applications of statistics (Proceedings of the Symposium Held at Wright State University, Dayton, Ohio, 1976)* (pp. 1–82). Academic Press.

Crandall, S. H. (1979b). Random vibration of vehicles and structures. In *Proceedings of the Seventh Canadian Congress of Applied Mechanics* (pp. 1–12). Sherbrooke.

Crandall, S. H., & Wittig, L. E. (1972). Chladni's patterns for random vibration of a plate. In G. Herrmann & N. Perrone (Eds.), *Dynamic response of structures* (pp. 35–71). New York: Pergamon Press.

Davis, R. E. (1966). Statistical dependence effect on normal mode response. *AIAA Journal, 4*, 2033–2034.

Elishakoff, I. (1975a). Distribution of natural frequencies in certain structural elements. *The Journal of the Acoustical Society of America, 57*(2), 361–369.

Elishakoff, I. (1975b). Turbulent flow excited vibration of a shallow cylindrical shell. *AIAA Journal, 13*(11), 09–1118.

Elishakoff, I. (1977a). Random vibrations of orthotropic plates clamped or simply supported all round. *Acta Mechanica, 28*(1–4), 165–176.

Elishakoff, I. (1977b). On the role of cross–correlations in the random vibrations of shells. *Journal of Sound and Vibration, 50*(2), 239–252.

Elishakoff, I. (1977c). Random vibrations of orthotropic plates with all edges clamped or simply supported. *Acta Mechanica, 28,* 165–176.

Elishakoff, I. (1977d). Flutter and random vibration in plates. In B. L. Clarkson (Ed.), *Stochastic problems in dynamics* (pp. 390–410). London: Pitman.

Elishakoff, I. (1999). *Probabilistic theory of structures* (pp. 317–380). Wiley: New York (second edition, Dover Publications, Mineola, 1999; third edition, World Scientific, Singapore, 2017).

Elishakoff, I., Van Zanten, A. Th., & Crandall, S. H. (1979). Wide-band random axisymmetric vibration of cylindrical shells. *Journal of Applied Mechanics, 46*(2), 417–423.

Garrelick, J. M., & Chayes, L. (1978). Comment on the contribution of cross-modal terms to the local response of structures to random pressure fields. *The Journal of the Acoustical Society of America, 63*(5), 1626–1628.

Hobeck, J. D., & Inman, D. J. (2012). Artificial piezoelectric grass for energy harvesting from turbulence-induced vibration. *Smart Materials and Structures, 21*, Article 105024.

Jullien, Y. G. (1966). Vibrations aléatoires des calottes sphériques élastiques. *Journal de Mécanique, 5*(4), 419–438. (in French).

Khabbaz, G. R. (1964). Significance of the cross-correlation between the modes of a structure on its response. *AIAA Journal, 2*, 2211–2217.

Langley, A. J., & Taylor, P. H. (1979). Chladni patterns in random vibration. *International Journal of Engineering Science, 17*, 1039–1047.

Li, P., Gao, S., & Cai, H. (2015). Modeling and analysis of hybrid piezoelectric and electromagnetic energy harvesting from random vibrations. *Microsystem Technologies, 21*(2), 401–414.

Lin, Y. K. (1967). *Probabilistic theory of structural dynamics* (Chap. 7, pp. 203–252). New York: McGraw-Hill.

Lin, J. T., & Alphenaar, B. (2010). Enhancement of energy harvested from a random vibration source by magnetic coupling of a piezoelectric cantilever. *Journal of Intelligent Material Systems and Structures, 21*(13), 1337–1341.

Litak, G., Friswell, M. I., & Adhikari, S. (2010). Magnetopiezoelastic energy harvesting driven by random excitations. *Applied Physics Letters, 96*(21), Article 214103.

Luan, X., Wang, Y., Jin, X., & Huang, Z. (2018). Optimal locations of piezoelectric patch on wideband random point-driven beam for energy harvesting. *Journal of Vibration and Acoustics, 140*(1), Article 011014.

Mercer, C. A. (1965). Response of a multi-supported beam to a random pressure field. *Journal of Sound and Vibration, 2*, 293–306.

Palmov, V. A. (1965). Thin plates under action of wide-band random loading. In *Proceedings of the Leningrad Polytechnics Institute Named After M.I. Kalinin* (Vol. 252, pp. 97–106). (in Rusian).

Powell, A. (1968). On the fatigue failure of structures due to vibrations excited by random pressure fields. *The Journal of the Acoustical Society of America, 30*(12), 1130–1135.

Tang, X.-D., & Zuo, L. (2012). Vibration energy harvesting from random force and motion excitations. *Smart Materials and Structures, 21*(7), Article 075025.

Yoon, H., & Youn, B. D. (2014). Stochastic quantification of the electric power generated by a piezoelectric energy harvester using a time–frequency analysis under non-stationary random vibrations. *Smart Materials and Structures, 23*(4), Article 045035.

Zhao, S., & Erturk, A. (2013). Electroelastic modeling and experimental validations of piezoelectric energy harvesting from broadband random vibrations of cantilevered bimorphs. *Smart Materials and Structures, 22*, Article 015002.

Appendix
The Effect of the Eigenvectors Associated with the Repeated Frequencies

In systems with repeated frequencies, the associated modes are not unique as was demonstrated by Li et al. (2012) and Xu and Liu (2008), among others. Indeed, let us take

$$\mathbf{U} = \mathbf{V}\boldsymbol{\alpha} \tag{A.1}$$

where $\boldsymbol{\alpha}$ is the constant matrix satisfying $\boldsymbol{\alpha}^T\boldsymbol{\alpha} = \mathbf{I}$, $\mathbf{I}$ is an identity matrix with the dimension equal to the number of repeated frequencies. Then $\mathbf{U}$ satisfies the following equation

$$\mathbf{KU} = \boldsymbol{\omega}\mathbf{MU} \tag{A.2}$$

as well as the modal orthogonality condition

$$\mathbf{U}^T\mathbf{MU} = \mathbf{I} \tag{A.3}$$

where $\boldsymbol{\omega} = \mathrm{diag}\left[\omega_1^2, \omega_2^2, \ldots, \omega_n^2\right]$. Since $\boldsymbol{\alpha}$ can be selected in numerous ways, the modes $\mathbf{U}$ are not unique. Therefore, there arises a problem of whether different modes would lead to the same probabilistic characteristics of responses. To address this problem, we consider Elishakoff's (1983, 1999) system as shown in Fig. 3.1, but with the following parameters:

$$m_i = 1\,\mathrm{kg}, \quad k_i = 1\,\mathrm{N/m}, \quad c_i = 1\,\mathrm{Ns/m}, \quad i = 1, 2, 3 \tag{A.4}$$

Then the mass, stiffness, and damping stiffness are changed into

$$\mathbf{M} = \begin{bmatrix} 1\,0\,0 \\ 0\,1\,0 \\ 0\,0\,1 \end{bmatrix}, \mathbf{K} = \begin{bmatrix} 3 & -1 & -1 \\ -1 & 3 & -1 \\ -1 & -1 & 3 \end{bmatrix}, \mathbf{C} = \begin{bmatrix} 1\,0\,0 \\ 0\,1\,0 \\ 0\,0\,1 \end{bmatrix} \tag{A.5}$$

I. Elishakoff, *Dramatic Effect of Cross-Correlations in Random Vibrations of Discrete Systems, Beams, Plates, and Shells*,
https://doi.org/10.1007/978-3-030-40394-2

The natural frequencies are obtained as

$$\omega_1^2 = 1\,\text{rad}^2/\text{s}^2, \omega_2^2 = \omega_3^2 = 4\,\text{rad}^2/\text{s}^2 \tag{A.6}$$

The modal matrix reads

$$\mathbf{V} = \begin{bmatrix} -0.5774 & -0.1543 & 0.8018 \\ -0.5774 & 0.7715 & -0.2673 \\ -0.5774 & -0.6172 & -0.5345 \end{bmatrix} \tag{A.7}$$

The correlation matrix of generalized velocities $\dot{Q}_1$, $\dot{Q}_2$, and $\dot{Q}_3$ are obtained as

$$\mathbf{R}_{\dot{q}\dot{q}}(0) = \begin{bmatrix} 0.3333 & 0.0117 & -0.0608 \\ 0.0117 & 0.0238 & -0.1237 \\ -0.0608 & -0.1237 & 0.6429 \end{bmatrix} \pi S_0 \tag{A.8}$$

As mentioned above, the modes associated with the repeated frequencies are not unique. Now let us suppose

$$\boldsymbol{\alpha}_1 = \begin{bmatrix} 0.7559 & -0.6547 \\ 0.6547 & 0.7559 \end{bmatrix} \tag{A.9}$$

We get

$$\mathbf{U}_1 = \begin{bmatrix} -0.5774 & 0.4082 & 0.7071 \\ -0.5774 & 0.4082 & -0.7071 \\ -0.5774 & -0.8165 & 0 \end{bmatrix} \tag{A.10}$$

The above modes correspond with the ones in by Elishakoff (1983). Let us suppose yet another choice for $\boldsymbol{\alpha}$, namely

$$\boldsymbol{\alpha}_2 = \begin{bmatrix} 0.8887 & -0.4586 \\ 0.4586 & 0.8887 \end{bmatrix} \tag{A.11}$$

We get

$$\mathbf{U}_2 = \begin{bmatrix} -0.5774 & 0.2306 & 0.7833 \\ -0.5774 & 0.5631 & -0.5913 \\ -0.5774 & -0.7936 & -0.1920 \end{bmatrix} \tag{A.12}$$

Using the relationship $\mathbf{R}_{\dot{x}\dot{x}}(0) = \mathbf{V}\mathbf{R}_{\dot{q}\dot{q}}(0)\mathbf{V}^T$, $\mathbf{R}_{\dot{x}\dot{x}}(0) = \mathbf{U}_1\mathbf{R}_{\dot{q}\dot{q}}(0)\mathbf{U}_1^T$, and $\mathbf{R}_{\dot{x}\dot{x}}(0) = \mathbf{U}_2\mathbf{R}_{\dot{q}\dot{q}}(0)\mathbf{U}_2^T$ in Eq. (3.9), the mean-square velocities corresponding to eigenvectors $\mathbf{V}$, $\mathbf{U}_1$ and $\mathbf{U}_2$ can be obtained, respectively. The results are given in

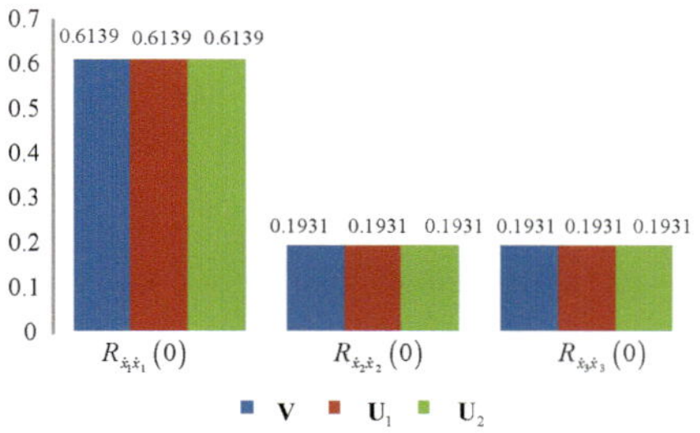

(a) Exact results (πS_0)

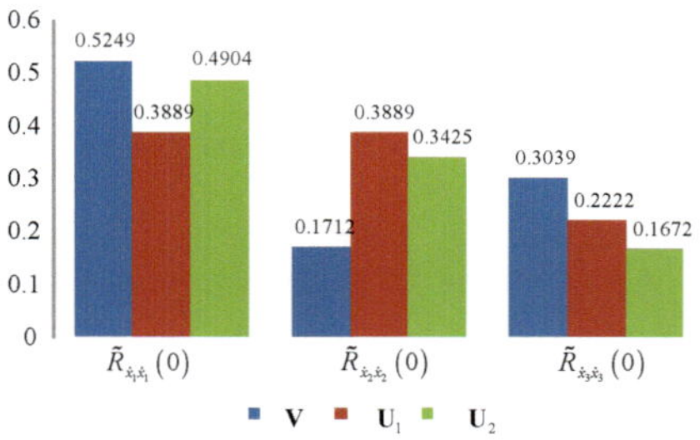

(b) Estimated results (πS_0)

Fig. A.1 The effects of selected eigenvectors associated with repeated frequencies on the calculated probabilistic characteristics of responses

Fig. A.1a. Similarly, if the cross-correlation terms have been omitted, the estimated mean-square velocities are calculated as shown in Fig. A.1b.

As is seen, the estimated results when the eigenvectors selected as $\mathbf{V}$, $\mathbf{U}_1$ and $\mathbf{U}_2$ are all *different* from each other, but the exact results are the *same*. This means that different modes associated with repeated frequencies can lead to different estimate probabilistic characteristics of response when the cross-correlations have been omitted. It should be stressed that if the above neglect is not conducted, i.e. if both direct and cross-correlations are taken into account, the same results are obtained, irrespective of the choice of modes.

If we select the modes as $\mathbf{U}_1$, the associated errors of mean-square velocity due to omitted modal cross-correlations reads

$$\tilde{\eta}_1 = \left| \frac{\widetilde{R}_{\dot{x}_1\dot{x}_1}(0) - R_{\dot{x}_1\dot{x}_1}(0)}{R_{\dot{x}_1\dot{x}_1}(0)} \right| \times 100\% = 36.65\% \tag{A.13}$$

$$\tilde{\eta}_2 = \left| \frac{\widetilde{R}_{\dot{x}_2\dot{x}_2}(0) - R_{\dot{x}_2\dot{x}_2}(0)}{R_{\dot{x}_2\dot{x}_2}(0)} \right| \times 100\% = 101.44\% \tag{A.14}$$

$$\tilde{\eta}_3 = \left| \frac{\widetilde{R}_{\dot{x}_3\dot{x}_3}(0) - R_{\dot{x}_3\dot{x}_3}(0)}{R_{\dot{x}_3\dot{x}_3}(0)} \right| \times 100\% = 15.11\% \tag{A.15}$$

If we omit terms $R_{\dot{q}_2\dot{q}_3}(0)$ and $R_{\dot{q}_3\dot{q}_2}(0)$, the error of $\widetilde{R}_{\dot{x}_1\dot{x}_1}(0)$ turn out to be 30% and the error in evaluation of $\widetilde{R}_{\dot{x}_2\dot{x}_2}(0)$ is 75%, which are exactly the same values as reported by Elishakoff (1983); and Lubliner and Elishakoff (1986).

Therefore, we conclude that if the cross-correlations are omitted, the following two features will be exhibited by the analytical and numerical derivations: (a) the results will not be unique as they will depend on the chosen eigenvectors, and (b) the results will lead to large numerical errors.

References

Elishakoff, I. (1983). *Probabilistic theory of structures*. New York-Wiley (Dover, 1999; World Scientific, 2017).

Lubliner, E., & Elishakoff, I. (1986). Random vibration of system with finitely many degrees of freedom and several coalescent natural frequencies. *International Journal of Engineering Science, 24*(4), 461–470.

Li, J., Yan, Q., & Chen, J. B. (2012). Stochastic modeling of engineering dynamic excitations for stochastic dynamics of structures. *Probabilistic Engineering Mechanics, 27*(1), 19–28.

Bibliography

Abdelnaser, A. S., & Singh, M. P. (1993a). Random response of antisymmetric angle-ply composite plates with levy boundary conditions. *Composites Engineering, 3*(9), 817–833.
Abdelnaser, A. S., & Singh, M. P. (1993b). Random vibrations of cantilevered composite beams with torsion-bending coupling. *Probabilistic Engineering Mechanics, 8*(3–4), 143–152.
Abe, M. (1998). Vibration control of structures with closely spaced frequencies by a single actuator. *Journal of Vibration and Acoustics, 120*, 120–124.
Abe, M., & Igusa, T. (1995). Tund Mass Dampers for structures with closely spaced natural frequencies. *Earthquake Engineering and Structural Dynamics, 24*(2), 247–261.
Abhinav, S., & Manohar, C. S. (2016). Global Response Sensitivity analysis of randomly excited dynamic systems. *Journal of Engineering Mechanics, 142*(3), article 04015094.
Abu-Hilal, M. (2003). Vibration of Beam With general boundary conditions due to moving random load. *Archive of Applied Mechanics, 72*(9), 637–650.
Accardi, L., Hida, T., Ohya, M., Si, S., & Watanabe, N. (2017). *White noise analysis and quantum information*. Singapore: World Scientific.
Adhikari, S. (2006). Random eigenvalue problems revisited. *Sadhana, 31*(4), 293–314.
Adhikari, S. (2007). Matrix variate distributions for probabilistic structural dynamics, *A1AA Journal, 45*(7), 1–148.
Adhikari, S. (2010). Generalized Wishart distribution for probsbilistic structural dynamics. *Computational Mechanics, 45*(5), 495–511.
Adhikari, S. (2011). Doubly spectral stochastic finite-element method for linear structural dynamics. *Journal of Aerospace Engineering, 24*(3), 264–276.
Adhikari, S., & Friswell, M. I. (2004). Random Eiglnvalue problems in structural dynamics. In *Proceedings of the 45th AIAA/ASME/ASCE/AHS/ASC Structures, Structural Dynamics and Materials Conference,* Palm springs, CA.
Adhikari, S., Friswell, M. I., & Inman, D. J. (2009). Piezoelectric Energy harvesting from broadband random vibrations. *Smart Materials and Structures, 18*, article 115005.
Adhikari, S., & Manohar, C. S., (1999). Dynamic analysis of framed structures with statistical uncertainties. *International Journal for Numerical Methods in Engineering, 44*(8), 1157–1178.
Adhikari, S., & Manohar, C. S. (2000). Transient dynamics of stochastically parametrized beams. *Journal of Engineering Mechanics, 126*(11), 1131–1140.
Adhikari, S., & Sarkar, A. (2009). Uncertainty in sturctural dynamics: Experimental validation of a Wishart random matrix model. *Journal of Sound and Vibration, 323*(3), 802–825.
Adrezin, R., & Benaroya, H. (1999). Response of a tension leg platform to stochastic wave forces. *Probabilistic Engineering Mechanics, 14*(1), 3–17.
Agrawal, O. P. (2003). Response of a diffusion-wave system subjected to deterministic and stochastic fields. *ZAMM: Zeitschrift Für angewandte Mathematik und Mechanik, 4*, 265–274.

I. Elishakoff, *Dramatic Effect of Cross-Correlations in Random Vibrations of Discrete Systems, Beams, Plates, and Shells*,
https://doi.org/10.1007/978-3-030-40394-2

Agrawal, O. P. (2004). Analytical Solution for stochastic response of a fractionally damped beam. *Journal of Vibration and Acoustics*, *126*(4), 561–566.

Ahmadi, G. (1978). Earthquake response of linear continuous systems. *Nuclear Engineering and Design*. *50*(2), 327–345.

Ahmadi G. (1979). Generation of artificial time histories compatibale with given resoponse spectrum—A review. *Solid Mechanics Archives*, *4*(3), 207–239.

Ahmadi, G., & Hashemi, J. (1973). The random vibration of a nonuniform cantilever beam with concentrated mass. *Vehicle System Dynamics*, *2*(4), 225–233.

Ahmadi, G., & Hashemi, J. (1974). Response of an infinitely long timoshenko beam to random loads. *The Journal of the Industrial Mathematics Society*, *24*(1), 29–40.

Ahmadi, G., & Mavni, R. (1972). On the dynamic response of an infinite Bernoulli-Euler beam to random load. *Industrial Mathematics*, *22*, Part 2, 65–72.

Ahmadi, G., & Satter, M. A. (1974). On the random vibration of a damped simply supported beam varrying concentrated masses. *The Journal of the Industrial Mathematics Society*, *24*, Part 1, 19–40.

Ahmadi, G., & Satter, M. A. (1975). Mean-square response of Beams to nonstationary random excitation. *AIAA Journal*, *13*, 1097–1100.

Akhmanov, S., Diakov, Yu., & Chirkin, A. (2017). *Statistical radiophysics and optics. Random vibrations and waves in linear systems*. Moscow: "Fizmatlit" Publishing House (in Russian).

Alavi, S. R., Rahmati, M., & Ziaei-Rad, S. (2017). A new approach to design safe-supported hard disk drives against random excitation by using optimiztion of rubbers spatial parameters. *Microsystems Technology*, *23*, 2023–2032.

Alderucci, T., & Muscolino, G. (2016). Fully Nonstationary analysis of linear structural systems subjected to multicorrelated stochastic excitation. *Journal of Risk and Uncertainty in Engineering Systems, Part A: Civil Engineering*, *2*(2), article 4015007.

Aldrich, F.J. (2005). Predicting turbulent boundary layer wall pressure fluctuations in adverse pressure gradients. In *ASME 2005 International Mechanical Engineering Congress and Exposition* (pp. 99–108). New York: ASME Press.

Allemang, R., & Brown, D. L. (1982). A correlation coefficiernt for modal vector analysis. In *Proceedings of the 1st International Modal Analysis Conference* (Vol. 1, pp. 110–116), Orlando, FL: Union College Press.

Alkhfaji, S. S., & Garvey, S. D. (2011). Modal correlation approaches for general second-order systems: Matching mode pairs and an application to Campbell diagrams. *Journal of Sound and Vibration*, *330*(23), 5615–5627.

Alotta, G., Failla, G., & Pinnola, F. P. (2017). Stochastic analysis of a Nonlocal fractional viscoelesatic bar forcced by Gaussian white noise. *Journal of Risk and Uncertainty in Engineering Systems*, *3*, article 030904.

Alwan, A., & Aluru, N. R. (2015). A nonstationary covariance function model for spatial uncertainties in electrostatically actuated microsystems. *International Journal for Uncertainty Quantification*, *5*(2), 99–121.

Amabili, M., Sakar, A., & Païdoussis, M. P. (2006). Chaotic vibrations of circular cylindrical shells: Galerkin versus reduced-order modes via the proper orthogonal decomposition method. *Journal of Sound and Vibration*, *290*(3–5), 736–762.

Amin, M., & Ang, A. H. S. (1968). Nonstationary stochastic model of earthquake motions. *Journal of Engineering Mechanics*, *94*, 559–583.

Amin, M., Tsao, H. S., & Ang, A. H. S. (1969). Significance of nonstationarity of earthquake motions. In *Proceedings, Fourth World Conference on Earthquake Engineering*, Santiago, Chile (Vol. 1, pp. 97–113).

Amini, Y., Fatehi, P., Heshmati, M., & Parandvar, H. (2016). Time domain and frequency domain analysis of functionally graded piezoelectric harvesters subjected to random vibration: Finite element modeling. *Composite Structures*, *136*, 384–393.

Ammon, D. (1990). Approximation and generation of Gaussian and non-Gaussian stationary process. *Structural Safety*, *8*(1–4), 153–160.

Andersen, L., & Nielsen, S. R. K. (2001). Vehicle moving along an infinite beam with random surface irregularities on a Kelvin foundation. *Journal of Applied Mechanisms, 69*(1), 69–75.
Andrianov, I. V., & Iskra, V. S. (1991). Application of Bolotin's asymptotic method in problems of vibraion control. *Problems of Mechanical Engineering, 36*, 79–82 (in Russian).
Ang, A. H. S. (2017). An Intuitive Basis of the probability density evolution method (PDEM) for stochastic dynamics. In *Risk and Reliability Analysis: Theory and Applications* (pp. 99–108), Springer, Cham.
Antonov, I. L. (1979). About Properties with probability one of linear oscillator's trajectories subjected to random load. *Herald of Moscow University, Series 1, Mathematics, Mechanics*, (4), 84–91 (in Russian).
Antonov, I. L. (1993). *Random Vibrations properties of trajectories*. Moscow: Moscow State University Press (in Russian).
Anuta, A. E. Jr. (1996). The model equations and equal frequencies in definite systems. *Computers and Structures, 60*(3), 357–367.
Aoki, S., & Hiroi T. (2005). Reduction method for residual stress of welded joint using random vibration. *Nuclear Engineering and Design, 235*(14), 1441–1445.
Aoki, S., Nishimura, T., & Hiroi, T. (2005). Reduction method for residual stress of welded joint using random vibration. *Nuclear Engineering and Design, 235*(14), 1441–1445.
Arcas, N. R. (1969). Prediction of stresses and fatigue life of acoustically excited aircraft structures. *Shock and Vibration Bulletin, 39*(2).
Aridogan, U., Basdogan, I., & Erturk, A. (2016). Random vibration energy harvesting on their plates using multiple piezopatches. *Journal of Intelligent Material Systems and Structures, 27*(20), 2744–2756.
Arnold, L. (1998). *Random Dynamical systems*. Berlin: Springer.
Arnoux ,A., Batou, A., Soize, C., & Cagliardini, L. (2012). Stochastic reduced-order model for dynamical structures having a high modal density in the low frequency range. In *International Symposium on Computational Modelling and Analysis of Vehicle Body Noix and Vibration*, University of Sussex, (pp. 1–10).
Arsalanov, A. M. (1987). *Design of strucural elements with preassigned reliability, subjected to random action*. Moscow: "Mashinostroenie" Publishing House (in Russian).
Arseniev, D. G., Kotov, V. V., Polianskii, V. A., & Smirnova, N. A. Biomorphic control is the problem of vibroisolation of random vibrations, *Scientific-Technical Transactions St. Petersburg State Polytechnic University*, (2), [169, 112–116, 201] (in Russian).
Ashlock, J. C., & Pak, R.Y. S. (2010). Application of random vibration techniques to resonant column testing. *Advances in Analysis Modeling and Design*, 2651–2661.
Aslanyan, A. G., & Lidsky, V. B. (1972). Natural frequency distribution of thin elastic shell. *Mechanics of Solids*, (3), 155–170 (in Russian).
Aslanyan, A. G., & Lidsky, V. B. (1974). *Distribution of eigenfrequencies of thin elastic shells*. Moscow: "Nauka" Publishing House (in Russian).
Aslanyan, A. G., & Lidsky, V. B. (1974). A formula for the number of frequencies of axisymmetric vibrations of a shell of resolution. *Soviet Physics Doklady, 20*, 461–470, 1975.
Aslanyan, A. G., & Tulovskii, V. N. (1973). Distribution of the natural frequencies of elastic shells. *Doklady Akademii Nauk SSSR, 208*(4), 801–804 (in Russian).
Aslanyan, A. G., Kuzina, Z. N., Lidsky, V. B., & Tulovskii, V. N. (1973). Distribution of the natural frequencies of a thin elastic shell of arbitrary outline. *PMM: Journal of Applied Mathematics and Mechanics, 37*(4), 604–617.
Atluri, S. N., Cho, J. Y., & Kim, H. G. (1999). Analysis of thin beams using the meshless local Petrov-Galerkin method, With generalized moving leant squares interpolations. *Computational Mechanics, 24*(5), 334–347.
Augusti, G., Baratta, A., & Casciati, F. (1983). *Probabilistic methods in structural engineering*. London,U.K.: Chapman and Hall.
Augustin, F., & Rentrop, P. (2012). Galerkin techniques for random ordinary differential equations. *Numerische Mathematik, 122*(3), 399–419.

Au, F. T. K., Cheng, Y. S., & Cheung, Y. K. (2001). Analysis of bridges under moving vehicles and trains: An overview. *Progress in Structural Engineering and Materials, 3* (3), 299–304.

Au-Yang, M. K. (2000). Joint and cross acceptances for cross-flow-induced vibration—Part I: Theoretical and finite element formations. *Journal of Pressure Vessel Technology, 122*(3), 349–354.

Awejcewicz, Y., & Krys'ko, A. V. (2003). Analysis of complex parametric vibrations of plates and shells using Bubnov-Galerkin approach. *Archive of Applied* Mechanics, *73*(7), 495–504.

Babitsky, V. I., & Kolovsky, M.Z. (1967). Vibration of a linear system with limiters due to random loads. *Mechanics of Solids* (3), 147–151 (in Russian).

Bäcker, A. (2007). Random waves and more: Eigenfunctions in chaos and mixed systems. *The European Physical Journal-Special Topics, 145*(1), 161–169.

Bachoo, R., & Bridge, J. (2013). The modal density distribution and density of fibre reinforced composite beam. *Journal of Sound and Vibration, 332*(8), 2000–2018.

Bachoo, R., & Bridge, J. (2018). The modal density of composite beams incorporating the effects of shear deformation and rotary inertia. *Journal of Sound and Vibration, 423*, 459–471.

Bagdasaryan, G. E. (1986). Application of V. V. Bolotin's asymptotic method for investigation of magnetoelastic vibrations of rectangular plates. *Problems Mashinostroeniya,* 25, 63–68 (In Russia).

Bailly, C., Lafon, P., & Candel, S. (1997). Subsonic and supersonic jet noise predictions from statistical source models. *AIAA Journal*, 35(11),1688–1696.

Balasubramonian, S., & Iyer, K.S. S. (1982). System optimization for random earthquake forces. *Computer Methods in Applied Mechanics and Engineering, 31*(2), 233–235.

Balazs, P., Kreuzer, W., & Waubke, H. (2007). A stochastic 2D-model for calculating vibrations in random layers. *Journal of Computational Acoustics, 15*(3), 271–283.

Banerjee, J. R., & Kennedy, D. (1985). Response of an axially loaded Timoshenko beam to random loads. *Journal of Sound and Vibration, 101*(4), 481–487.

Bardell, N. S. (1994). Chiladni Figures for completely free parallelogram plates: An analytical study. *Journal of Sound and Vibration, 174*(5), 655–676.

Barnoski, R. L., & Maurer, J. R. (1969). Mean-square response of simple mechanical systems to nonstationary random excitation. *Journal of Applied Mechanics, 36*, 221–227.

Barshtein, M. F. (1960). Application of probabilistic methods to analysis of structures to seismic excitation. *Structural Mechanics and Analysis of Structures*, (2) (in Russian).

Barszcz, T. (2007). Decomposition of vibration signals into deterministic and nondeterministic components and its capabilities of fault detection and identification. *International Journal of Applied Mathematics and Computer Science, 19*(2), 327–335.

Bartholomew, P., & Pitcher, N. (1984). Optimization of structures with repeated normal-mode frequencies. *Engineering Optimization, 7*(3), 195–208.

Baskin, I. L., Lizarev, A. D., & Rostanina, N. B. (1975). Theoretical and empirical natural frequency density of nonshallow spherical shells. In *Proceedings of the Tenth All-Union Conference in Theory of Shells and Plates* (pp. 45–54). Tbilisi: "Mezniereba" Publishing House (in Russian).

Batou, A., & Soize, C. (2013). Uncertainty quantification in low-frequency dynamics of complex beam-like structures having a high-modal density. *International Journal of Uncertainty Quantification, 3*(6), 475–485.

Batou, A., Soize, C., Arnoux, A., Cagliardini, L., & Brie, N. (2013). Reduced-order model in structural dynamics for high-model density in the LF range applications to automotive vehicle and fuel assemblies. In *DINAME 2013, XV International Sympossium on Dynamic Problems of Mechanics* (pp. 1–10).

Battaglia, F. (1979). Some Extensions in the evolutionary spectral analysis of a stochastic process. *Bollettino dell'Unione Matematica Italiana, 13*(5),1154–1166.

Bauer, L., Keller, H. B., & Reiss, E. L. (1975). Multiple eigenvalues lead to secondary bifurcation. *SIAM Review, 17*(1), 101–122.

Bazhenov, V. A., Dekhtyaruk, E. S., Zakharchenko, T. G., Lukyanchenko, O.A., & Imdukh, V. F. (2000). Forced Stationary vibration of flexible plates induced by a random narrow-band excitaion. *Strength of Materials*, *32*(2), 178–186.

Beck, T. W., Dillon, M. A., De Freitas, J., & Stock, M. S. (2009). Cross-correlation analysis of mechanomyographic signals detected in two axes. *Physiological Measurements*, *30*(12), 1465–1471.

Becker, G., Camarinopoulos, L., & Kabranis, D. (2002). Dynamic reliability under random shocks. *Reliability Engineering and System Safety*, *77*(3), 239–251.

Beer, F. P. (1961). On the response of linear systems to time-dependent, unidimensional loadings. *Journal of Applied Mechanics, 281*, 50–56.

Bellizzi, S., & Bouc, R. (1996). Spectral response of asymmetrical random oscillators. *Probabilistic Engineering Mechanics*, *11*, 51–59.

Balmés, E. (1993). High Modal density, curve veering, localization: A Different perspective on the structural response. *Journal of Sound and Vibration, 161*, 358–363.

Belozerkovskii, A. B., Zovdak, V. A., Chernykh, O. V., & Yaroshok, Yu. A. (2003). Random vibrations of frames of carts of underground station car with road imperfections taken into account. *Dynamics and Strength of Machines*, *1*(12), 10–17 (in Russian).

Belyaev, A. K. (1990). On the application of the locality principle in structural dynamics. *Acta Mechanics, 83*(3–4), 213–222.

Belyaev, A. K. (1991a). Vibration State of complex mechanical structures under wide-band excitation. *International Journal of Solids and Structures, 27*, 811–823.

Belyaev, A. K. (1991b). Theory of vibrational conductivity. *ZAMM: Zeiitschrift fuer angewandte mathematick und Mechanik, 71*, 127–129.

Belyaev, A. K. (1992). Dynamic Simulation of high-frequency vibration of extended complex structures. *Interational Journal of Mechanics of Structures and Machines, 20*, 155–168.

Belyaev, A. K. (1993). High-Frequency Vibration of extended complex structures. *Probabilistic Engineering Mechanics, 8*(1), 15–24.

Belyaev, A. K. (1994). Vibrational conductivity approach to high-frequency dynamics. *Nuclear Engineering and Design, 150*(2–3), 281–286.

Belyaev, A. K. (1999). On combination of continuous and discrete energy approaches to high frequency dynamics of structres. In *Proceedigns of the 6th International Congress of Sound and Vibration* (pp. 2341–2348), International Institute of Acoustics and Vibration.

Belyaev, A. K. (2003). Combining Continuous and discrete energy approaches to high frequency dynamics of structures. In A. Belyaev & A. Guran (Eds.), *Selected topics in structonics and mechatronic system* (pp. 221–273), Singapore: World Scientific.

Belyaev, A. K. (2003). High-frequency dynamics: The missing link in the evolution of dynamics. In A. Nilsson & H. Boden (Eds.), *Proceedings of the 10th International Congress on Sound and Vibration* (pp. 589–600).

Belyaev, A. K. (2015). Thermodynamic analogy in problems of high-frequency vibration propagation in complex system. *Mechanics of Solids, 50*(2), 218–226.

Belyaev, A. K., & Krutzik, N. J. (1994). Localization of high-frequency vibrations of secondary systems of power plates. *Acta Mechanica, 102*(1), 1–10.

Belyaev, A. K., & Palmov, V.A. (1980). Theory of vibroconductivity. In M.Petyt (Ed.), *Proceedings of the 1st Conference on Recent Advances in Structural Dynamics* (pp. 157–168). University of Southampton.

Belyaev, A. K., & Palmov, V. A. (1986). Integral theories of random vibration of complex structures, In I. Elishakoff & R. H. Lyon (Eds.), *Random vibration—Status and recent developments* (pp. 19–38). Amsterdam: Elsevier.

Belyaev, A. K., & Pradlwarter, H. J., Wide-band vibration of driven components of complex structures. In (G.I.Schüller, M. Shinozuka, & J.P.Yao (Eds.), *Proceedings of COSSAR'93* (Vol. 1, pp. 85–91). Rotterdam: Balkema.

Belyaev, A. K., & Pradlwarter, H. J. (1995). Wide-band random vibration in members of complex structures. *International Journal of Solids and Structures, 32*(24), 3629–3641.

Benaroya, H. (1992). Random eigenvalues, algebraic methods and structural dynamic models. *Applied Mathematics and Computation, 52*(1), 37–66.
Benaroya, H., & Han, S. M. (2005). *Probability models in engineering and science.* Boca Raton: Taylor and Francis.
Benaroya, H., & Rehak, M. (1998). Finite element methods in probabilistic structural analysis: A selective review. *Applied Mechanics Reviews, 41*(5), 201–213.
Benaroya, H., & Nagurka, M. (2010). *Mechanical vibration: Analysis, uncertainties, and control* (3rd edition). Boca Raton: CRC Press.
Benasciutti, D. (2014). Some Analytical expressions to measure the accuracy of the "Equivalent von Mises Stress" in vibration multiaxial fatigue. *Journal of Sound and Vibration, 333*, 4326–4340.
Bendat, J. S. (1976). Solutions for the multiple input/output problem. *Journal of Sound and Vibration, 44*(3), 311–325.
Bendat, J. S., & Piersol, A. G. (1966). *Measurement and analysis of random data.* New York: Wiley.
Bendat, J. S., & Piersol, A. G. (1980). *Engineering applications of correlation and spectral analysis.* New York: Wiley-Interscience.
Bendat, J. S., & Pearsol, A. G. (1989). *Random data analysis and measurement procedures.* New-York: Wiley.
Bendat, Y. S., & Piersol, A. G. (1971). *Random data: Analysis and measurement procedures.* New York: Wiley.
Beranek, L. L. (1960). *Noise reduction.* New York: McGraw-Hill.
Beranek, L. L., & Ver, L. (1992). *Noise and vibration control engineering.* New York: Wiley.
Beranek, L. L., & Work, G. A. (1949). Sound Transmission through mutliple structures containing flexible blankets. *The Journal of the Acoustical Society of America, 21*(4), 419–428.
Bergman, L. A. (1979). *Moments of time to first passage of the linear oscillator.* Doctoral Thesis, Department of Civil Engineering, Case Western Reserve University, Cleveland, OH.
Bergman, L. A. (2001). Uncertainty Modelling in dynamical systems: A perspective. In D. J. Ewins & D. J. Inman (Eds.), *Structural dynamics @ 2000: Current status and future directions* (pp. 9–16). Baldock, UK: Research Saludes Press Ltd.
Bergman, L. A., & Heinrich, J. C. (1980). Solution of the Pontriagin–Vitt equation for the moments of time to first passage of the randomly acclerated particle by the FEM. *International Journal for Numerical Methods in Engineering, 14*, 1408–1412.
Bergman, L. A., & Henrich, J. C. (1980). On the reliability of the linear oscillator and systems of coupled oscillators. *International Journal for Numerical Methods in Engineering, 18*, 1271–1295.
Bergman, L. A., & McFarland, D. M. (1989). Wide-band random excitation of beams and rectangular plates coupled to discrete substructures. *Structural Safety, 6*(2–4), 87–97.
Bergman, L. A., & Nicholson, J. W. (1985). Stationary response of combined linear dynamic systems to stationary excitation. *Journal of Sound and Vibration, 103*(2), 225–236.
Bernal, D. (2005). Closely spaced roots and defectiveness in second-order systems. *Journal of Engineering Mechanics, 131*, 276–281.
Bernard, M. L. (1989). Optimization of precision of structures with repeated roots. In *Computer Aided Optimum Design of Structures: Recent Advances*, (Proceedings of the OPTI—89 Conference). Boston: Computational Mechanics Publications.
Bernard, M. L., & Bronowicki, A. J. (1994). Modal expansion method for eigensensitivity with repeated roots. *AIAA Journal, 32*(7), 1500–1506.
Bessey, R. L., Chu, W. H., & Kana D.D., The Response of Cylindrical Sheels to Random Acoustic Excitation Over Broad Frequency Ranges, Southwest Research Institute, 1970.
Bhat, R. B., Rao, B. V., & Wagner, H. (1973). Structural response to random acoustic excitation. *Earthquake Engineering and Structural Dynamics*, 2, 117–132.
Bhat, R. B., & Sankar, T. S. (1981). Effect of stiffener arrangement on the random response of a flat panel. *The Shock and Vibration Bulletin, 51*, Part 3, 81–87.
Bhat, W. V., & Wilby, J. F. (1971). Interior Noise radiated by an airplane fuselage subjected to turbulent layer excitation and evaluation of noise reduction treatments. *Journal of Sound and Vibration, 18*(4), 449–464.

Bhattacharyya, B., & Chakraborty, S. (2002). Stochastic dynamic sensitivity of uncertain structures subjected to random earthquake loading, *249*(3), 543–556.
Biagini, P., Borri, C., & Facchini, L. (2007). Wind response of large roofs of stadiums and arena. *Journal of Wind Energy and Industiral Aerodynamics, 95*, 871–887.
Biedermann, J., Winterb, R., Wandel, M., & Böswald, M. (2017). Energy based correlation criteria in the mid-frequency range. *Journal of Sound and Vibration, 400*, 457–480.
Bienek, M. P. (1965). Creep under random loading. *AIAA Journal, 2*, 1559–1561.
Bies, D. A., & Hausen, C. H. (2009). *Engineering noise control: Theory and practice* (Fourth Edition). London: Span Press.
Birgersson, F. (2003). *Prediction of random vibration using spectral methods*. Ph.D. Dissertation, Department of Aeronautical and Vehicle Engineering, Stockholm: Royal Institute of Technology.
Birgersson, F., Ferguson, N. S., & Finnveden, S. (2003). Application of the spectral finite element method of turbulent boundary layer induced vibration of plates. *Journal of Sound and Vibration, 259*(4), 873–891.
Bischof, C. H., & Sun, X. (1996). On Tridiagonalizing and diagonalizing symmetric matrices with repeated eigenvalues. *SIAM Journal on Matrix Analysis and Applications, 17*, 869–885.
Bishop, R.E. D., & Price, W. G. (1974). *Probabilistic theory of ship dynamics*. London: Chapman and Hall.
Björkenstam, U. (1973)., *Random loading on structures under creep*. Ph.D. Dissretaion, Division of Solid Mechanics, Chalmers University of Technology, Gothenburg.
Björkenstam, U. (1974). *Random Loading on Hooke-Norton Structures. ZAMP-Zeitschrift für angewandte Mathematik und Physik, 25*(3), 319–330.
Blake, W. K. (2012). *Mechanics of flow-induced sound and vibration, 2: Complex flow-structure interactions*. Amsterdam: Elsevier.
Blanzé, C., & Champaney, L. (2004). A computational strategy for the random response of assemblies of structures. *International Journal of Solids and Structures, 41*(22–23), 6383–6405.
Blevins, R. D. (1989). An Approximate method for sonic fatigue analysis of plates and shells. *Journal of Sound and Vibration, 129*, 51–71.
Blevins, R. D. (2006). Modal density of rectagular volumes, areas and lines. *The Journal of the Acoustical Society of America, 119*(2), 788–791.
Bliven, D. O., & Soong, T. T. (1969). On Frequencies of elastic beams with random imperfections. *Journal of the Franklin Institute*, 287(4), 297–304.
Bobby, S., Suksuwan, A., Spence, S.M. J., & Kareem, A., Reliability-based topology optimization of uncertain building systems, subject to stochastic excitation. *Strucutral Safety, 66*(1–16).
Boccotti, P. (1983). On wind wave kinematics. *Meccanica, 18*(4), 205–216.
Bofilios, D. A., & Vaicaitis, R. (1987). Response of double-wall composite shells to random point loads. *Journal of Aircraft, 24*(4), 268–273.
Bogdanoff, J. L., & Chenea, P. F. (1961). Dynamics of some disordered linear system. *International Journal of Mechanical Sciences, 3*, 157–169.
Bogdanoff, J. L., & Goldberg, J. E. (1959). On the Transient behavior of a system under a random disturbance. In *Fourth Mid-Western Conference on Solid Mechanics* (pp. 488–496). The University of Texas.
Bogdanoff, J. L., & Goldberg, J. E. (1960). On the Bernoulli-Euler beam theory with random excitation. *Journal of Aero/Space ScienceS, 27*, 371–376.
Bogdanoff, J. L., Goldberg, J. E., & Bernard, M. C., Response of a simple structure to a random earthquake-type disturbance. *Bulletin of the Seismological Society of America, 52*(2), 293–310.
Bogdanoff, J. L., & Kozin, F. (1961). On the Statistical analysis of linear vehicle dynamics. In *Mechanics of Soil-Vehicle Systems, Proceedings of the 1st International Conference on the Mechanics of Soil-vehicle Systems* (pp. 195–223).
Bogomolov, S. I., & Khodyrev, V. I. (1967). On analysis of the probabilistic spectrum of free vibrations of turbine blades. *Civil Engineering and Architecture*, (11), 47–52 (in Russian).

Bogomolov, S. I., & Khvorost, V. A. (1971). Towards determination of lifetime of beams under random loadings. *Dynamics and Strength of Machines, 13*, 53–57. Kharkov State University Press (in Russian).

Boily, S., & Charron, F. (1999). The vibroacoustic response of a cylindrical shell structure with viscoelastic and poroelastic materials. *Applied Acoustics*, 58(2), 131–152.

Bolotin, V.V. (1960). The edge effect in the oscillations of elastic shells. *PMM: Journal of Applied Mathematical Mechanics, 24*, 1257–1272.

Bolotin, V. V. (1963a). On the density of natural oscillation frequencies in thin elastic shells. *Journal of Applied Mathematics and Mechanics, 27*(2), 538–543.

Bolotin, V. V. (1963b). On elastic vibrations excited by broadband forces. *Izvestiya Vyschikh Uchebnykh Zavedenii* N3 (in Russian).

Bolotin, V. V. (1964). Application of probability theory methods in the theory of plates and shells. In *Proceedings of Fourth AU-Union Conference on Theory of Plates and Shells* (pp. 15–63). Erevan: Armenia's Academy of Sciences.

Bolotin, V. V. (1965). *Statistical methods in structural mechanics* (first edition). Moscow: State Publishing House for Building, Architecture and Building Materials (in Russian).

Bolotin, V. V. (1966a). Broadband random vibrations of elastic systems. *International Journal of Solids and Structures, 2*, 105–124.

Bolotin, V. V. (1966b). Stochastic boundary-value problems in the theory of plates and shells. In *Proceedings of the 6th All-Union Conference on Theory of Shells and Plates*, Moscow, Nauka (in Russian).

Bolotin, V. V. (1966c). On the broad-band random vibration of elastic systems. In *Proceedings, 11th International Congress of Applied Mechanics* (pp. 233–238). Berlin: Springer.

Bolotin, V. V. (1980). On the parametric resonance density. *PMM: Journal of Applied Mathematics and Mechanics, 44*, 776–781.

Bolotin, V. V. (ed.). (1999). *Vibrations in technical engineering*, Vol. 1, Vibrations of Linear Systems, Moscow: "Mashinostroenie" Publishing House (in Russian).

Bolotin, V. V., & Elishakoff, I. (1971). Random Oscillation in elastic shells containing an acoustic medium. *Mechanics of Solids*, 6(5), 99–106.

Bolotin, V. V., Elishakoff, I., Efimtsov, B. M., Moskalenko, V. N., & Sharii, N. V. (1969). Method of analysis of acoustic fields inside stiffened shells. In V.V.Bolotin (Ed.), *Proceedings of the Scientific Technical Conference Based on Results of Scientific Investigations in Years 1968–1969, Section of Energetic Machinery, Subsection of Dynamics and Strength of Machines* (pp. 5–26). Moscow: Power Engineering Institute, Press (in Russian).

Bolotin, V. V., Makarov, B. P., Mishenkov, G. V., & Shveiko Yu.Yu. (1960). Asymptotic method of investiating spectrum of free vibrations of plates. In *Raschety na Prochnost (Strength Analysis)* (Vol. 6, pp. 231–253) (in Russian).

Bolton, J. S., & Green, E. R. (1993). Smart foams for active absorption of souind. In *Second Conference on Recent Advances in Active Control of Sound and Vibration*, Blacksburg, VA (pp.130–149).

Bonilha, M. W., & Fahy, F. J. (1996). On the Vibration field correlation of randomly excited flat plate structures, I: Thoery. *Joined of Sound and Vibration, 214*(3), 443.

Bonte, M.H. A., De Boer, A., & Liebregts, R., Determining the Von Mises stress power spectral density for frequency domain fatigue analysis including out-of-phase stress components. *Journal of Sound and Vibration, 302*, 379–386 (2007).

Borboni, A., & Faglia, R.(2013). Stochastic Evaluation and analysis of free vibrations in simply supported piezoelectric bimorphs. *Journal of Applied Mechanics, 80*(2), article 021003.

Borgman, L. E. (1967). Spectral analysis of ocean wave forces on piling. *Journal of Waterways and Harbors, 2*, 129–156.

Borgman, L. E. (1969). Ocean Wave simulation for engineering design. *Journal of Wateways and Harbors, 95*, 557–583.

Borino, G., Di Paola, M., & Muscolino, G. (1988). Non-stationary spectral moments of base excited MDOF systems. *Earthquake Engineering and Dynamics, 16*, 745–756.

Borisyuk, A. O., & Grinchenko, V.T. (1997). Vibration and noise generation by elastic elements excited by a turbulent flow. *Journal of Sound and Vibration, 204*(2), 213–237.
Boyce, W. E. (1962). Random Vibration of elastic strings and bars. In *Proceedings of the Fourth, U. S. National Congress of Applied Mechanics*, Berkeley, CA (Vol. 1, pp. 77–85).
Boyce, W. E. (1966). Stochastic Nonhomogeneous Sturm-Liouville problems. *Journal of Franklin Institute, 282*, 206–212.
Boyce, W. E. (1968). Random eigenvalue problems. In *Probabilistic Methods in Applied Mathematics* (Vol. 1, pp. 7–23). New York: Academic Press.
Boyce, W. E., & Goodwin, B. E. (1964). Random transverse vibrations of elastic beams. *Journal of the Society for Industrial and Applied Mathematics, 12*(3), 613–629.
Bozich, P. J., & White, R. W. (1970). A study of the vibration responses of shells and plates to fluctuating pressure environment, NASA, Vol. 1515.
Brandt, A. (2011). *Noise and vibration analysis: Signal analysis and experimental procedures*. Chichester,U.K.: Wiley.
Bratt, J. F. (1984). Standing waves in thin plates tensioned by in-plane loads subject to high frequency excitation. *Journal of the Franklin Institute, 318*(6), 431–438.
Braun, S. (2001). Correlation functions. In S. Braun (Ed.), *Encyclopedia of vibration* (pp. 294–300). New York: Academic Press.
Bräutigam, H., & Wedig, W. (1978). Breitbandtilger für stochastisch erregte Schwingungssysteme. *VDI-Berichte, 320*, 195–204 (in German).
Brekhovkikh, L. M. (1960). *Waves in layered media*. New York: Academic Press.
Brenneman, B., & Talley, J. Q. (1986). RMS fatigue curves to random vibrations. *Journal of Pressure Vessel Technology, 68*(4), 538–542.
Brenner, C. E., & Bucher, C.G. (1992). Stochastic response of uncertain systems. *Archive of Applied Mechanics, 62*(8), 507–516.
Brinkmeier, M., Dörre, K., Stephen, J., & Eigen, M. (1999). Two-beam cross-Correlation: A method to characterize Transport phenomena in micrometer-sized structures. *Analytical Chemistry, 71*(3), 609–616.
Brown, D. L., Allemang, R. J., Zimmerman, R., & Mergeay, M. (1979). *Parameter estimation techniques for modal analysis*. SAE Technical Paper 790221.
Brown, K. T. (1984). Measurement of modal density: An improved technique for use on lightly damped structure. *Journal of Sound and Vibration, 96*(1), 127–132.
Brownell, F. H. (1955). An extension of Weyl's asymptotic law for eigenvalues. *Pacific Journal of Mathematics, 5*(4), 483–490.
Brownell, F. (1957). Extended asymptotic eigenvalue distribution for bounded domain in n-space. *Journal of Mathematics and Mechanics, 6*, 119–166.
Bryja, D., & Sniady, P. (1988). Random vibration of a Suspension bridge due to highway traffic. *Journal of Sound and Vibration, 125*(2), 379–387.
Bryja, D., & Sniady P. (1991). Spatially coupled vibrations of a suspension bridge under random highway traffic. *Earthquake Engineering and Structural Dynamics, 20*, 999–1010.
Bucher, C.G. (1988). Approximate nonstationary random vibration analysis for MDOF systems. *Journal of Applied Mechanics*, 55, 197–200, 1988.
Bucher, C.G. (2009). *Computational analysis of randomness in structural mechanics*. Boca Raton: CRC Press.
Bucher, C. G., & Brenner, C. E. (1992). Stochastic response of uncertain system. *Archive of applied machines, 62*, 507–516.
Bucher, C. G., & Lin, Y. K. (1988). A note on spectral moments in nonstationary random vibration. *Probabilistic Engineering Mechanics, 3*(1), 53–55.
Bucher, C. G., & Schuëller, G. I. (1991). Non-Gaussian response of linear systems. In G. I. Schuëller (Ed.), *Structural dynamics*. Berlin: Springer (pp. 103–127).
Buciarelli, L. L. Jr., & Kuo, C. (1970). Mean-square response of a second-order system to nonstationary random excitation. *Journal of Applied Mechanics, 37*, 612–616.

Bullmore, A. J., Nelson, P. A., & Elliott, S. J. (1990). Theoretical studies of the active control of propeller-induced cabin noise. *Journal of Sound and Vibration, 140*, 191–217.

Bukowski, J., Korn, L., & Watrenberg, D. (1995). Correlated inputs in quantitative risk assessment: The effects of distributional shape. *Risk Analysis, 15*(2), 215–219.

Bull, M. K. (1967). Wall pressure fluctuations associated with subsonic turbulent boundary layer flow. *Journal of Fluid Mechanics, 28*, 719–754.

Bull, M. K. (1996). Wall pressure fluctuation beneath turbulent boundary layers: Some reflections on forty years of research. *Journal of Sound and Vibration, 190*(3), 299–315.

Bunke, H. (1972). *Geowöhnliche Differential Gleichungen mit zufälligen Parametern*. Berlin: Akademie Verlag (in German).

Burgos, R. B., Santos, M.A. C., & Silva, R. R. (2015). Analysis of beams and thin plates using the wavelet-Galerkin method. *International Journal of Engineering and Technology, 7*(4), 261–266.

Burlon, A., Failla, G., & Arena, F. (2018). Exact stochastic analysis of coupled bending–torsion beams with in-span supports and masses. *Probabilistic Engineering Mechanics, 54*, 53–64.

Burroughs, C. B., Fischer, R. W., & Kern, F. K. (1997). An introduction to statistical energy analysis. *The Journal of the acoustical Siociety of America, 101*(4), 1779–1789.

Byrcroft, G. N. (1960). White noise representation of earthquakes. *Journal of Engineering Mechanics, 86*, 1–16.

Cacciola, P., Colajanni, P., & Muscolino, G. (2005). A modal approach for the evaluation of the response sensitivity of structural systems subjected to non-stationary random processes. *Computer Methods in Applied Mechanics and Engineering, 194*(42),4344–4361.

Caddemi, S. (1994). On the approximation of covariance kernels. *Probabilistic Engineering Mechanics, 9*, 245–254.

Cai, G. Q., & Zhu, W. Q. (2017). *Elements of stochastic dynamics*. Boca Raton: CRC Press.

Campbell, G. W. (1972). *The use of an absorber as a vibration reduction device for improved reliability of seismic structures*. Master's Thesis, Aerospace and Mechanical Engineering Department, The University of Arizona.

Camperi, S., Ghanchi-Tehrani, M., Zilletti, M., & Elliott, S. J. (2016). Active vibration control of an inertial actuator subject to broadband excitation. *Journal of Physics: Conference Series, 744*(1), article 012038.

Carcatterra, A., Akay, A., & Koc, I. M. (2006). Near-irreversibility in a conservative linear structure with singularity point in its modal density. *The Journal of the Acoustical Society of America, 119*, 2141–2149.

Carkovs, J., & Matvejevs, A. (2017). On the stability of pin-joined beam affected by random pulsating load. *Proceedings of the Latvian Academy of Sciences, Series B, 21*, 63–68.

Casciati, F., & Faravelli, L. (1991) *Fragility analysis of complex structural systems*. Taunton: Research Studies Press.

Casciati, F., Grigoriu, M., Der Kiureghian, A., Wen, Y. K., & Vrouwenvelder, T. (1997). Report on the working group on dynamics. *Structural Safety, 19*(3), 271–282.

Caughey, T. K. (1963). Nonstationary Random inputs and response, Chapter 3. In S. H. Crandall (Ed.), *Random vibration* (Vol. 2). Cambridge, MA: MIT Press.

Caughey, T. K., & Dienes, J. K. (1962). The behavior of linear systems with random parametric excitation. *Journal of Mathematical Physics, 41*, 300–318.

Caughey, T. K., & Gray, A. H., Jr. (1963). Discussion of distribution of structural response to earthquakes. *Journal of Engineering Mechanics, 89*(2), 159–168.

Caughey, T. K., & Stumpf, H. J. (1961). Transient response of a dynamic system under random excitation, *Journal of Applied Mechanics, 28*, 563–566.

Ceberio, M., Ferson, S., Kreinovich, V., Chopra, S., Xiang, G., Murguia, A., & Santillan, J. (2017). Now to take into account dependence between the inputs. *Journal of Uncertain Systems, 1*, 14–37.

Cederbaum, G. (1990). Random vibrations of viscoelastic laminated plates. *Journal of Applied Mechanics, 57*, 688–693.

Cederbaum, G., Elishakoff, I., Aboudi, J., & Librescu, L. (1992). Dynamic response of moderately thick laminated composite shells to random excitation, Ch. 3. In *Random vibration and reliability of composite structures* (pp. 55–72). Lancaster, PA: Technomic.
Cederbaum, G., Elishakoff, I., & Librescu, L. (1989). Random vibration of composite plates within the first order shear deformation theory. *Journal of Composite Structures, 12*, 97–111.
Cederbaum, G., Librescu, L., & Elishakoff, I. (1988). Dynamic response of flat panels made of advanced composite materials subjected to random excitations, *The Journal of the Acoustical Society of America, 84*, 660–666.
Cederbaum, G., Librescu, L., & Elishakoff, I. (1989a). Remarks on the dynamics of a higher-order laminated plate theory and its application in random vibration response. *International Journal of Solids and Structures, 25*, 515–526.
Cederbaum, G., Librescu, L., & Elishakoff, I. (1989b). Response of laminated plates to nonstationary random excitation. *Journal of Structural Safety, 6*, 99–113.
Cederkvist, J. (1982). Design of beams subjected to random loads. *Journal of Structural Mechanics, 10*, 49–65.
Chang, C. C., & Yang, H. T. Y. (1991). Random vibration of flexible, uncertain beam element. *Journal of Engineering Mechanics, 117*(10), 2329–2350.
Chang, T.-P. (1993a). Forced vibration of a mass-loaded beam with a heavy tip body. *Journal of Sound and Vibration, 164*(3), 471–489.
Chang, T.-P. (1993b). Dynamic finite element analysis of a beam on random foundation. *Computers and Strucutres, 48*(4), 583–589.
Chang, T.-P. (1994). Deterministic and random vibration of an axially loaded Timoshenko beam resting on an elastic foundation. *Journal of Sound and Vibration, 178*(1), 55–66.
Chang, T.-P. (2013). Deterministic and random vibration analysis of fluid-contacting transversely isotropic magneto-electro-elastic plates. *Computers and Fluids, 84*, 247–254.
Chang, T.-P., & Chang, H.-C. (1997). Perturbation method for probabilistic dynamic finite element analysis of a rectangular plate, *Journal of Structural Mechanics, 25*(4), 397–415.
Chang, T.-P., Chang, H.-C., & Liu, M.-F. (2006). A finite element analysis on random vibration of nonlinear shell structures. *Journal of Sound and Vibration, 291*(1), 240–257.
Chang, T.-P., Lin, G.-L., & Chang, E. (2006). Vibration analysis of a beam with an internal hinge subjected to a random moving oscillator, *International Journal of Solids and Structures, 43*(21), 6398–6412.
Chang, T.-P., & Liu, M. F. (2009). Vibration Analysis of a uniform beam traversed by a moving vehicle with random mass and random velocity. *Structural Engineering and Mechanics, 31*(6), 737–749.
Chase, D. M. (1980). Modeling the wavevector-frequency spectrum of turbulent boundary layer wall pressure. *Journal of Ssound and Vibration, 70*(1), 29–67.
Chen, T.-U. (1993). Design sensitivity analysis for repeated eigenvalues in structural design. *AIAA Journal, 31*(12), 2347–2350.
Chen, B., Zhong, P., Cheng, W., Chen, X., & Yang, Q. (2017). Correlation and combination factors of wind forces on cylindrical roof structures. *International Journal of Structural Stability and Dynamics, 17*(9), article 1750104.
Chen, C. L., & Chen, C. W. (2001). Random response of a rotating composite blade with flexure-torsion coupling effect by the finite element method. *Composite Structures, 54*(4), 407–415.
Chen, C. L., & Chen, L. W. (2001). Random vibration by a damped thick rotating blade of generally orthotropic material, *Composite Structures, 53*, 365–377.
Chen, C. L., & Chen, L. W. (2001). Random response of rotating corporate blade with flexure-torsion coupling effect by the FEM. *Composite Structures, 54*, 407–415.
Chen, C. L., & Chen, L. W. (2002). Random vibration of a rotating blade with external and internal damping by the FEM. *Journal of Sound and Vibration, 252*, 697–715.
Chen, C. C. T., & Yang, T. Y. (1991). Flexible beam and frame elements under nonwhite and nonzero mean loads. *Journal of Engineering Mechanics, 117*(6), 1346–1364.

Chen, G., Coleman, M. P., & Zhou, J. (1992). The equivalence between the wave propagation method and Bolotin's method in the asymptotic estimation of eigenfrequencies of a rectangular plate. *Wave Motion*, 16, 285–297.
Chen, G., Coleman, M. P., & Zhou, J. (1991). Analysis of vibration eigenfrequencies of a thin plate by the Keller-Rubinow wave method I: Clamped boundary conditions with rectangular or circular geometry. *SIAM Journal of Applied Mathematics, 51*, 967–983.
Chen, G.-H., Meng, Z., & Yang D.-X. (2018). Exact nonstationary responses of thin rectangular thin plate on Pasternak foundation excited by stochastic moving loads. *Journal of Sound and Vibration, 412*, 166–183.
Chen, G.-H., Zhou, J., & Yang D.-X. (2017). Benchmark solutions of stationary vibration of rectangular thin plate based on discrete analytical method. *Probabilistic Engineering Mechanics, 50*, 17–24.
Chen, J. B. (2008). strategy for selecting representative points via tangent spheres in the probability density evolution method. *International Journal for Numerical Methods in Engineering, 74*(13), 1988–2014.
Chen, J. B., & Li, J. (2009). A note on the principle of preservation of probability and probability density evolution equation. *Probabilistic Engineering Mechanics, 24*(1), 51–59.
Chen, J. J., Che, J. W., Ma, H. B., & Dai, J. (2001). Dynamic optimum design Based on reliability for trans structures. *Acta Mechanica Solida Sinica, 22*(1), 54–60.
Chen, J. J., Che, J. W., Sun, H. A., Ma, H. B., & Cui, M. T. (2002). Probabilistic dynamic analysis of truss structures. *Structural Engineering and Mechanics, 13*(2), 231–239.
Chen, K. F., & Jiao, Q. Y. (2009). A parsimonious mass-spring system capable of being tuned for arbitary repeated natural frequencies. *Mechanics Research Communications, 36*(2), 149–153.
Chen, K. F., & Jiao, Q. Y. (1995). Distinction of natural frequencies of chain structure vibration system. *Journal of Beijing Agricultural Engineering University, 15*, 24–28.
Chen, K. F., & Jiao, Q. Y. (1999). On the repeated natural frequencies for torsional vibration of shafts. *Applied Mathematics and Mechanics, 20*(11), 1279–1285.
Chen, K. K., & Soong, T. T. (1971). Covariance properties of waves propagation in random medium. *The Journal of Acoustical Society of America, 49*, 1639–1642.
Chen, L., & Letchford, C. W. (2004). A determinstic-stochastic hybrid model of downbursts and its impact on a cantilevered structure. *Engineering Structures, 26*(5), 619–620.
Chen, M. T., & Harichandran, R. S. (2001). Response of an earth dam to spatially varying earthquake ground motion. *Journal of Engineering Mechanics, 127*(9), 932–939.
Chen, M.-T., & Ali, A. (1996). A Robust integration technique for general-purpose random vibration analysis. In *Proceedings of the 67th Shock and Vibration Symposium* (pp. 479–488). Monterey, CA.
Chen, S.-H. (1992). Random vibration analysis of large-scale structures with random parameters. *Computers and Structures, 43*(4), 681–685.
Chen, S.-H., Liu, Z.-S., & Zhang, Z.-F. (1992). Random vibration analysis for large-scale structures with random parameters. *Computers and Structures, 43*(4), 681–685.
Chen, S.-H., Liu, Z.-S., Shao, C. S., & Zhao, Y. Q. (1993). Perturbation analysis of vibration modes with close frequencies. *Communications in Numerical Methods in Engineering, 9*(5), 427–438.
Chen, Y. (1991). *Stochastic response of secondary systems in base-isolated structures*. Master's Thesis, Clarkson University, Potsdam, N.Y.
Chen, Y., & Soong, T. T. (1988a). State-of-the-art review: Seismic response of secondary systems. *Engineering Structures, 10*, 218–228.
Chen, Y., & Soong, T. T. (1988b). Seismic response of secondary system. *Engineering Structures, 10*(3), 218–228.
Chen, Y., & Ahmadi, G. (1992). Stochastic earthquake response of secondary systems in base-isolated structures. *Earthquake Engineering and Structural Dynamics, 21*, 1039–1057.
Chen, Y., & Soong, T. T. (1988). State-of-the art review: Seismic response of secondary systems. *Enginering Structures, 10*(3), 218–228.

Chen, Z., Sun, Y., Zhang, H., & Zhang, Z. (2016). A study on the fluctuating pressure correlation of a cantilevered elliptic annular roof structure. *International Journal of Structural Stability and Dynamics, 16*(1), article 1640013.

Cheng, Y., & Du X. (2016). System reliability analysis with dependent component failures during early design stage—A feasibility study. Journal of Mechnical Design, *138*(5), article 051405.

Chen, X. (2015). Revisiting combination rules for estimating extremes of linearly combined correlated wind load effects. *Journal of Wind Engineering and Industrial Aerodynamics, 141*, 1–11.

Cheung, Y. L., Wong, W. O., & Cheng, L. (2012). Minimization of the mean square velocity response of dynamic structures using an active—passive dynamic vibration absorber. *The Journal of the Acoustical Society of America, 132*(1), 197–207.

Chgu, W. J., Au-Yang, M. K. (1972). Random response of rectangular panels to the pressure field beneath a turbulent boundary layer in subsonic flows, *NASA TN-D-6970*

Chien, C. F., & Soroka, W. W. (1976). Spatial correlation of acoustic pressures in steady and decaying reverberant sound fields. *Journal of Sound and Vibration, 48*(2), 235–242.

Chirkov, V. P. (1975). Random vibrations of a thin-walled structures carrying concentrated masses. *Mechanics of Solids, 10*(3), 141–145.

Chladni, E. F. F. (1786). *Über die Longitudinal Schwingungen der Seiten und Stäbe*. Erfurt: George Adam Keyser (in German).

Chladni, E. F. F. (1787). *Entdeckungen über die Theorie des Klanges*. Leipzig: Weidmanns, Erben and Reich (in German).

Chladni, E. F. F. (1802). *Die Akustik*. Leipzig: Breitkopf und Härtel (in German).

Chladni, E. F. F. (1809). *Traité d'Acoustique*. Paris: Courcier (in French).

Chladni, E. F. F. (1826). Über meine Aufnahme bei Napoleon und sonst in Paris. *Caecilic, 5*, 137–144 (in German).

Chladni, E. F. F. (2015). *Treatise on acoustics: The first comprehensive english translation of E.F. F. Chladni's Traité d'Acoustique* (R. T. Beyer, Trans.). Cham: Springer.

Chladni, E. (1802). *Die Akustik*. Leipzig: Breitkopf and Härtels (in German). Retrieved December 1, 2017, from https:// vlp.mpiwg-berlin.mpg.de/refrences?id=lit29494.

Chladni, E. F. F. (1817). *Neue Beiträge Zur Akustik*. Leipzig: Breitkopf und Härtel (in German).

Chmielewski, T. (1982). *Metody Probabilistyczne n Dynamice Konstrukcji (Probabilistic methods in structural dynamics)*. Opole (in Polish).

Cho, D. S., & Lee, W. K. (2000). Modal interactions of a randomly excited hinged clamped beam. *Journal of Sound and Vibration, 237*(3), 377–393.

Choi, K. K., & Haug, E. J. (2012). Optimization of structures with repeated eigenvalues. In E. J. Haug & J. Cea (Eds.), *Optimization of distributed parameter structures* (pp. 219–277). Amsterdam: Sijthoff and Noordhoff.

Choi, K. K., Haug, E. J., & Lam, H. L. (1982). A numerical method for distributed parameter structural optimization problems with repeated eigenvalues. *Journal of Structural Mechanics, 10*, 191–207.

Chonan, S. (1978). Random vibration of an elastically supported circular plate with an elastically restrained edge and an initial tension. *Journal of Sound and Vibration, 58*(3), 443–454.

Chonan, S. (1980). Random vibration of an initailly stressed thick plate on elastic foundation. *Journal of Sound and Vibration, 71*(1), 117–127.

Chonan, S. (1981). Random vibration of an annular plate. *Journal of Sound and Vibration,78*(1), 1–13.

Chonan, S. (1985). Random vibration of a prestressed, orthotropic, thick rectangular plate on generalized flexible foundation. *The Journal of the Acoustical Society of America, 78*, 598–604.

Chonan, S. (1988). Random vibration of prestressed, orthotropic, thick rectangular plate on a generalized flexible foundation. *The Journal of the Acoustical Society of America, 28*, 598–604.

Christakos, G. (2017). *Spatiotemporal random fields* (2nd ed.). Amsterdam: Elsevier.

Chrukiuri, A., & Barbone, P. E. (1998). High modal density approximations for equipment in the time domain. *The Journal of Acoustical Society of America, 104*(4), 2048–2053.

Chu, S. L., Amin, M., & Singh, S. (1972). Spectral treatment of actions of three earthquake components on structures. *Nuclear Engineering and Design, 21*, 126–136.

Ciappi, E., De Rosa, S., Franco, F., Guyader, J. L., Hambric, S. A., Leung, R. C. K., & Hanford, A. D. (2018). *Flinovia-flow induced noise and vibration issues and aspects-II*. Springer.

Ciappi, E., & Magionesi, F. (2009). Full scale analysis of the response of an elastic ship panel excited by turbulent boundary layers. *Noise Control Engineering Journal, 57*(3),179–192.

Ciappi, E., Magionesi, F., De Rosa, S., & Franco, F. (2009). Hydrodynamic and hydroelastic analyses of a plate excited by the turbulent boundary layer. *Journal of Fluids and Structures, 25*(2), 321–42.

Ciappi, E., Magionesi, F., De Rosa, S., & Franco, F. (2012). Analysis of the scaling laws for the turbulence driven panel responses. *Journal of Fluids and Structures, 32*, 90–103.

Clark, C. (1967). The asymptotic distribution of eigenvalues and eigenfunctions for elliptic boundary value problems. *SIAM Review, 9*(4), 627–646.

Clark, C. (1970). On the asymptotic formula for the eigenvalues of membranes and plates. *Unione Matematica Italiana, 3*(2), 191–196.

Clark, D. L., Bae, H., & Forster, E. E. (2018). Non-deterministic metamodeling for correlated and uncorrelated random variable. In *Paper AIAA-2018–0667, SciTech Conference*.

Clarkson, B. L. (1959). The effect of jet noise on aircraft structures. *Aeronautical Quarterly, 10*, 103–126.

Clarkson, B. L. (1962). The Design of structures to resist noise fatigue. *Journal of the Royal Aeronautical Soc*iety, *66*, 603–616.

Clarkson, B. L. (1981). The derivation of modal densities from point impedances. *Journal of Sound and Vibration, 77*(4), 583–584.

Clarkson, B. L. (1981). A structural dynamicist looks at statistical energy analysis. *Proceedings of the Institute of Acoustics*, 1–8.

Clarkson, B. L. (1986). Experimental determination of modal density. In I. Elishakoff & R. H. Lyon (Eds.), *Random Vibration—Status and Recent Developments* (pp. 59–86). Amsterdam: Elsevier.

Clarkson, B. L., & Ford, R. D. (1962). The response of a typical aircraft structure to jet noise. *Journal of the Royal Aeronatuical Society, 66*, 31–40.

Clarkson, B. L., & Mercer, C. A. (1965). Use of cross-correlation in studying the response of lightly damped structures to random forces. *AIAA Journal, 3*(12), 2287–2291.

Clarkson, B. L., & Pope, R. J. (1981). Experimental determination of modal densities and loss factors of flat plates and cylinders. *Journal of Sound and Vibration, 77*(4), 535–549.

Clarkson, B. L., & Ransky, M. F. (1983). Modal density of honeycomb plates. *Journal of Sound and Vibration, 91*(1), 103–118.

Coe, C. F., & Chyu, W. J. (1972). Pressure-fluctuation inputs and response of panels underlying attached and separated supersonic turbulent boundary layers, Conference Preprint No. 113. In *Symposium on Acoustic Fatigue* (pp. 5.1–5.20). AGARD.

Collins, J. D., & Thomson, W. T. (1969). The eigenvalue problem for structural systems with statistical properties. *AIAA Journal, 7*(4), 642–648.

Coleman, M. P. (1997). Analysis of vibration by the wave propagation method and Bolotin's method for a rectangular thin plate with at least one side roller-supported. *Wave Motion, 25*, 169–180.

Coleman, M. P. (2003). Analysis and computation of teh vibration spectrum of the shallow cicular cylindrical shell with rectangular domain. *Journal of Sound and Vibration, 263*, 335–356.

Colwell, R. C. (1933). Diagonal Symmetry in chladni plates. *Journal of Franklin Institute, 215*(2),169–177.

Constantinou, M. C., & Tadjbakhsh, I. G. (1984). Response of a sliding structure to filtered random excitations. *Journal of Structural Mechanics, 12*, 401–418.

Conte, J. P., & Peng, B. F. (1996). An explicit closed-form solution for linear systems subjected to nonstationary random excitation. *Probablistic Engineering Mechanics, 11*(1), 37–50.

Cook, R. G. (1964). Digital *Simulation of random vibration*. Ph.D. Thesis, Department of Mechanical Engineering, Massachusetts Institute of Technology.

Cook, R. K., Waterhouse, R. V., Berendt, R. D. (1955). Measurement of correlation coefficients in reverberant sound fields. *The Journal of the Acoustical Society of America, 27*(6), 1072–1077.

Corcos, G. M. (1967). The structure of the turbulent pressure at the wall of a boundary layer. *Journal of Sound and Vibration, 6*, 59–70.
Cornell, C. A., & Winterstein, S. R. (1987). Temporal and magnitude dependence in earthquake recurrence models. In Y. K. Lin & R. Minai (Eds.), *Stochastic approaches in earthquake engineering* (pp. 18–39). Berlin: Springer.
Corotis, R. B., & Marshall, J. A. (1977). Oscillator response to modulated random excitation. *Journal of Engineering Mechanics, 103*, 501–513.
Corotis, R. B., & Vanmarcke, E. H. (1975). Time-dependent spectral contents of system response. *Journal of Engineering Mechanics, 101*(5), 623–637.
Corotis, R. B., Vanmarcke, E. H., & Cornell, C. A. (1972). First passage of nonstationary random process. *Journal of Engineering Mechanics, 98*(2), 401–412.
Cortinez, V. H., Piovan, M. T., & Rossi, R. E. (1999). A consistent derivation of the Timoshenko's beam theory. *Structural Engineering and Mechanics, 7*(5), 527–532.
Cottis, M. G. (1968). On the dynamic response of an orthopic finite cylindrical shell to an arbitrary pressure field. *Journal of Sound and Vibration, 7*(1), 31–38.
Cottis, M. G., & Jasonides, J. G. (1964). The response of a finite thin cylindrical shell to random of a pressure field. In *Proceedings of the Second International Conference on Acoustical Fatigue in Aerospace Structures*, Dayton, OH (pp. 185–212).
Cottone, G., & Di Paola, M. (2009). On the use of fractional calculus for the probabilistic characterization of random variables. *Probabilistic Engineering Mechanics, 24*(3), 321–330.
Coughlin, A. M. (2000). Patterns in the sand: Mathematical exploration of chladni patterns. *Journal of Mathematics and Science: Collaborative Explorations, 3*(1), 131–146.
Courant, R. (1919). Über die Abhängigkeit der Schwingungszahlen einer Member von Ihrer Begrenzung and über asymptotische Eigenwerteilung. *Göttinger Nachrichten*, 255–264 (in German).
Courant, R. (1920). Über die Eigenwerte bei den Dibberrentialglerichungen der mathematischen Physik. *Mathematische Zeitung, 7*, 1–57 (in German).
Courant, R. (1922). Über die Schwingungen eigenspannter Platten. *Mathematische Zeitung, 15*, 195–200 (in German).
Courant, R., & Hilbert, D. (1953). *Methods of mathematical physics*. New York: Wiley Interscience.
Craig, R. R., Jr., Chung, Y.-T., & Blair, M. A. (1985). Modal vector for closely-spaced frequency modes. In *Proceedings of the International Modal Analysis Conference*, Orlando, FL (pp. 672–678).
Cramer, H., & Leadbetter, M. R. (1967). *Stationary and related stochastic processes*. New York: Wiley.
Crandall, S. H. (Ed.). (1958). *Random vibration*. New York: MIT Press/Wiley.
Crandell, S. H. (1959). Random vibration. *Applied Mechanics Reviews, 12*, 739–745.
Crandell, S. H. (1966). Random vibration. In H. N. Abramson, H. Liebowitz, J. M. Crowley, & S. Juhasz (Eds.), *Applied mechanics surveys* (pp. 681–689). Washington, D.C.: Spartan books; (Vol. 12, pp. 739–745) (1959).
Crandall, S. H. (1970). First crossing probabilistic of the linear oscillator. *Journal of Sound and Vibration, 12*, 285–299.
Crandall, S. H. Introduction to random processes. In I. Holand, D. Kavlie, G. Moe, & R. Sigbjornsson (Eds.), *Safety of structures under dynamic loading* (pp. 19–34).
Crandall, S. H. (1980a). Localized response reductions in wide-band random vibration of uniform structures. *Archives of Applied Mechanics, 49*(5–6), 347–359.
Crandall, S. H. (1980b, July 3). Personal Communication to I.E.
Crandall, S. H. (1982, November 29). Personal Communication to I.E.
Crandall, S. H. (1980, July 3). Private Communication.
Crandall, S. H. (1982, November 29). Private Communication.
Crandall, S. H. (1983, April 19). Private Communication.
Crandall, S. H. (1986). Image sum procedure for estimating wide-band response of plates. In *23rd Annual Technical Meeting, Society of Engineering Science*. University of Buffalo, NY.

Crandall, S. H. (1988). Spatial correlation in structural response in wide-band exciatation. In N. Sri Namachchivaya, H. Hilton, & Y. K. Wen (Eds.), *Stochastic structural dynamics* (pp. 213–234). University of Illinois at Urbana-Champaign.

Crandall, S. H., Chandiramani, K. L., & Cook, R. G. (1966). Some first-passage problems in random vibration. *Journal of Applied Mechanics, 33*, 532–538.

Crandall, S. H., Lee, S. S., & Williams, J. H. (1974). Accumulated slip of a friction controlled mass excited by earthquake motions. *Journal of Applied Mechanics, 41*, 1094–1098.

Crandall, S. H., & Weiss, G. (1959). Random vibration. *Physics Today, 12*, 49.

Crandall, S. H., & Zhu, W. Q. (1983a). Wide band random excitation of square plates. In K. Hennig (Ed.), *Random vibrations and reliability* (pp. 231–244). Berlin: Akademie-Verlag.

Crandall, S. H., & Zhu, W. Q. (1983b). Random vibration: A survey of recent developments. *Journal of Applied Mechanics, 50*(46), 953–962.

Crandall, S. H., & Zhu, W. Q. (1983c). Wide band random vibration of square plates. In K. Hennig (Ed.), *Random vibrations and reliability* (pp. 231–243). Berlin: Akademie-Verlag.

Crandall, S. H., & Zhu, W. Q. (1984). Wide Band random vibration of an equilateral triangular plate. In T. C. Huang & P. D. Spanos (Eds.), *Random vibrations* (pp. 35–50). New York: ASME Press.

Crandall, S. H., & Zhu, W. Q. (1986). Wide band random vibration of an equilateral triangular plate. *Probabilistic Engineering Mechanics, 1*, 5–12.

Crandall, S. H., & Yildiz, A. (1962). Random vibrations of beams. *Journal of Applied Mechanics, 29*, 267–275.

Crauel, H., & Gundlach, M. (Eds.) (2007). *Stochastic dynamics*. Berlin: Springer.

Cremer, L., Heckl, M., & Ungar, E. E. (1975). *Structure—Borne sound*. Berlin: Springer.

Crocker, M. J., & White, R. W. (1967). Response of a simply supported beam to a three-dimensional reverberant acoustic field. *The Journal of the Acoustical Society of America, 41*, 1064–1065.

Crowson, H. L. (1963). Generalization of the note "an error analysis on the digital computation of the autocorrelation function". *AIAA Journal, 1*(8), 1968–1969.

Crowson, H. L. (1963). An error analysis in the digital computation of the autocorrelation function. *AIAA Journal, 1*, 488–489.

Cumming, I. G. (1967). Derivation of the moments of a continuous stochastic system. *International Journal of Control, 5*, 85–90.

Cunningham, P. R. (2001). *Response prediction of acoustically-excited composite honeycomb sandwich structures with double curvature*. Ph.D. Thesis, University of Southampton.

Cupial, P. (1997). Calculation of the natural frequencies of composite plates by the Rayleigh-Ritz method with orthogonal polynomials. *Journal of Sound and Vibration, 201*(3), 385–387.

Curling, L. R., & Païdoussis, M. P. (2003). Analysis for random flow-induced vibration of cylindrical structures, subjected to turbulent axial flow. *Journal of Sound and Vibration, 264*, 795–833.

Curtis, A. F., & Boykin, T. R. (1961). Response of two degree of freedom systems to white noise base excitation. *The Journal of the Acoustical Society of America, 33*, 655–663.

Czogala, E. (1976). Random vibration of a cylindrical shell due to an excitation with uniformly varying frequency. *Bulletin de L' Academic Polonaise des Sciences, 24*, 71–78.

Daborn, P. M., Roberts, C., Ewins, D. J., & Ind, P. (2014). Next generation random vibration testing. In *Topics in Modal Analysis II, Vol 8: Proceeding of the 32nd IMAC, Conference Proceedings of the Society for experimental Mechanics*.

Dahlberg, T. (1978). Ride comfort and road holding of a two_DOF vehicle traveling on a randomly profiled road. *Journal of Sound and Vibration, 58*(2), 179–187.

Dai, X. (2009). *Random vibration of piping systems containing a turbulent fluid*. M. S. Thesis, University of Nebraska, Lincoln, Nebraska.

Dailey, R. L. (1989). Eigenvector derivatives with repeated eigenvalues. *AIAA Journal, 27*(4), 486–491.

Das, A. K., & Dey, S. S. (1992). Effects of tuned mass dampers on random responses of bridges. *Computers and Structures, 43*(4), 745–750.

Das, A. K., & Dey, S. S. (1994). Random vibration of beams with localized region of damage. *Computers and Structures, 51*(1), 33–38.
Davenport, A. G. (1961). The application of statistical concepts in the wind loading of structures. In *Proceedings, Institute of Civil Engineers* (Vol. 19, pp. 187–196).
Davenport, A. G. (1962). Response of slender, line-like structures to gusty winds. *Journal of the Institution of Civil Engineers, 23*, 389–408.
Davenport, A. G. (1964). A note on distribution of largest value of random function with application to gust loading. In *Proceedings, Institute of Civil Engineers* (Vol. 28, pp. 187–196).
Davidson, J. W. (1974). *Optimum design of structures with random parameters*. Ph.D. Thesis, University of California, Los Angeles.
Davidson, J. W., Felton, L. P., & Hart, G. C. (1976). Reliability-based optimization for dynamic loads. *Journal of Structural Engineering, 103*(10), 2021–2035.
Davidson, J. W., Felton, L. P., & Hart, G. C. (1977). Optimum design, of structures with random parameters. *Computers and Structures, 7*, 481–486.
Davies, H. G. (1971). Low Frequency random excitation of water-loaded rectangular plates. *Journal of Sound and Vibration, 15*(1), 107–126.
Davies, H. G. (1971). Sound from turbulent boundary layer excited panels. *The Journal of the Acoustical Society of America, 49*(3), Part 2, 878–889.
Davies, M. G. (1963). Factoring the spectral matrix. *IEEE Transactions on Automatic Control, AC-8*, 296–305.
Davies, P., & Hammond, J. (1984). Envelope modelling of impulse response functions of systems having a high modal density. In *ICASSP'84. IEEE International Conference on Acoustics, Speech, and Signal Processing* (Vol. 9, pp. 634–637). IEEE Press.
Davis, R. B. (2017). Spatial Correlation function of cylinders in difference acoustic fields. *AIAA Journal, 55*(11), 4005–4010.
Davis, W. R., Jr. (1973). *Nonstationary spectral analysis for linear dynamic systems*. Sc. D. Thesis, Massachusetts Institute of Technology.
Davis, H.G., & Bhandsker, S.L. (1982). Random point excitation of coupled beams. *Journal of Sound and Vibration, 84*(4), 557–562.
Davis, W. R., Jr., & Bucciarelli, L. L., Jr. (1975). Nonstationary spectral analysis for linear dynamical systems. *AIAA Journal, 13*(1), 25–31.
De Rosa, S., Franco, F., Romano, G., & Scaramuzzino, F. (2003). The random structural response due to a turbulent boundary layer excitation. *Wind and Structure, 6*(6), 437–450.
Debchaudury, A., & Gasparini, D. A. (1982). Response of MDOF systems to vector random excitation. *Journal of Engineering Mechanics, 108*(2), 367–385.
Dehasa, J. S. (1980). The eigenvalue density of rational Jacobi matrices II. *Linear Algebra and Its Applications, 33*, 41–55.
De La Fuente, E. (2009). Von Mises stresses in random vibration of linear structures. *Computers and Structures, 87*, 1253–1262.
De La Fuente, E., & San Millián, J. (1998). An exact solution for random analysis of linear structural systems. In S. Idelsohn, E. Oñate, & E. Dvorkin (Eds.), *Computational mechanics: new trends and applications*. Barcelona: CIMNE.
De Langhe, K., Sas, P., & Dayuan, X. (1992). Random response analysis of a composite aeroplane test panel subjected to acoustic pressure loading. In *Proceedings of the International Conference on Noise and Vibration* (pp. 998–1007).
Demse, T. P. (2011). The great L's of French mathematics. *Journal of Mathematical Sciences: Advances and Applications, 12*(1), 1–39.
Deodatis G. (1991). Weighted Integral method II: Response variability and reliability. *Journal of Engineering Mechanics, 117*(8), 1865–1877.
Deodatis, G. (1996). Simulation of ergodic multivariate stochastic processes. *Journal of Engineering Mechanics, 122*(8), 778–787.
Deodatis, G., & Shinozuka, M. (1988). Auto-regressive model for nonstationary stochastic processes. *Journal of Engineering Mechanics, 114*(11), 1995–2012.

Demic, M. (1984). Assesment of random vertical vibration of human body fatigue using physiological approach. *IMeCHE Engineers*, 153/84.

Demic, M., Lukic, J., & Milic, Z. (2002). Some aspects of the investigation of random vibration influence on ride comfort. *Journal of Sound and Vibration, 253* (1), 109–129.

Demirdövan, N., Khalil, M., & Golonzka, O. (2001). Correaltion effects in two-dimensional vibrational spectroscopy of compleded vibrations. *Journal of Physical Chemistry, 105*(34), 8025–8030.

Der Kiureghian, A. (1991). *CQC modal combination rule for high-frequency modes*. In *Transactions, 11th International Conference on Structural Mechanics in Reactor Technology, K1* (pp. 231–236). Tokyo, Japan.

Der Kiureghian, A. (2000). The geometry of random vibrations and solutions by FORM and SORM. *Probabilistic Engineering Mechanics, 15*(1), 81–90.

Der Kiureghian, A. (2008). Analysis of structural reliability under parameter uncertainty. *Probabilistic Engineering Mechanics, 23*, 351–358.

Der Kiureghian, A., & Crempien, J. (1989). An evolutionary model for earthquake ground motion. *Structural Safety, 6*(2–4), 235–246.

Der Kiureghian, A., & Weng C.-D. (1992). A computer assisted learning system for random vibratina. *Computer and Structres, 43*(5), 975–993.

Der Kiureghian, A., & Neuenhofer, A. (1992). Response spectrum method for multi-support seismic excitations. *Earthequake Engineering and Structural Dynamics, 21*, 713–740.

Der Kiureghian, A., Sackman, J. L., & Nour-Omid, B. (1983). Dynamic analysis of Light equipment in structures: Response of stochastic input. *Journal of Engineering Mechanics, 109*(1), 90–110.

Der Kiureghian, A., & Wung, C.-D. (1992). A computer-assisted learning system for random vibrations. *Computers and Strcutures, 43*(5), 975–993.

De Rosa, S., & Franco, F. (2008). Exact and numerical responses of a plate under turbulent boundary excitation. *Journal of Fluids and Structures, 24*(2), 212–230.

De Rosa, S., & Fraco, F. (2008). A scaling Procedure for the Response of an isolated system with high modal overlap factor. *Mechanical Systems and Signal Processing, 22*(7), 1549–1565.

De Rosa, S., Franco, F., & Ciappi, E. (2015). A simplified method for the analysis of the stochastic response in discrete coordinates. *Journal of Sound and Vibration, 339*, 359–375.

De Silva, C. W. (2005). *Vibration and shock handbook*. Boca Raton: CRC Press.

Dey, S., Mukhopadhyay, T., Spickenheuer, A., Adhikari, S., & Heinrich, G. (2016). Bottom up surrogate based Approah for stochastic frequency response analysis of laminated composite plates. *Composite Structures, 140*, 712–727.

Dey, S. S. (1979). Finite element method for random response of structures due to stochastic excitation. *Computer Methods in Applied Mechanics and Engineering, 20*, 173–194.

Diederich, F. W. (1956). The dynamic response of a large airplane to continuous random atmosphere disturbances. *The Journal of the Acoustical Society of America, 23*(10), 917–930.

Dhanaut, J. M., Cheng G.-F., & Mei, C. (2014). Response of plates under statistically unsynchronized uniform random loads using Monte-Carlo simulation. *Journal of Computational and Applied Mechanics, 9*(1), 8–18.

Dhillep, M., & Bose, P. R. (2008). A comparative study of 'missing' correlation methods for response spectrum method of seismic analysis. *Computers and Structures, 86*, 2087–2094.

Dietrich, F. W. (1956). The dynamic response of a large airplane to continuous atmospheric disturbances. *Journal of Aeronautical Sciences, 23*, 917–930.

Di Lorenzo, S., Di Paola, M., Pinnola, F. P., & Pirrotta, A. (2014). Stochastic response of fractionally damped Beams. *Probabilistic Engineering Mechanics, 35*, 37–43.

Di Paola, M. (1984). A method for the analysis of vibrating systems subjected to random excitation. *Meccanica, 19*, 61–67.

Di Paola, M. (1985). Transient spectral moments of linear systems. *Solid Mechanics Archives, 10*, 225–243.

Di Paola, M. (1993). Stochastic differential calculus. In *Dynamic motion: Chaotic and stochastic behaviour* (pp. 29–92). Vienna: Springer.

Di Paola, M. (1998). Digital simulation of wind field velocity. *Journal of Wind Engineering and Industrial Aerodynamics, 74*, 91–109.
Di Paola, M., & Elishakoff, I. (1996). Non-stationary response of linear systems under stochastic Gaussian and Non-Gaussian excitation: A brief overview of react results. *Chaos, Solutions, and Fractals, 7*(7), 961–971.
Di Paola, M., Failla, G., & Pirrotta, A. (2012). Stationary and non-stationary stochastic response of linear fractional viscoelastic systems. *Probabilistic Engineering Mechanics, 28*, 85–90.
Di Paola, M., & Falsone, G. (1993). Itô and stratonovich integrals for delta-correlated processes. *Probabilistic Engineering Mechanics, 8*, 197–208.
Di Paola, M., & Falsone, G. (1994). Stochastic dynamics of MDOF structural systems under non-normal filtered inputs. *Probabilistic Engineering Mechanics, 9*(4), 265–272.
Di Paola, M., & Falsone, G. (1997). Higher order statistics of the response of MDOF linear systems excited by linearly parametric white noises and external excitations. *Probabilistic Engineering Mechanics, 12*(3), 179–188.
Di Paola, M., Falsone, G., Muscolino, G., & Ricciardi, G. (1991). Modal analysis for random response of MDOF systems. In Y. K. Lin & I. Elishakoff (Eds.), *Stochastic structural dynamics 1: New theoretical developments* (pp. 63–82). Berlin: Springer.
Di Paola, M., Ioppolo, M., & Muscolino, G. (1984). Stochastic seismic analysis of MDOF linear systems. *Engineering Structures, 6*, 113–118.
Di Paola, M., & Muscolino, G. (1985). Response maxima of SDOF system under seismic action. *Journal of Structural Division, 111*, 2033–2045.
Di Paola, M., & Muscolino, G. (1986). On the convergent part of high spectral moments. *Journal of Sound and Vibration*, 110, 233–245.
Di Paola, M., & Muscolino, G. (1988). Analytic evaluation of spectral moments. *Journal of Sound and Vibration*, 124, 479–488.
Di Paola, M., & Ricciardi G. (1992). Vibration of a bridge under a random train of moving loads. In *Probabilistic mechanics and structural and geotechnical reliability* (pp. 136–139). New York: ASCE Press.
Dickinson, S. M. (1975a). Bolotin's method Applied to the buckling and lateral vibration of stressed plates. *AIAA Journal, 13*(1), 109–110.
Dickinson, S. M. (1975b). Modified Bolotin method to buckling and vibration of stressed plates. *AIAA Journal, 13*(12), 1672–1673.
Dickinson, S. M. (1975c). Reply by author to W. W. King. *AIAA Journal*, 13(11), 1544.
Dickinson, S. M., & Warburton, G. W. (1967). Natural frequencies of Plate systems using the edge effect method. *Journal of Mechanical Engineering Science*, 9(12), 318–324.
Dickow, K. A., Brunskog, J., & Ohlrich, M. (2013). Modal density and modal distribution of bending wave vibration fields in ribbed plates. *The Journal of the Acoustical Society of America, 134*(4), 2719–2729.
Dimentberg, M. F. (1962). Forced vibrations of panels under random loading. In *Theory of plates and shells, Proceedings of the 2nd all-union conference on theory of plates and shells* (pp. 270–273). Kiev: Ukranian Academy of Sciences Press (in Russian).
Dimentberg, M. F. (1967). Determining the statistical characteristics of a linear system from measurements of its motion. *PMM: Journal of Applied Mathematics and Mechanics, 31*(6), 1089–1094.
Dimentberg, M. F. (1989). *Random process in dynamic system with variable parameter*. Moscow: "Nauka" Publishing House (in Russian).
Ditlevsen, O. (1985). Fatigue model for elastic bars in turbulent wind. *Journal of Wind Engineering and Industrial Aerodynamics, 18*(1), 27–52.
Ditolla, R. J. (1986). Random *Vibration analysis by the power spectrum and response spectrum methods*. Ph.D. Thesis, The Univesity of Arizona.
Dogan, V. (2007). Vibration of anti-symmetric angle-ply composite plates under random excitation. *Aircraft Engineering and Aerospace Technology, 79*(5), 475–484.

Dogan, V. (2014). Random vibration of shear deformable FGM plates. *Journal of Mechanics of Materials and Structures, 9*(4), 365–384.

Dokainish, M. A., & Elmadang, M. M. (1980). Random response of tractor-semitrailer system. *Vehicle System Dynamics, 9*(2), 87–112.

Dokunin, A. V., Krasnikov Yu. D., & Khurgin, A. Ya. (1978). *Statistical dynamics of mountain machines*. Moscow: "Mashinostroere" Publishing House (in Russian).

Donnati, P., Grosjean, P. A., Mistrott, P., & Roure, L. (1983). The subjective equivalence on sinusodial and random whole body vibration in sitting position. *Ergonomics, 26*, 251–273.

Dong, R. G., McDowell, T. W., & Welcome, D.E. (2005). Biodynamic response at the palm of the human hand subjected to a random vibration. *Industrial Health,43*(1), 241–255.

Dong, R. G., Welcome, D. E., McDowell, T. W., & Wu J. Z. (2004). Biodynamic response of human fingers in a power grip subjected to a random vibration. *Journal of Biomechanical Engineering, 126*(4), 447–457.

Dowell, E. H. (1969a). Noise or flutter or both?. *Journal of Sound and Vibration, 11*, 159–179.

Dowell, E. H. (1969b). Transmission of Noise from a turbulent boundary layer through a flexible plate into a cavity. *The Journal of the Acoustical Society of America, 46*(1), Part 2, 238–252.

Dowell, E. H. (1980a). Master plan for prediction of vehicle interior noise. *AIAA Journal, 18*(4), 353–366.

Dowell, E. H. (1980b). Interior Noise studies for single and double-walled cylindrical shells. *Journal of Aircraft, 17*(9), 690–699.

Dowell, E. H. (1984). On Asymptotic Approximations to beam modal shapes. *Journal of Applied Mechanics, 51*, 439.

Dowell, E. H., Gorman, G. F., & Smith, D. A. (1977). Acoustoelasticity: General theory, acoustic natural modes and forced response to sinusoidal excitation, including comparison with experiment. *Journal of Sound and Vibration, 52*(4), 519–542.

Dowell, E. H., & Kubota, Y. (1985). Asymptotic modal analysis and statistical energy analysis of dynamical systems. *Journal of Applied Mechanics, 52*(4), 949–957.

Dowell, E. H., Nefske, D. J., Sung, S. H., & Culver, D. R. (2017). *Asymptotic model analysis of structural and acoustic systems.* San Rafael, CA: Morgan, R. Claypool Publishers.

Dowell, E. H., & Vaicaitis, R. (1975). A primer for structural response to random pressure fluctuations. *NASA-CR-142671*.

Druesne, F., Hamdaoui, M., & Lardeur, P. (2016). Variability of dynamic responses of frequency dependent visco-elastic sandwich beams with material and physical properties modeled by spatial random fields. *Composite Structures, 152*, 316–323.

Duan, J.-Q. (2015). *An introduction to stachastic dynamics.* Cambridge, U.K.: Cambridge University Press.

Duan, J.-Q., Luo, S.-L., & Wang, C. (Eds.) (2010). *Recent developments in stochastic dynamics aand stochastic analysis*. Singapore: World Scientific.

Dunn, F., Hartmann, W. H., Campbell, D. M., & Fletcher, N.H. (2015). *Springer handbook of acosutics*. Berch: Springer.

Dzhankulaev, A. Ya. (2004). *Forced vibrations of a one-dimensional elastic continuous-discrete systems under harmonic and random excitations*. Ph.D. Dissertation, Nalchik (in Russian).

El Hami, A., & Radi, B. (2016). *Stochastic dynamics of structures*. London: ISTE-Wiley.

Elie, B., Gautier, F., & David, B. (2013). Estimation of mechanical properties of Panels Based on modal density and mean mobility measurements. *Mechanical Systems and Signal Processing, 40*(2), 328–344.

Elishakoff, I. (1970). Analysis of sound isolation properties of some types of multilayered structures. In *Proceedings of the Moscow power engineering institute, dynamics and strength of machines*, (Vol. 74, pp. 48–54). Moscow: Power Engineering Institute Presss (in Russian).

Elishakoff, I. (1987). Generalized Eringen problem: Influence of axial force on random vibration response of simply supported beam. *Structural Safety, 4*(4), 255–265.

Elishakoff, I. (1988). Wide-band random vibration of continuous structures with associated effect of cross-correlations. In W. Schiehlen & W. Wedig (Eds.), *Analysis and estimation of stochastic mechanical systems* (pp. 32–42). Vienna: Springer-Verlag.
Elishakoff, I. (1994). Comment on "Random vibration of externally damped viscoelastic timosheno beams with general boundary conditions" by M. P. Singh and A. S. Abdelnaser (*Journal of Applied Mechanics, 60*, 149–156, 1991); *Journal of Applied Mechanics, 61*, 224–225.
Elishakoff, I. (1995). Random Vibration of structures: A personal perspective. *Applied Mechanics Review, 48*(12), 809–825.
Elishakoff, I., & Ajenjo, A. (2019). Closed-form random vibration response for columns on elastic foundation and subjected to axial force. *Engineering Structures, 194*, 431–440.
Elishakoff, I., Cederbaum, G., & Librescu, L. (1989). Random vibration of moderately thick composite cylindrical shells excited by a random loading. *American Institute of Aeronautical and Astronautics Journal, 27*, 975–981.
Elishakoff, I., & Colombi, P. (1993). Combination of probabilistic and convex models of uncertainty when scarce knowledge is present on acoustic excitation parameters. *Computer Methods in Applied Mechanics and Engineering, 104*(2), 187–209.
Elishakoff, I., & Ducreux, B. (2014). Dramatic effect of cross correlations in random vibration of point-driven spherically curved panel. *Archive of Applied Mechanics, 84*, 473–490.
Elishakoff, I., & Hettema, Ch. (1987). Nonstationary random vibrations With reduce. In P. D. Spanos (Ed.), *Probabilistic methods in civil engineering* (pp. 529–532). New York.
Elishakoff, I., Librescu, L., & Cederbaum, G. (1990). Stationary and nonstationary random vibration of laminated plates via higher order theory. In *Proceedings of the AIAA/ASME/ASCE/ASH/ASC 31st Structures, Structural Dynamics and Materials Conference* (Vol. 2, pp. 1857–1864). Long Beach, CA.
Elishakoff, I., Lin, Y. K., & Zhu, L. P. (1995). Random vibration of uniform beams with varying boundary conditions by the dynamic-edge-effect method. *Computer Methods in Applied Mechanics and Engineering, 121*(1), 59–76.
Elishakoff, I., & Livshits, D. (1984). Some closed-form solutions in random vibrations of Bernoulli-Euler beams. *International Journal of Engineering Sciences, 22*(11/12), 1291–1302.
Elishakoff, I., & Livshits, D. (1989). Some closed-form solutions in random vibrations of Timoshenko beams. *Probabilistic Engineering Mechanics, 4*, 47–54.
Elishakoff, I., & Livshits, D. (2011). Comparison of asymptotic method With explicit solutions in random vibration. *Memoirs on Differential Equations and Mathematical Physics, 52*, 109–121.
Elishakoff, I., & Santoro, R. (2014). Random vibration of a point-driven two-span beam on an elastic foundation. *Archive of Applied Mechanics, 84*, 355–374.
Elishakoff, I., & Soize, C. (2012). *Nondeterministic mechanics*. Vienna: Springer.
Elishakoff, I., & Spencer, B. F., Jr. (1987). Reliability of an uncertain sliding structure . *Journal of Sound and Vibration, 114*, 399–404.
Elishakoff, I., & Zhu, L. P. (1993). Random vibration of structures by the finite element method. *Computer Methods in Applied Mechanics and Engineering, 105*(3), 359–373.
Ellingwood, B. (1994). Probability-based codified design for earthquakes. *Engineering Structures, 16*(7), 498–506.
Elliott, G. H. (1988). The Evaluation o the modal density of paraboloidal and similar shells. *Journal of Sound and Vibration, 126*(3), 477–483.
Elliott, S. J., Nelson, P. A., Smothers, I. M., & Boucher, C. C. (1971). In-flight experiments on the active control of propeller-induced cabin noise. *Journal of Sound and Vibration, 140*(2), 219–238.
Ellyin, F., & Chandrasekhar, P. (1977). Probabilistic dynamic response of beams and frames. *Journal of Engineering Mechanics, 103*, 411–421.
El-Raheb, M. (2011). Modal properties of a cyclic symmetric hexagon lattice. *Computers and Structures, 89*(23–24), 2249–2260.
Erickson, L. L. (1969). Modal densities of sandwich panels: Theory and experiment. *The Shock and Vibration Bulletin, 39*, Part 3, 1–16.

Erickson, L. L. (1970). Modal density estimates for sandwich panels: Theory and experiment. NASA Technical Note, *NASA TN D-5771*.

Eringen, A. C. (1957). Response of beams and plates to random loads. *Journal of Applied Mechanics, 24*, 46–52.

Eringen, A. C. (1958). Response of Tall buildings to random earthquakes. In R. M. Haythornthwaite, L. H. Donnell, & M. M. Frocht (Eds.), *Proceedings 3rd U. S. National Congress of Applied Mechanics,* Brown University, Rhode Island (pp. 141–151).

Eringen, A. C. (1962). Stochastic loads. In W. Flügge (Ed.), *Handbook of engineering mechanics* (pp. 66.13–66.16). New York: McGraw-Hill.

Eringen, A. C., & Dunkin, J. W. (1960). The elastic half plane subjected to surface tractions with random magnitude or separtation. *Journal of Applied Mechanics, 27*, 701–709.

Ermolenko, E. M. (1980). Effect of orthotropy parameters of orthotropy on the spectrum in problems of vibrations of shell. *Journal of Applied Mechanics and Technical Physics, 21*(1), 153–160.

Ermolenko, V. M., & Kornev, V. M. (1979). Spectrum singularities in cylindrical orthotropic shell stability problems. *Soviet Applied Mechanics, 15*(8), 758–761.

Esadze, S.Yu. (2000). Analysis of structures to random seismic excitations with subsequent statistical analysis of results, *Proceedings of the Georgia's State Technical University*, (1) (429), 63–68 (in Georgian).

Esadze, S. Yu., & Berdzenishvili, B. (1998). Abiout a model obtained by taking into account joint random vibrations of earthquake foundation's horizontal and vertical vibrations. *Proceedings of the Georgia's State Technical University*, (1) (417), 45–53 (in Georgian).

Eslimy-Isfahany, S.H. R., & Banerjee, J. R. (1996). Response of composite beams to deterministic and random loads. *AIAA Paper 96–1411-CP*.

Eslimy-Isfahany, S.H. R., Banerjee, J. R., & Sobey, A. J. (1996). Response of a bending-torsion coupled beam to deterministic and random loads. *Journal of Sound and Vibration, 195*(2), 267–283.

Esmailzadeh, M., Lakis, A. A., Thomas, M., & Marcouiller, L. (2009). Prediction of the response of a thin structure subjected to a turbulent boundary layer induced random pressure field. *Journal of Sound and Vibration, 328*, 109–128.

Etkin, B. (1966). Theory of the response of a slender vertical strucutre to a turbulent wind with shear. In *Proceedings, NASA Conference on Wind Load Problems of Launch Vehicles*, Langley Field, VA (pp. 21.1–21.15).

Etkin, B., & Hansen, J. S. (1984). Effect of a Damper on the wind-induced oscillations of a tall mast. *Journal of Wind Engineering and Industrial Aerodynamics, 17*, 11–29.

Etaix, N., Dubois, J., Fink, M., & Ing, R. K. (2013). Increasing the modal density in plates for mono-element focusing in air. *The Journal of the Acoustical Society of America, 134*(2), 1049–1054.

Evdokimenkov, V., Dineev, V., & Karp, K. (2017). *Engineering methods of probablistic analysis of aviation and cosmic system*. Moscow: "Fizmatlit" Publishing House (in Russian).

Ewins, D. J. (2016). Exciting vibrations: The role of testing in the era of supercomputers and uncertainties. *Meccanica, 51*, 3241–3258.

Ezeh, J. C., Ibarugbulon, M. O., Opara, H. E., & Oguaghamba, O. A. (2015). Characteristic orthogonal polynomial application to Galerkin's indirect variational method for simply supported plate under implane loading. *International Journal of Research in Engineering and Technology, 34*(2), 232–239.

Ezekiel, M. (1930). *Methods of correlation analysis*. New York: Wiley.

Fabian, L. (1973). *Zufallsschwingungen und ihre Behandlung*, Berlin: Springer (in German).

Fahy, F. J. (1970). Response of a cylinder to random sound in contained fluid. *Journal of Sound and Vibration, 13*, 171–194.

Fahy, F. J. (1970). *Acoustically induced vibration of containing structures*. Ph.D. Thesis, University of Southampton, England.

Fahy, F. J. (1971). Statistics of acoustically induced vibrations. In *Proceedings of the 7th International Congress on Acostics,* Paper 19V5, Budapest, Hungary.

Fahy, F. J. (1974). Statistical energy analysis—A critical review. *Shock and Vibration Digest, 6*, 14–33.

Fahy, F. J. (1985). *Sound and structural vibration: Radiation, transmission and response*. San Diego: Academic Press.

Fahy, F. Y. (1993). Statistical energy analysis: A wolf in sheep's clothing?. In *Proceedings, Inter-Noise*, Belgium (Vol. 93, pp. 13–26).

Fahy, F. J. (1994). Statistical energy analysis: A critical overview. *Philosophical Transactions of the Royal Society of London A: Mathematical, Physical and Engineering Sciences, 346*(1681), 431–447.

Fahy, F. J., & Gardonio, P. (2007). *Sound and structural vibration: Random transmission and response*. San Diego, CA: Academic Press.

Fahy, F. Y., & Mohammed, A. D. (1992). A study of uncertainty in applications of statistical energy analysis to coupled beam and plate systems. Part I: Computational experiments. *Journal of sound and Vibration, 158*(1), 45–67.

Fahy, F. J., & Price, W. G. (Eds.) (1999). *IUTAM symposium on statistical energy analysis*. Dordrecht: Springer.

Fahy, F. J., & Walker, J. G. (1998). *Fundamentals of noise and vibration*. London: Span Press.

Falsone, G. (1994). Cumulants and correlations for linear systems under non-stationary delta-correlated processes. *Probabilistic Engineering Mechanics, 9*(3), 157–166.

Failla, G. (2017). Stationary response of beams and frames with fractional dampers via exact frequency response functions. *Journal of Engingeering Mechanics,143*(5), article D4016004.

Failla, G., Di Paola, M., Pirrotta, A., Burlon, A., & Dunn, I. (2019). Random vibration mitigation of beams via tuned mass dampers with spring inertia effects. *Meccanica, 54*(9),1365–1383.

Falsone, G. (1992). Stochastic response of combined primary-secondary structures under seismic input. *Earthequake Engineering and Structural Dynamics, 21*, 927–943.

Falsone, G. (2000). A new light on the response of linear systems subjected to random non-stationary filtered inputs. *Journal of Engineering Mechanics, 126*, 1273–1286.

Falsone, G., & Neri, F. (2000). Stochastic modelling of earthquake excitation following the EC8: Power spectrum and filtering equations. *European Earthquake Engineering, 1*, 3–12.

Falsone, G., & Neri, F. (2003). The non-stationary cross-correlation coefficients in the seismic analysis of MDOF structural systems. *Earthquake Engineering and Structural Dynamics, 32*, 2289–2299.

Falsone, G., & Settineri, D. (2011). New differential equations governing the response cross-correlations of linear systems subjected to coloured loads. *Journal of Sound and Vibration, 330*(12), 2910–2927.

Fan, F. G., & Ahmadi, G. (1990). Nonstationary Kanai-Tajimi models for El Centro 1940 and Mexico City 1985 earthquakes. *Probabilistic Engineering Mechanics, 5*, 171–181.

Fang, T. (1995). *Engineering random vibration*. Beijing: Industrial Press (in Chinese).

Fang, T. (1988). Calculating response spectral moments via complex modal analysis. In *Proceedings, 6th International Modal Analysis Conference*, Kissimmee, FL (pp.1321–1325).

Fang, T., Leng, X. L., & Song, C. Q. (2003). Chebyshev polynomial approximation for DYNAMICAL response problem of random system. *Journal of Sound and Vibration, 266*(1), 198–206.

Fang, T., & Sun, M. (1997). A unified approach to two types of evolutionary random response problem in engineering. *Archive in Applied Mechanics, 67*(7), 496–506.

Fang, T., Li, J. Q., & Sun, M. N. (2002). A universal solution for evolutionary random response problems. *Journal of Sound and Vibration, 253*(4), 909–916.

Fang, T., & Wang, Z. N. (1986a). Complex modal analysis of random vibration. *AIAA Journal, 24*(2), 342–344.

Fang, T., & Wang, Z. N. (1986b). Mean square response to band-limited white noise excitation. *AIAA Journal, 24*(5), 860–862.

Fang, T., & Wang, Z. N. (1986c). Stationary response to second order filtered white noise excitation. *AIAA Journal, 24*(6), 1048–1051.

Fang, T., & Wang, Z. N. (1986d). Complex modal analysis of random vibrations. *AIAA Journal, 24*(2), 342–344.
Fang, T., & Wang, Z. N. (1986e). Time domain modal analysis of random vibrations. In *Proceedeings of the 4th International Conference on Modal Analysis* (pp. 270–275).
Fang, T., & Wang, Z. N. (1987a). Stationary response due to exponential cosine random excitations. In *5th International Modal Analysis Conference*. London (pp. 574–579).
Fang, T., & Wang, Z. N. (1987b). Stationary response due to arbitrarily filtered white noise excitation. *Journal of Northwestern Polytechnic University*, English extra issue, 41–48.
Fang, T., & Zhang, T. S. (1989). Calculating response spectral moments by complex modal analysis. *Acta Mechanica Sinica, 5*(2), 159–166.
Fang, T., & Zhang, T.S. (1993). Complex modal analysis of structural random vibrations. In E. J. Cui (Ed.), *Aero-Hydroelasticity: Developments and Applications* (pp. 491–497). Beijing: Seismological Press.
Fang, T., & Zhang, T. S. (1995). Time domain complex modal analysis of random vibration. *The Progress in Natural Science, 5*(1), 113–118.
Farabee, T. M., & Casarella, M. J. (1991). Spectral features of wall pressue fluctuations beneath turbulent boundary layers. *Physics of Fluids, 3*, 2410–2420.
Faraday, M. (1831a). Acoustic streaming over vibrating plates. *Philosphical transactions of the royal scoiety, 121*, 299–318.
Faraday, M. (1831b). On a peculiar class of acoustical figures, and on certain forms assumed by groups of particles upon vibrating elastic surfaces. *Philosophical Transactions of the Royal Society of London, 121*, 299–340.
Fareed, A. M., Schmidt, G., & Wahl, F. (2001). Experimental identification of modal density parameters of light weight structures. *Technische Mechanik, 2*(3), 215–225.
Farooq, U., & Feeny, B. F. (2008). Smooth orthongonal decomposition for modal analysis of randomly excited systems. *Journal of Sound and Vibration, 316*(1), 137–146.
Farshidianfar, M. H., & Farshidianfar, A. (2013). Mode count and modal density of acoustical casities bars, plates and cylindrical shells. *ICSV20*, 172–175.
Farshidianfar, M. H., Farshidianfar, A., & Moghadam, M. M. (2014). Mode count and modal density of Isotropic circular cylindrical shells using a modified wavenumber space integration method. *Journal of Vibration and Acoustics, 136*(4), article 041004.
Ferguson, N. S., & Clarkson, B. L. (1986). The modal density of honeycomb shells. *Journal of Vibration, Acoustics, Stress, and Reliability in Design, 108*(4), 399–404.
Ferson, S., & Hajagos, J. G. (2006). Varying correlation coefficients can underestimate uncertainty in probablistic models. *Reliability Engineering and System Safety, 91*, 1461–1467.
Field, R. V., & Grigoriu, M. D. (2017). *Random vibration of complex systems: A unified approach.* Academic Press.
Fisher, R. A. (1921). On the probable error of a coefficient of correlation deduced from a small sample. *Metron, 1*, 3–32.
Fischer, P., Belyaev, A. K., & Pradlwarter, H. J. (1995). Combined integral and FE analysis of broad-band random vibration in structural members. *Probabilistic Engineering Mechanics, 10*, 241–250.
Field, C. F. (1969). Cross-correlation and milling machine dynamics—A case study. *International Journal of Mechanical Tool Design, 9*, 81–96.
Field, J. R., Richard, V., Red-horse, J. R., & Paez, T. L. (2000). *A nondeterministic shock and vibration application using polynomialchaos expansion*. Report NO. SAND 2000–0800C, Sandia National Labouratory, Albuquerque NM.
Finnveden, S. (2004). Evaluation of modal density and group velocity by a FEM. *Journal of Sound and Vibration, 273*(1), 51–75.
Finnveden, S., Birgersson, F., Ross, U., & Kremer, T. (2006). A model of wall pressure correlation for prediction of turbulence-induced vibration. *Journal of Fluids and Strucutres, 20*(8), 1127–1143.
Flax, A. H. (1977). Comment on "Modal coupling in lightly damped structures" by T. K. Hasselman. *AIAA Journal, 15*(11), 1662.

Fletcher, N. H., & Rossing, T. D. (1998). Two-dimensional systems: Membranes plates and shells. *The Physics of Musical Instruments*, 70–101.

Florence, S.J. K., Renji, K., & Subramanian, K. (2018). Modal density of honeycomb sandwich composite cylindrical shells considering transverse deformation. *International Jounral of Acoustical Vibration, 23*(1).

Foster, E. T. (1968). Semilinear random vibrations in discrete systems. *Journal of Applied Mechanics, 35*(3), 560–564.

Fragkoulis, V. C. (2017). *Random vibration of systems with singular matrices*. Ph.D. Thesis, University of Liverpool.

Fragkoulis, V. C., Kougioumtzoglou, I. A., & Pantelous, A. A. (2016). Linear random vibration of structural systems systems with singular matrices. *Journal of Engineering Mechanics, 142*(2), article 04015081.

Franca, L. P., Do Carmo, E.G.D. (1989). The Galerkin gradient least-squares method. *Computer Methods in Applied Mechanics and Engineering, 74* (1), 41–54.

Franco, F., De Rosa, S., & Ciappi E. (2013). Numerical approximations on the predictive responses of plates under stochastic and convective loads. *Journal of Fluids and Structures, 42*, 296–312.

Frendi, A. (1997). Coupling between a supersonic turbulent boundary layer and a flexible structure. *AIAA Journal, 35*(1), 58–66.

Friedrich, H., & Hennig, K. (1975). Bestimmung des Verhaltens von linearen systemes mit *n* Freihetsgraden bei Einwirkung einer stochastischen Fremderreggung. *Maschinenbau technik, 21*, 163–165 (in German).

Friswell, M. I. (1996). The derivatives of repeated eigenvalues and their associated eigenvectors. *Journal of Vibration and Acoustics, 118*, 390–397.

Frýba, L. (1968). Non-stationary vibrations of bridges under random moving load. In *Eighth Congress of the International Association for Bridge and structural Engineering (IABSE)*, Ne York (pp. 1223–1236).

Frýba, L. (1972). *Vibrations of solids and structures under moving loads*. Groningen: Noordhoff International Publishing.

Frýba, L. (1976). Non-stationary response of a beam to a moving random force, *Journal of Sound and Vibration, 46*, 323–338.

Frýba, L. (1986). Random vibration of a beam on elastic foundation under moving force. In I. Elishakoff & R. H. Lyon (Eds.), *Random vibration—Status and recent developments* (pp. 127–148). Amsterdam: Elsevier.

de la Fuente E. (2009). Von Mises stresses in random vibration of linear structures. *Computers and Structures, 87*(21–22), 1253–1262.

Fujimori, Y. (1970a). Response analysis of the plate to boundary layer pressure fluctuations. In *Proceedings of the Nineteenth Japan National Congress for Applied Mechanics* (pp. 121–127).

Fujimori, Y. (1970b). Response analysis of a simply-supported plate to boundary layer pressure fluctuations, NAL TR-207.

Fujimori, Y. (1970c). Response analysis of a clamped plate to boundary layer pressure fluctuations, NAL TR-401.

Fujimori, Y. (1973). Spectrum analysis of the response of multimodal kinear system to nonstationary input. In *Proceedings of the 23rd Japan National Congress for Applied Mechanics* (pp.467–475). University of Tokyo Press.

Fujimori, Y. (1974). Shear and moment response of the airplane wing to nonstationary turbulence. *AIAA Journal, 12*(11), 1459–1460.

Fujimori, Y. (1988). Stochastic aspects of structural dynamics in aerospece vehicles, in N. Sri Namachchivaya, H. Hilton & Y. K. Wen (Eds.), *Stochastic structural dynamics* (pp. 235–254). University of Illinois at Urbana-Champaign.

Fujimori, Y., & Lin, Y. K. (1973a). Analysis of airplane response to nonstationary turbulence including wing bending flexibility. *AIAA Journal, 11*(3), 334–339.

Fujimori, Y., & Lin, Y. K. (1973b). Analysis of airplane response to nonstationary turbulence including wing bending flexibility II. *AIAA Journal, 11*(9), 1343–1345.

Fujino, Y., & Pheinsunsom, P. (1987). Modal interaction in dynamical system with closely-spaced natural frequencies under self-excited and random forces. In Y. K. Lin & R. Minai (Eds.), *Stochastic approaches in earthquake engineering* (pp. 52–67). Berlin: Springer.

Fujiwara, N., & Mirotsu, Y. (1973). Optimum configurations of vibration reducers for beam systems with random excitation. *Bulletin of University of Osaka*, Perfect Se.A, *22*(1), 23–33.

Fung, Y. C. (1953). Statistical aspects of dynamic loads. *Journal of the Aeronautical Sciences, 20*(5), 317–322.

Fung, Y. C. (1955). The analysis of dynamic stresses in aircraft structures during landing as nonstationary random process. *Journal of Applied Mechanics*, *22*, 449–457.

Gader, R.-L. (1997). *Stability and random vibrations of compressed-twisted straight columns, interacting with liquid flow*. Ph.D. Dissertation. Moscow (in Russian).

Ganesh Raja, M., & Navayaman S. (2007). Active control of tensegrity structures under random excitation. *Smart Materials and Structures, 16*(3), 809–815.

Gao, J., Du F., & Ding, S. (2013). Global sensitivity analysis for dynamic systems with stochastic input process. *Reliability Engineering and System Safety, 118*, 106–107.

Gao, W., Chen, Y. Y., Ma J., & Liang, Z. T. (2004). Dynamic response analysis of stochastic frame structure under nonstationary random excitation. *AIAA Journal, 42*(9), 1818–1822.

Gao, W., Chen, J., Cui, M., & Cheng, Y. (2005). Dynamic response analysis of linear stochastic truss structures under stationary random excitation. *Journal of Sound and Vibration*, *281*(1–2), 311–321.

Gaonkar, G. H. (1971). Linear Systems with non-stationary random inputs. *International Journal of Control, 14* (1), 161–174.

Gaonkar, G. H., & Hohenemser, K. H. (1969). Flapping response of lifting rotor blades to atmospheric turbulence. *Journal of Aircraft, 6*(6), 496–503.

Gaonkar, G. H., & Hohenemser, K. H. (1971). Stochatic properties of turbulence excited rotor blade vibrations. *AIAA Journal, 9*(3), 419–424.

Gaonkar, G. H., Hohenemser, K. H., & Lin, Y. K. (1972). Random gust response statistics for coupled torsion-flapping rotor blade vibrations. *Journal of Aircraft*, *9*, 726–729.

Gardner, C. (1985). *Handbook of stochastic methods for physics, chemistry and the natural sciences*. Berlin: Springer.

Gardner, C. (2009). *Stochastic methods: A handbook for the natural and social sciences*. Berlin: Springer.

Garrison, M. R., Miles, R. N., Sun, J. Q., & Bao, W. (1994). Random response of a plate partially covered by a constrained layer damper. *Journal of Sound and Vibration, 172*(2), 231–245.

Garcia, D. A., & Rosales, M. B. (2016). Deflection in sawn timber beams with stochastic properties. *European Journal of Wood and Wood Production*, 1–17.

Gasparini, D. A. (1979). Response of MDOF systems to nonstationary random excitation. *Journal of Engineering Mechanics*, *105*(1), 13–27.

Gasparini, D. A., & DebChaudbury, A. (1980). Dynamic response to nonstationary nonwhite excitation. *Journal of Engineering Mechanics, 106*(6), 1233–1248.

Gasparini, D. A., DebChaudhury, A., & Gazetas, G. (1981). Random vibration of cantilever shear beams. *Earthquake Engineering and Structural Dynamics, 9*, 599–612.

Gawronski, W., & Sawicki, J. T. (1997). Response errors of non-proportionally lightly damped structures. *Journal of Sound and Vibration*, *200*(4), 543–550.

Gaylard, M. E. (1982). Distribution of the natural frequencis in matrix-characterized structures.

Gbur, G. Physics demonstrtaions: Chladni patterns. Retrieved July 29, 2019, from https://tenyurl.com/Chladni-demonstrations.

Gentry, C. A., Guigou, C., & Fuller, C. R. (1997). Smart foam for applications in passive/active noise radiation control. *The Journal of the Acoustical Society of America*, *101*(4), 1771–1778.

Gershgorin, A. D. (1984). Random vibration of aircraft engine blades. *Strength of Materials, 16*(1), 77–80.

Ghanem, R., & Ghosh, D. (2007). Efficient characterization of the random eigenvalue problem in a polynomial chaos decomposition. *International Journal for Numerical Methods in Engineering, 72*(4), 486–504.

Ghienne, M., & Blanzé, C. (2016). A simplified method for random vibration analysis of structures with random parameters. *Journal of Physics: Conference Series, 744*(1), article 012174.

Ghosh, D. (2013). Application of random eighvalue problem in forced response analysis of a linear stochastic structure. *Archive of Applied Mechanics, 83*, 1341–1357.

Ghosh, D., Ghanem, R., & Red-Horse, J. (2005). Analysis of equivalence and modal interaction of stochastic system. *AIAA Journal, 43*(10), 196–2201.

Gilbert, G. (1973). Cabin noise levels. *Business Commercial Aviation, 39*, 80–86.

Gil-Santos, E., Ramos, D., Jana, A., Calleja, M., Raman, A., & Tamayo, J. (2009). Mass sensing based on deterministic and stochastic responses of elastically coupled nanocantilevers. *Nano Letters, 9*(12), 4122–4127.

Gikhman, I. I., & Skorokhod, A. V. (1972). *Stochastic differential equations*. New York: Springer.

Gikhman, I. I., & Skorokhod, A. V. (1977). Controllable random process,. Kiev: "Naukova" Dumka Publishers (in Russian).

Gladky, V. F. (1975a). *Dynamics of flying vehicle*. Moscow: "Nauka" Publishing House (in Russian).

Gladky, V. F. (1975b). *Strength, vibration and reliability of flying vehicle structures*. Moscow: "Nauka" Publishing House (in Russian).

Gladwell, G.M. L., & Bishop, R.E. D. (1991). Dynamicist and communicator. *Journal of Sound and Vibration, 151*(2), 181–191.

Gladysz, M., & Sniady, P. (1984). Dynamic response of a discrete system to a poissonian stream of forces. *Archives of Civil Engineering, 30*, 37–51 (in Polish).

Gladysz, M., & Sniady, P. (2009). Spectral density of the bridge beam response with uncertain parameters under a random train of moving forces. *Archive of Cave and Mechanical Engineering, 9*, 31–47.

Gloyna, F. L., Jacobs, L. D., & Lagerquiest, D. R. (1970). Response of complex structures of turbulent boundary layers. *Journal of Aircraft, 7*, 210–219.

Gnedenko, B. B., & Rozanov Yu. A. (1989). *Random process, mathematical statistics and their applications*. Moscow: Moscow State University Press (in Russian).

Gnutzmann, S., & Lois, S. (2013). On the modal count statistics for separation systems in any dimensions. *Journal of Physics A: Mathematical and Theoretical, 46* (4).

Goda, Y. (2010). *Random seas and design of maritaime structurers* (2nd Edition). Singapore: World Scientific.

Goldenveizer, A. L. (1970). On the frequency distribution density of oscillations of a thin shell. *PMM: Journal of Applied Mathematics and Mechanics, 34*(5), 910–915.

Goldenveizer, A. L., Lidsky, V. B., & Tovstik, P. E. (1979). *Free oscillations of thin elastic shells*. Moscow: "Nauka" Publishing House (in Russian).

Goller, G., Pradlwarter, H. J., & Schuëller, G. I. (2013). Reliability assessment in structural dynamics. *Journal of Sound and Vibration, 332*, 2488–2499.

Gordon, V. A. (2003). Random longitudinal vibrations of in homogenous bars. In *Fundamental and applied technological problems of machine building* (pp. 238–241). Oreal: Oreal State Technical University Press (in Russian).

Goβmann, E., & Waller, H. (1983). Analysis of multicorrelated wind-excited vibrations of structures using the covariance method. *Engineering Structurers, 5*, 264–272.

Goto, H., & Kameda, H. (1969). Statistical influence of the future earthquake ground motion. In *Proceedings, 4th World Conference of Earthquake Engineering*, Chile, (Vol. 1, pp. 39–54).

Gouedard, P., Stehly, L., Brenguier, F. (2008). Cross-correlaion of random fields: Mathematical approach and application. *Geophysical Prospecting, 56*(3), 375–393.

Gouedard, P., Stehly, L., Brenguier, F., Campillo, M., Colin de Verdiere, Y., Larose, E., et al. (2008). Cross-correlation of random Fields: mathematical approach and applications. *Geophysical Prospecting, 56*(3), 375–393.

Graham, L. L., & Deodatis, G. (2001). Response and eigenvalue analysis of stochastic finite element systems with multiple correlated material and geometric properties. *Probabilistic Engineering Mechanics, 16*(1), 11–29.
Graham, W. R. (1996). Boundary layer induced noise in aircraft. Part 1: The flat plate model. *Journal of Sound and Vibration, 192*(1), 101–120.
Graham, W. R. (1997). A comparison of models for the wavenumber–frequency spectrum of turbulent boundaryllayer pressures. *Journal of Sound and Vibration, 206*(4), 541–565.
Greenslade, T. (2010). On the Web: http:// physics.kenyon.edu/EerlyApparatus/Acoustics/Chladni.html. Kenyon College, Gambier, Ohio.
Greenspon, J. E. (1960). Axially symmetric vibration of a thick cylindrical shell in an acoustic medium. *The Journal of Acoustical Society of America, 32*, 1017–1025.
Greenspon, J. E. (1967). Far-field radiation from randomly vibrating structures. *The Journal of the Acoustical Society of America, 41*(5), 1201–1207.
Greenspon, J. E. (1970). *Random structural acoustics* (Vol. 1). JG Engineering Research Associates.
Greenspon, J. E. (2011). *Structural acoustics: Deterministic and random phenomena.* Boca Raton: CRC Press.
Grewal, A., Zimcik, D. G., & Lapointe, R. (1997). Vibro-acoustic modelling in aircraft cabin noise transmission and control. In AIAA 97–1101, *38th AIAA/ASME/ASCE/AHS/ASC Structures, Structural Dynamics, and Materials Conference and Exhibit, AIAA/ASME/AHS Adaptive Structures Forum,* Kissimmee, FL.
Grigoriu, M. (1981). Mean-squire structural response to stationary ground acceleration. *Journal of Engineering Mechanics, 107*, 969–986.
Grigoriu, M. (1986a). Response of linear systems to quadratic Gaussian excitation. *Journal of Engineering Mechanics, 112*(6), 523–535.
Grigoriu, M. (1986b). Structural response to uncertain seismic excitations. *Journal of Structural Engineering, 112*, 1355–1365.
Grigoriu, M. (1987). White noise processes. *Journal of Engineering Mechanics, 113*(5), 757–765.
Grigoriu, M. (1992). Transient response of linear systems to stationary Gaussian inputs. *Probabilistic Engineering Mechanics, 7*, 159–164.
Grigoriu, M. (1995a). *Applied non-Gaussian processes: Examples, theory, simulation, linear random vibration, and MATLAB solutions.* Englewood Cliffs, N.J.: Prentice-Hall.
Grigoriu, M. (1995b). Linear system subject to non-Gaussian α-stable processes. *Probabilistic Engineering Mechanics, 10*, 23–34.
Grigoriu, M. (2000). Stochastic mechanics. *International Journal of Solids and Structures, 37*, 197–214.
Grigoriu, M. (2002). *Stochastic calculus: Applications in science and engineering.* Basel: Birkhäuser.
Grigoriu, M. (2010). Linear random vibration by stochastic reduced-order models. *International Journal for Numerical Methods in Engineering, 82*(12), 1537–1559.
Grigoriu, M. (2012). *Stochastic systems: Uncertainty quantification and propagation.* Berlin: Springer.
Grigoriu, M., & Alibe, B. (1986). Response of offshore structures to random waves. *Journal of Engineering Mechanics, 112*(8), 729–744.
Grigoriu, M., & Ariaratnam, S. T. (1988). Response of linear systems to polynomials of Gaussian process. *Journal of Applied Mechanics, 55*, 905–910.
Grigoriu, M., & Bunu, A. (1986). Response of offshore structures to random waves. *Journal of Engineering Mechanics, 112*(8), 729–744.
Grigoriu, M., Ruiz, S. E., & Rosenblueth, E. (1988). The Mexico earthquake of September 19, 1985—Non-stationary model of seismic ground acceleration. *Earthquake Spectra, 4*, 551–568.
Grinchenko, N. I., & Pavlyuk Yu. S. (1969). Dynamical analysis of a system with elastic liquid links under random excitation. *Mechanics of Solids,* (6), 139–145 (in Russian).
Grinchenko, N. I., & Shefer, L. A. (1976). Spectral method estimating the fatigue life under random excitation. *Problems of Strength,* (1), 18–22 (in Russian).

Gruner, J. (1997). *Diskrete Schwingungssysteme mit stochastischer Fremderregung*. Doctoral Thesis, Technical University Chemnitz (in German).

Gudkov, A. I., & Leshakov, P. S. (1968). *External loads and strength of flying vehicles* (2nd Edition). Moscow: "Mashinostroenie" Publishing House (in Russian).

Guillemin, V. (1985). A new proof of Weyl's formula on the asymptotic distribution of eigenvalues. *Advances in Mathematics, 55*(2), 131–160.

Gulgazaryan, G. R., & Lidskii, V. L. (1982). Density of free vibration frequencies for a thin anisotropic shell made of anisotropic layers. *Mechanics of Solids, 17*(3), 159–162.

Gunda, R., Vijayakumar, S. M., & Singh, R. (1995). Method of images for the harmonic response of beams and rectangular plates. *Journal of Sound and Vibration, 185*(5), 791–808.

Gupta, A. K., & Singh, M.P. (1977). Design of column sections subjected to three components of earthquake. *Nuclear Engineering and Design, 41*, 129–133.

Gupta, A. K. (1990). *Response spectrum method in seismic analysis and design of structures*. Cambridge, MA: Blackwell.

Gupta, A., & Manohar, C. S. (2002). Dynamic stiffness method for circular stochastic Timoshenko beams: Reponse variability and reliability analyzes. *Jounal of Sound and Vibrations, 253*(5), 1051–1085.

Gupta, A. K., Hassan, T., & Gupta, A. (1996). Correlation coefficients for modal response combination of non-classically damped systems. *Nuclear Engineering and Design, 165*, 67–80.

Gupta, A. K., Mori, T. F., & Szekely, G. J. (2000). How to transform correlated random variables into uncorrelated ones. *Applied Mathematical Letters, 13*(6), 31–33.

Gupta, S., & Manohar, C. S. (2005). Multivariate extreme value distributions for random vibration applications. *Journal of Engineering Mechanics, 131*(7), 712–720.

Gurley, K. R., & Kareem A. (1999). Applications of wavelet transforms in earthquake, wind and ocean engineering. *Engineering Structures, 21*(2), 149–167.

Gurley, K. R., Kareem, A., & Tognarelli M. A. (1996). Simulation of a class of non-normal random processes. *International Journal of Non-Linear Mechanics, 31*(5), 601–617.

Gusev, A. A. (1972). Towards analysis of transient regimes of random vibrations. *Herald of Institutions of Higher Learning, Mechanical Engineering,* (1), 10–14 (in Russian).

Gusev, A. S. (1989). *Fatigue strength and life of strucures*. Moscow: "Mashinostroenie", Publishing House (in Russian).

Gusev, A. S. (2009). *Probabilistic methods for machinery and structural mechanics*. Moscow: "MGTU" Press (in Russian).

Gusev, A. S., Grigoriev Yu. B., Starodubzeva, S. A., & Fomicheva, V. F. (2016). Random vibrations of beam partially submerged into the liquid. *Vestnik MGTU Imeni, N. E. Baumana, Transactions of the Moscow State Technical University,* (2), 81–91 (in Russian).

Gusev, A. S., & SvetlitskyV. A. (1984). *Analysis of structures under random excitations*. Moscow: "Mashinostroenie" Publishing House (in Russian).

Guyader, J. L. (2009). Characterization and reduction of dynamic models of vibrating systems with high modal density. *Journal of Sound and Vibration, 328*(4), 488–506.

Haas, C. N. (1999). On modeling correlated random variables in risk assessment. *Risk Analysis, 19*(6), 1205–1214.

Hagedorn, P., & Nascimento, N. (1985). Structural vibration excited by random field process: The effect of the crosscorrelation of the excitation. In *Proceedings, 4th International Conference on Strucutral Safety and Reliability* (pp. II-547—II-551).

Hache, F., Elishakoff, I., & Challamel, N. (2018). Critical comparison of exact solutions in Random vibration of beams using three versions of Bresse–Timoshenko theory. *Probabilistic Engineering Mechanics, 53*, 95–108.

Haines, C. W. (1967). Hiearchy methods for random vibrations of elastic strings and beams. *Journal of Engineering Mathematics, 1*(4), 293–305.

Halehouse, I. (1971). Somic fatigue of aircraft structures due to jet engine fan noise. *Journal of Sound and Vibration, 17*(3), 287–298.

Hambric, S. A., Hwang, Y. F., & Bonness, W. K. (2004). Vibration of plates with clamped and free edges excited by low-speed turbulent boundary layer flow. *Journal of Fluids and Structures, 19*(1), 93–110.

Hambric, S. A., Hwang, Y. F., & Bonness, W.K. (2002). Vibrations of plates with clamped and free edges excited by highly subsonic turbulent boundary layer flow. In *ASME 2002 International Mechanical Engineering Congress and Exposition* (pp. 1179–1189). ASME Digital Collection, New York.

Hambric, S. A., Shaw, M., Campbell, R. L., & Conlon, S. C. (2015). Calculating structural vibration and stress from turbulent flow induced forces. In *Flinovia-flow induced noise and vibration issues and aspects* (pp. 343–356). Cham: Springer.

Hambric, S. A., Sung, S. H., & Nefske, D.J. (2016). *Engineering vibroacoustic analysis: Methods and applications*. Chichester,U.K.: Wiley.

Hammond, J. K. (1968). On the response of simple and multidegree of freedom systems to non-stationary random excitation. *Journal of Sound and Vibration, 7*(3), 393–416.

Hammond, J.K.(1973a) Evolutionary spectra in random vibrations. *Journal of the Royal Statistical Society, 35*(2), 167–179.

Hammond, J. K. (1973b). Evolutionary spectra in random vibrations. *Journal of the Royal Statistical Society, 1335*, 167–188.

Hammond, J. K., & Harrison, R. F. (1981). Nonstationary response of vehicles on rough ground—A state space approach. *Journal of Dynamic System, Measurement and Control, 103*(3), 245–250.

Handa, K. N. (1970). *The response of tall structures to atmospheric turbulence*, Ph.D. Thesis, University of Southampton, Southampton, U.K.

Harichandran, R. S. (1987). Correlation analysis in space-time modeling of strong ground motion. *Journal of Engineering Mechanics, 113*(4), 629–634.

Harichandran, R. S. (1993). An Efficient adaptive algorithm for large-scale random vibration analysis. *Earthquake Engineering and Structure Dynamics, 22*(2), 151–165.

Harichandran, R. S. (1995). Random vibration under propagating excitation: Closed-form solutions. *Journal of Engineering Mechanics, 118*(3), 575–586.

Harichandran, R. S., & Chen, M. T. (1996). Statistical moments of principal stress related quantities in random vibration analysis. In *Probabilistic Mechanics and Structural Reliability* (pp. 962–965). New York: ASCE Press.

Harichandran, R. S., & Naja, M. K. (1997). Random vibration of laminated composite plates with material non-linearity. *International Journal of Non-Linear Mechanics, 32*(4), 707–720.

Harichandran, R. S., & Wang, W. (1988). Response of simple beam to spatially varying earthquake excitation. *Journal of Engineering Mechanics, 114*(9), 1526–1541.

Harichandran, R. S., & Wang, W. (1990). Response of indeterminate two-span beam to spatially varying seismic excitation. *Earthquake Engineering and Structural Dynamics, 19*(2), 173–187.

Harichandran, R. S., Hawwari, A., & Sweidan, B. (1996). Response of Long-span bridges to spatially varying ground motion. *Journal of Structural Engineering, 122*(5), 476–484.

Harmoko, H., Yap, F. F., Vahdati, N., & Li C. (2009). Design and analysis of shock and random vibration isolation of operating hard disk drive in harsh environment. *Shock and Vibration, 16*(2), 143–154.

Harris, C. M. (1979). *Handbook of noise control*. New York: McGraw-Hill.

Harris, C. M., & Crede Ch. E. (1961). *Shock and vibration handbook*. New York: McGraw-Hill.

Harrison, R. F. (1983). *The nonstationary response of vehicles on rough ground*. Ph.D. Thesis, University of Southampton.

Hart, F. D., & Desai, V. D. (1967). Additive property of modal density for a composite structure. *The Journal of the Acoustical Society of America, 42*(5), 1203.

Hasenzagl, R., Irschik, H., & Ziegler, F. (1988). Design charts for random vibrations of elasto-plastic oscillators subjected to Kanai-Tajimi spectra. *Reliability Engineering and System Safety, 23*(2), 109–126.

Hasselman, T. K. (1962). Random excitation of continuous structures to random loading. *ZAMM: Zeiitschrift fuer angewandte mathematick und Mechanik, 42*, 475–476 (in German).

Hasselman, T. K. (1967). *Transient response of a single degree of freedom system to a nonstationary narrow- band process*. M.Sc. Thesis, University of California, Los Angeles.
Hasselman, T. K. (1969). An analytical basis for time-modulated random vibration testing, *NASA CR-66770*.
Hasselman, T. K. (1970). A comparison of solutions for the mean-square response of a single mechanical oscillator to nonstationary random excitation. *Journal of Applied Mechanics, 37*, 1187–1189.
Hasselman, T. K. (1972). Linear response to nonstationary random excitation. *Journal of Engineering Mechanics, 98*, 519–529.
Hasselman, T. K. (972). Study of modal coupling procedures for the shuttle: A matrix method for damping synthesis, *NASA CR -112253*.
Hasselman, T. K. (1977). Reply by author to A. H. Flax. *AIAA Journal, 15*(11), 1662–1663.
Hasselman, T. K. (1978). Modal coupling in lightly damped structures. *AIAA Journal, 14*, 1627–1628.
Hasselman, T. K., & Hart, G. C. (1972). Modal analysis of random structural systems. *Journal of Engineering* Mechanics, *98*(3), 561–579.
Haug, E. J. (1980). Optimization of distributed parameter structures with repeated eigenvalues. In P. J. Holmes (Ed.), *New approaches to nonlinear problems in dynamics* (pp. 511–520). SIAM Publications.
Haug, E. J., & Choi, K. K. (1982). Systematic occurrence of repeated eigenvalues in structural optimization. *Journal of Optimization Theiory and Applications, 38*, 251–274.
He, J. (2009). A reliability approximation for structures subjected to non-stationary random excitation. *Structural Safety, 31*(4), 268–274.
Healey, A. J., Nathman, E., & Smith, C. C. (1977). An analytical and experimental study of automobile dynamics with random railway inputs. *Journal of Dynamic Systems, Measurement, and Control, 99*(4), 284–292.
Heimann, B. (1982). Kovarianzanalyse von Schwingungssystemen unter gleichzeitiger periodischer zufalliger Fremderregung. *ZAMM: Zeitschrift für angewandte Mathematik und Mechanik, 62*(12), 667–679 (in German).
Heinkelè, C., Pemot, S., Sgard, F., & Lamarque, C.-H. (2006). Vibration of an oscillator With random damping: Analytical expression for the probability density function. *Journal of Sound and Vibration, 296*(1–2), 383–400.
Heinrich, W., & Hennig, K. (1977). *Zufallsschwingungen mechanischer Systeme*. Berlin: Academie-Verlag (in German).
Hennig, K. (1972). Zur Berechnung stationärer Zufallsschwingungen flacher Schalon. *ZAMM: Zeitschrift für angewandte Mathematik und Mechanik, 52*(10), T317–T320 (in German).
Hennig, K., Friedrich, H., & Knöfel, L. (1977). Berechnung von fremderregten Zufallsschwingungen Fester Kontinua. *Machinenbautechnik, 26*(11), 496–504 (in German).
Heredia-Zavoni, E., & Vanmarcke, E.H. (1994). Seismic random-vibration analyses of multisupport-structural systems. *Journal of Engineering Mechanics, 120*(5), 1107–1128.
Heredia-Zavoni, E., & Vanmarcke, E.H. (1995). Closure to seismic random-vibration analysis of multisupported-structural systems. *Journal of Engineering Mechanics, 121*(9), 1037–1038.
Heredia-Zavoni, E., & Vanmarcke, E. H. (1998). Random-vibration based response spectrum method for multi-supported structural systems. *Journal of Seismology and Earthquake Engineering, 1*(1), 35–50.
Hess, R. W., Herr, R. W., & Mayers, W. H. (1959). *A study of the acoustic fatigue characteristics of some flat and curved aluminum panels exposed to random and discrete noise*. NASA Technical Report TN D1.
Heuer, R., & Irschik, H. (1987). Time-dependent power spectral densities of randomly vibrating kerchhoff plates. In Y. K. Lin & G. J. Schuëller (Eds.), *Stochastic structural dynamics*.
Heuer, R., Irschik, H., & Ziegler, F. (1991). Thermally forced random vibrations of thick plates. *Computational stochastic mechanics*, 727–739. Springer.
Hida, T. (2017). *White noise: Functionals of Gaussian and other noises*. Singapore: World Scientific.

Hida, T., & Streit, L. (Eds.) (2017). *Let us use white noise*. Singapore: World Scientific.
Hien, T. D., & Kleiber, M. (1991). Stochastic design sensitivity in structural dynamics. *International Journal for Numerical Methods in Engineering, 32*(6), 1247–1265.
Ho, B. L., & Kalman, R. E. (1966). Effective construction of linear state-variable models from input-output descriptions. *Regeleungstechnik, 12*, 545–592.
Ho, D.U.E. N. (1974). Buckling load of non-linear systems with multiple eigenvalues. *International Journal of Solids and Structures, 10*(11), 1315–1330.
Hodges, C. H. (1986). Theories of noise and vibration transmission in complex structures. *Reports on Progress in Physics, 49*(2), 107.
Hodges, C. H., & Woodhouse, J. (1986). Theories of noise and vibration transmission in complex structures. *Reports in Progress in Physics, 49*, 107–170.
Höllinger, F. (1983). Time-Harmonic and nonstationary stochastic vibrations of arch dam-reservoir-systems. *Acta Mechanica, 49*(3–4), 153–167.
Holman, R. E., & Hart, G. C. (1972). Structural response to segmented nonstationary excitation. *AIAA Journal, 10*, 1473–1478.
Honerkamp, J. (1994). *Stochastic dynamic systems: Concepts , numerical methods, data analysis*. New York: Wiley-VCH.
Hopkins, C. (2002). Statistical energy analysis coupled plate systems with low modal density and low modal overlap. *Journal of Sound and Vibrations, 251*(2), 193–214.
Horacek, J. (1986a). The comutation of the dynamic response of cylindrical shells in a turbulent flow. *Acta Technica CSAV,* (2), 214–229.
Horacek, J. (1986b). The estimation of the dynamic response of cylindrical shell to a random pressure field. *Acta Technica CSAV, 33*(1), 708–716.
Hornick, R. J., & Lefritz, N. M. (1966). A study and review of human response to prolonged random vibration. *Human Factors, 8*(6), 481–492.
Hoshiya, M. (1972). Simulation of multi-correlated random process and application to random process and application structural vibration problems. *Proceedings of the Japan Society of Civil Engineers, 204*.
Hoshiya, M. (1974). *Random vibration analysis*. Tokyo (in Japanese).
Hoshiya, M. (1986). Response covariance to multiple excitations. *Journal of Engineering Mechanics, 112*(4), 412–421.
Hoshiya, M., & Chiba, T. (1980). Simulation methods of multi-dimensional nonstationary stochastic processes by time domain models. *Proceedings of Japan Society of Civil Engineering, 296*.
Hoshiya, M., Ishii, K., & Nagata, S. (1984). Recursive covariance of sturctural responses. *Journal of Engineering Mechanics, 110*(12), 1743–1755.
Hoshiya, M., & Shah, H. (1969). Free vibration of a beam-column with stochastic properties. In *Proceedings of the ASME-EMD Specialty Conference*, Purdue University, Lafayette, IN (pp. 107–111).
Hoshiya, M., & Shah, H. (1971). Free vibration of stochastic beam-column. *Journal of Engineering Mechanics, 97*(4), 1239–1255.
Hosseini, S.A. A., & Khadem, S. E. (2005). Free vibration analysis of rotating beams with random properties. *Strcutural Engineering and Mechanics, 20*(3), 293–312.
Hosseini, S.A. A., & Khadem, S. E. (2007). Vibration and reliability of a rotating beam with random properties under random excitation. *International Journal of Mechanical Sciences, 49*(12), 1377–1388.
Hosseinloo, A. H., Yap, F. F., & Chua, E. T. (2014). Random vibration protection of a double-chamber submerged jet impingement cooling system: A continous model. *Aerospace Sience Technology, 35*, 29–38.
Hosseinloo, A. H., Yap, F. F., & Vahdati, N. (2013). Analytical random vibration analysis of boundary-excited thin recatangular plates. *International Journal of Structural Stability and Dynamics, 13*, article 1250062.
Hou, G.J. W., & Kenny, S. P. (1989). Eigenvalue and eigenvector approximate analysis for repeated eigenvalue problem. *AIAA Journal, 30*, 2317–2324.

Hou, Z.-K., & Iwan, W. D. (1991). Nonstationary response of Linear systems under uncorrelated parametric and external excitations. *Probabilistic Engineering Mechanics*, *6*(2), 74–82.

Hou, Z.-K., Zhou Y-S., Dimentberg, M. F., & Noori, M. (1996). Random response to periodic excitation with correlated disturbances. *Journal of Engineering Mechanics, 122*(11), 1101–1108.

Houdijk, A. (1928). Le mouvement Brownien d'un fil, (Brownian Motion of a Wire). *Archives Neerlandaises des Science Exactes et Naturelles,* Series IIIA, *11*, 212–217.

Houbolt, J., & Sen, A. (1972). Cross-spectral functions based on von Kármán's spectral equation, *NASA CR-2011*.

Housner, G. W. (1955). Properties of strong ground motion earthquake. *Bulletin of the Seismological Society of America, 45*, 197–218.

Housner, G. W. (1970). Design spectrum, in R. L. Wiegel (Ed.), *Earthquake engineering* (pp. 93–106). Englewood Cliffs, NY: Prentice-Hall.

Hover, F. S., Techet, A. H., & Trianfylou, M. S. (1998). Forces on oscillating uniform and tapered cylinders in cross flow. *Journal of Fluid Mechanics, 363*, 97–114.

Howart, H.V. C., & Griffin, M. J. (1990). Subjective response to combined noise and vibration: Summation and interaction effects. *Journal of Sound and Vibration, 143*(3), 443–454.

Howell, L. J., & Lin, Y. K. (1971). Response of flight vehicle to nonstationary atmospheric turbulence. *AIAA Journal*, *9*(11), 2201–2207.

Hu, B., & Schiehlen, W. O. (1997). On the simulation of sochastic process by spectral representation. *Probabilitstic Engineering Mechanics, 12*(2), 105–113.

Hu, P. Y. (1969). Response of linear systems to magnitude limited random excitation. *Journal of Engineering for Industry, 991*–996.

Hu, S.L. J. (1993). Responses of dynamic systems excited by non-Gaussian oulse processes. *Journal of Engineering and Mechanics, 119*(9), 1818–1827.

Huang, M. F., Chn, C. M., Kwok, K. C., & Hitchcock, P. A. (2009). Cross correlations of modal responses of tall buildings in wind-induced lateral-torsional motion. *Journal of Engineering Mechanics, 135*(8), 802–812.

Huang, W., & Dececioglu, D. B. (2000). A simplified random vibration analysis on portable electronic products. *IEEE Transactions on Components and Packaging Techniques, 23*(3), 505–515.

Huang, Z. L., Liu, Z. H., & Zhu, W.Q. (2004). Stationary response of multi-degree-of-freedom vibro-impact systems under white noise excitations. *Journal of Sound and Vibration*, *275*(1), 223–240.

Hughes, T.Y. R., Engel, G., Mazzel, L., & Larson, M. G. (2000). The continuous Galerkin method is locally conservative. *Journal of Computational Physics, 163*(2), 467–488.

Hunter, R. Creating digital chladni patterns. Retrieved July 8, 2019, from https://thelig.ht/chladni/.

Hurtado, J. E., & Barbat, A. H. (1998). Monte carb techniques in computational stochastic mechanics. *Archives of Computational Methods in Engineering, 5*(1), 3–30.

Hwang, J. H., & Kim, J. S. (2000). On the approximate solution of aircraft landing gear under nonstationary random excitations. *Journal of Mechanical Science and Technology, 14*(9), 968–977.

Hwang, Y. F., Bonness, W. K., & Hambric, S. A. (2009). Comparison of semi-empirical MODELS for turbulent boundary layer wall pressure spectra. *Journal of Sound and Vibration*, *9*, 319(1–2), 199–217.

Ibragimov, I., & Rozanov Yu. A. (1970). *Gaussian random process*. Moscow: "Nauka" Publishing House (in Russian).

Ibrahim, H. H., Tawfik, M., & Negm, H. M. (2008). Thermoelastic random response of shape memory alloy hybrid composite plates. *Journal of Aircraft, 45*(3), 962–970.

Ibrahim, R. (2007). *Parametric random vibration*. Mineola, New York: Dover.

Ibrahim, R. A. (1987). Structural dynamics with parameter uncertainties. *Applied Mechanics Reviews, 40*(3), 309–328.

Ibrahim, R. A., & Roberts, J. W. (1978). Parametric vibration, Part V: Stochastic problems. *Shock and Vibration Digest*, *10*(12), 17–38.

Ibrahim, S. R. (1977). Random decrement technique for modal identification of structures. *Journal of Spacecraft, 14*(11), 696–700.

Ibrahim, Y. (1991). Covariances function of non-stationary response of linear systems. *Journal of Sound and Vibration, 86*, 303–317.

Ichchou, M. N., Hiverniau, B., & Troclet, B. (2009). Equivalent 'Rain on the Roof' loads for random spatially correlated excitations in the mid–high frequency range. *Journal of Sound and Vibration, 322*(4–5), 926–940.

Igumentsev, E. A. (1985). Probability characteristics of the frequency spectra of natural vibrations of turbocompressor blades. *Strength of Materials, 17*(4), 549–554.

Igusa, T. (1989). Characteristics of response to nonstationary white noise: Theory. *Journal of Engineering Mechanics, 115*(9), 1904–1918.

Igusa, T. (1989). Characteristics of response to nonstationary white noise: Application. *Journal of Engineering Mechanics, 115*(9), 1919–1934.

Igusa, T., & Der Kiureghian, A. (1988). Response of uncertain systems to stochastic excitation. *Journal of Engineering Mechanics, 114*(5), 812–832.

Igusa, T., Der Kiureghian, A., & Sackman, J. L. (1984). Modal decomposition method for stationary response of non-classically damped systems. *Earthquake Engineering and Structural Dynamics, 12*(1), 121–136.

Igusa, T., & Xu K. (1991). Wide-band response of multiple subsystems with high modal density. In Y. K. Lin & I. Elishakoff (Eds.), *Stochastic structural dynamics 1: New theoretical developments* (pp. 131–146). Berlin: Springer.

Imenitov, L. B. (1973). Investigation of natural transverse vibrations of plates without using Kirchoff-Love hypothesis. In *Theory of Shells and Plates* (pp. 479–482). Moscow: "Nauka" Publishing House (in Russian).

Impollonia, N., Failla, I., & Ricciardi, G. (2016). Parametric statistical moment method for damage detection and health monitoring. *Journal of Risk and Uncertainty in Engineering Systems, Part A: Civil Engineering, 2*(2), article (4016001).

Irschik, H., & Ziegler, F. (1985). Nonstationary random vibrations of yielding frames. *Nuclear Engineering and Design, 90*(3), 357–364.

Irschik, H., & Ziegler, F. (1991). Nonstationary random vibrations: Approximate formulation versus exact theory. *Probabilistic Engineering Mechanics, 6*(2), 83–91.

Itô, K. (1942). Differential equations determining a Markoff Process. *Journal of Pan-Japan Mathematical Colloquium, 1077*, 1352–400.

Ivankina, O. P., & Lebedev, B. S. (2016). Optimization of the driving mode of the vehicle on bumpy dirt road for safety the transported cargo. *EuroAsian Scientific Journal*, (1) (in Russian).

Iwan, W. D., & Hou, Z. K. (1989). Explicit solutions for the response of simple systems subjected to nonstationary random excitation. *Structural Safety, 6*, 77–86.

Iwan, W. D., & Jensen, H. (1993). On the dynamic response of continuous systems including model uncertainties. *Journal of Applied Mechanics, 60*, 484–490.

Iwankiewicz, R. (). Vibration of a beam under stochastic impulses excitation. *Engineering Reports, 30*, 405–417 (in Polish).

Iwankiewicz, R. (1985a). Dynamic response of linear structures to random trains of impulses with random spatial shapes. *Journal of Theoretical and Applied Mechanics, 4*(4), 485–504.

Iwankiewicz, R. (1985b). Contribution of mode shapes and cross-correlations in the dynamic response of a spherical shell to random impulses. In *Proceedings of the 2nd Conference on Random Problems in Structural Mechanics*, Gdansk, Poland (pp. 97–104) (in Polish).

Iwankiewicz, R. (1986). Response of vibratory systems to general random pulses. *Journal of Sound and Vibration, 108*, 429–446.

Iwankiewicz, R. (1989). On the probability of response of a linear oscillator to a random pulse train. *Rozprawy Inzynierskie, 37*, 247–268.

Iwankiewicz, R. (1990). Response of linear vibratory systems driven by renewal point processes. *Probabilistic Engineering Mechanics, 5*, 111–121.

Iwankiewicz, R. (1995). *Dynamical mechanical systems under random impulses*. Singapore: World Scientific.

Iwankiewicz, R. (2003). Dynamic systems under random impulses driven by a generalized erlang renewal process. In *Proceedings of 10th IFIP WG 7.5 Working Conference* (pp. 103–110), Kansai University, Osaka, Japan.

Iwankiewicz, R., & Sniady, P. (1984). Vibration of a beam under a random stream of moving forces. *Journal of Structural Mechanics, 12*(1), 13–26.

Iwankiewicz, R., & Sobczyk, K. (1983). Dynamic response of linear structures to correlated random impulses. *Journal of Sound and Vibration, 86*(3), 303–317.

Iwankiewicz, R., & Witt, M. (1983). Analysis of random response of a shallow cylindrical shell. *Rozprawy Inzynierskie, 31*(1), 103–114 (in Polish).

Iyengar, R. N., & Iyengar, K. T. (1969). A nonstationary random process for earthquake accelerations. *Bulletin of Seismological Society of America, 59*, 1163–1188.

Iyengar, R. N., & Jaiswal, O. R. (1993). A new model for non-Gaussian random excitations. *Probabilistic Engineering Mechanics, 8*(3–4), 281–287.

Iyengar, R. N., & Jaiswal, O. R. (1995). Rnadom field modeling of railway track irregularities. *Journal of Transportation Engineering, 121*(4), 303–308.

Iyengar, R. N., & Dash, P. K. (1976). Random vibration analysis of stochastic time-varying systems. *Journal of Sound and Vibration, 45*(1), 69–89.

Jacobs, L. D., & Lagerquist, D. R. (1968). Finite element analysis of complex panel to random loads, *AFFDL-TR-68–44*, Wright-Patterson Air Force Base.

Jacobs, L. D., Lagerquist, D. R., & Gloyna, F. L. (1969). Response of complex structures to turbulent boundary layers. *AIAA Paper No. 69–20.*

Jacobson, M. J. (1968). Stress and deflection of honeycomb panels loaded by spatially uniform white noise. *AIAA Journal, 6*, 1503–1510.

Jacquelin, E., Adhikari, S., Sinou J-J., & Friswell, M. I. (2015). Polynomial chaos expansion and steady state response of a class of random dynamical systems. *Journal of Engineering Mechanics, 141*(4), article 04014145.

Jacquelin, E., Adhikari, S., Friswell, M. I., & Sinou, J.-J. (2016). Role of roots of orthogonal polynomials in the dynamic response of stochastic systems. *Journal of Engineering Mechanics, 142*(8), article 06016004.

Jacquot, R. G. (1972). Optimal random vibration absorbers. *Journal of Engineering Mechanics, 99*(3), 612–616.

Jacquot, R. G. (2000). Random vibration of damped modified beam systems. *Journal of Sound and Vibration, 234*(3), 441–454.

Jacquot, R. G. (2001). Suppression of random vibration in plates using vibration absorbers. *Journal of Sound and Vibration, 248*(4), 585–596.

Jacquot, R. G. (2003). The spatial average mean square motion as an objective function for optimizing damping in damped modified systems. *Journal of Sound and Vibration, 259*(4), 955–965.

Jacquot, R. G. (2004). Optimal damper location for randomly forced cantilever beams. *Journal of Sound and Vibration, 269*, 623–632.

Jarczewska, K., Koszela, P., Sniady, P., & Korzec, A. (2011). Identification of the structure parametes using short-time non-stationary stochastic excitation. *Journal of Sound and Vibration, 330*(14), 3352–3367.

Jenkins, G. M., & Watts, D. G. (1968). *Spectral analysis and its applications.* San Francisco: Holden Day.

Jensen, H., & Iwan, W. D. (1992). Response of systems with uncertain parameters to stochastic excitation. *Journal of Engineering Mechanics, 118*(5), 1012–1015.

Jensen, H. A. (2005). Structural optimization of linear dynamical systems under stochastic excitation: A moving reliability database approach. *Computer Methods in Applied Mechanics and Engineering, 194*(12–16), 1757–1758.

Jensen, R. A., & Lambert, R. G. (1967). Numerical calculation of some response statistics for a linear oscillator under impulsive noise excitation. *The Journal of the Acoustical society of America, 41*, 827–835.

Jha, V. K., Sankar, T. S., & Bhat, R. B. (1983). Computer aided synthesis of a satellite antenna structure with probabilistic constraints. *The Shock and Vibration Bulletin, 53*, Part 3, 79–90.

Jha, V. K., Sankar, T. S., & Bhat, R. B. (1983). Dynamic behavior of a satellite antenna structure in random vibration environment. *The Shock and Vibration Bulletin, 53*, Part 3, 91–103.

Jiahao, L., & Wanxie, Z. (1998). Some notes on FEM and structural random response analysis. *Chinese Journal of Computational Mechanics, 2*.

Jin, X.-L., Chen, M. Z. Q., & Huang, Z. (2019). Suppressing random response of a regular structure by an inerter-based dynamic vibration absorber. *Journal of Vibration and Acoustics, 14*(4), article 041004.

Johnsen, T. L., & Dey, S. S. (1978). ASKA part II-linear dynamic analysis random response. Report: ASKA UM 218, Institut für Statik und Dynamik der Luft-und Raumfahrtkonstruktionen, University of Stuttgart.

Johnson, E. A., Proppe, C., Spencer, B. F., Bergman, L. A., & Schuëller, G. I. (2003). Parallel processing in computational stochastic dynamics. *Probabilistic Engineering Mechanics, 18*(1), 37–60.

Jonckheere, S., Vandepitte, D., & Desmet, W. (2015). A wave based approach for the dynamic bending analysis of Kirchhoff plates under distributed deterministic and random excitation. *Computers and Structures, 156*, 42–57.

Jones, A., & Beadle, C. (1970). Random vibration response of cantilever plates using the finite element method. *AIAA Journal, 8*, 1905–1907.

Jullien, Y. G. (1962). On the response of continuous structures to random excitation. In *Proceedings of the Fourth, U. S. National Congress of Applied Mechanics* (Vol. 1, pp. 219–224).

JullienY. G. (1963). The exact solution of the response of a damped rod to random excitation. *Journal of Applied Mechanics, 30*, 312.

Jun, L., & Jin X-D. (2005). Response of flexure-torison coupled composite thin-walled beams with closed cross-sections to random loads. *Mechanics Research Comunications, 32*(1), 25–41.

Jun, L., Rongying, S., Hongxing, H., & Jin X-D. (2004). Response of monosymmetric thin-walled Timoshenko beam to random excitation. *International Journal of Solids and Structures, 41*(22), 6023–6040.

Jun, L., Sh C-X., Kong X-S., Li X-B., & Wu W-G. (2014). Stochastic response of an axially loaded composite Timoshenko beam exhibiting bending-torsion coupling. Archive of *Applied Mechanics, 84*(1), 109–122.

Kako, T., Shinozuka, M., Hwang, H., & Reich, M. (1983). *FEM-based random-vibration analysis of nuclear structures under seismic loading*, Report No. BNL-NUREG-32G26, Toshiba Corporation.

Kalnins, A., & Dym, C. L. (1976). *Vibration: Beams, plates , and shells*. Dowden, Hutchinson and Ross.

Kaminski, M. (2010). Generalized stochastic pertubation technique in engineering computations. *Mathematical and Computational Modelling, 51*(3–4), 272–285.

Kaminski, M. (2013). *The stochastic perturbation method for computational mechanics*. Chichester, U.K.: Wiley.

Kaminski, M., & Szafran, J. (2009). Random eigenvibrations of elastic structures by the response functiion method and teh generalized stochastic perturbation technique. *Archives of Civil and Mechanical Engineering, 9*(4), 5–32.

Kanai, K. (1957). Semi-empirical formula for the seismic characteristics of the ground. *University of Tokyo Bulletin, Earthquake Research Institute, 35*, 309–325.

Kandianis, F. (1974). Notes on the properties of correlation functions in linear system analysis. *Journal of Sound and Vibration, 36*(2), 207–223.

Kandler, G., & Fūssl, J. (2017). A probabilistic approach for the linear behavior of fluid laminated timber. *Engineering Structures, 148*, 179–685.

Kang, W., Zhang, Z., Zhou, K., & Chen, Y. (2019). The Random vibration and force transmission characteristics of the elastic propeller-shafting system induced by inflow turbulence. *Ocean Engineering, 188*, article 106317.

Kapur, K. K., Ching-U. Ip. & Howard, E.-P. (1968). Random vibration using finite element approach. *Shock and Vibration Bulletin, 38*, 7–16.

Karczub, D. G., & Norton, M. P. (2000). Correlation between dynamic strain and velocity in randomly excited plates and cylindrical sheels with clamped boundaries. *Journal of Sound and Vibration, 230*(5), 1069–1101.

Kareem, A. (1987). Wind effects on structures: A probabilistic viewpoint. *Probabilistic Engineering Mechanics, 2*(4),166–200.

Kareem, A. (1992). Dynamic response of high-rise buildings to stochastic wind loads. *Journal of Wind Engineering and Industrial Aerodynamics, 42*(1–3), 1101–1112.

Kareem, A. (1998). Aerodynamic response of structures with parametric uncertainties. *Structural Safety, 5*(3), 205–225.

Kareem, A., & Gurley, K. (1996). Damping in structures: Its evaluation and treatment of uncertainty. *Journal of Wind Engineering and Industrial Aerodynamics, 59*(2), 131–157.

Kareem, A., & Kline, S. (1995). Performance of multiple mass dampers under random loading. *Journal of Structural Engineering, 121*(2), 348–361.

Kareem, A., & Li Y. (1991). FFT-based simulation of multi-variate nonstationary random processes. In I. Elishakoff & Y. K. Lin (Eds.), *Stochastic structural dynamics 2*, (pp. 73–90). Berlin: Springer.

Kareem, A., & Sun, W.-J. (1987). Stochastic response of structures with fluid-containing appendages. *Journal of Sound and Vibration, 119*(3), 389–408.

Karmalita, V. (2019). *Digital processing of random oscillations.* Berlin: de Gruyter.

Karsh, P. K., Mukhopadhyay Dey, S. (2018). Stochastic dynamic analyssis of twisted functionally graded plates. *Composites, Part B: Engineering*, in press.

Katafygiotis, L. S., & Beck, J. L. (1988). Treating model uncertainties in structural dynamics. In *Proceedings, 9th World Conference on Eaarthquake Engineering*, Tokyo (Vol. V, pp. 301–306).

Katafygiotis, L. S., & Beck, J. L. (1995). A very efficient moment calculation method for uncertain linear dynamic systems. *Probabilistic Engineering Mechanics, 10*(2), 117–128.

Kaufman, S., Lapinski, W. L., & Caa, R. C. (1961). Response of a single-degree-of-freedom isolation to a random disturbance. *The Journal of the Acoustical Society of America, 33*(8), 1108–1112.

Kaul, M. N. (1978). Stochastic characterization of earthequakes through their response spectrum. *Earthequake Engineering and Styructural Dynamics, 6*, 497–509.

Kaul, O. N. (1976). *Vibration response of multilayer structures to random excitation.* Ph.D. Thesis, Indian Institute of Technology, Delhi, India.

Kaul, O. N., Gupta, K. N., & Nakra, B. C. (1979). Response of constrained visco-elastic beams to random excitation. *Archiwum Mechaniki Stosowany, 31*(2), 875–895.

Kaul, O. N., Gupta, K. N., & Nakra, B. C. (1984). Random vibration of an unconstrained viscoelastically damped beam. *Technische Mechanik, 6*(3), 48–56.

Kazakov, I. E. (1977). *Statistical dynamics of systems of variable structure.* Moscow: "Nauka" Publishing House (in Russian).

Kaziev, A. M., Khuranov, V.Kh., & Kostenko, O. V. (2016). Investigation of vector excitations on beams. *Don's Engineering Herald, 46*(3) (in Russian).

Keane, A. J., & Price, W. G. (Eds.) (1997). *Statistical energy analysis: An overview, with applications in structural dynamics.* Cambridge University Press.

Keane, A., & Nair, P. B. (2005). *Computational approaches for aerospace design: The pursuit of excellence.* Hokoken, NJ: Wiley.

Keane, A. J., & Price, W. G. (1987). Statistical energy analysis of strongly coupled systems. *Journal of Sound and Vibration, 117*, 363–386.

Keller, J. B. (1959). Review of "Response of a Simply Supported Timoshenko Beam to a Purely Random Gaussian Prcoess" by J. C. Samuels and, A. C. Eringen. *Mathematical Review, 20*, 925.

Keller, J. B., & Rubinow, S. I. (1960). Asymptotioc solutions of eigenvalue problems. *Annals of Physics, 9*, 24–75.

Kelly, J. M., & Sackman, J. L. (1980). Consevatism in summation rules for closely spaced modes. *Earthquake Engineering and Structural Dynamics, 8*, 63–74.

Kessissoglou, X. Y., & Lucas, G. I. (2009). Gaussian orthogonal ensemble spacing Statistics and the statistical overlap factor applied to dynamic systems. *Journal of Sound and Vibration, 324*, 1039–1066.

Keswick, P. K., & Norton, M. P. (1987). A comparison of modal density measurement techniques. *Applied Acoustics, 20*(2), 137–153.

Khabbaz, G. R. (1970). Comparison of mechanical coupling factor by two methods. *The Journal of the Acoustical Society of America, 47*, 392–393.

Kharkevich, A. A. (1962). *Spectrum and analysis*. Moscow: "Fizmatgiz" Publishing House (in Russian).

Kharkova, N. V. (1971). On the lower portion of the spectrum of natural axisymmetric vibrations of a thin elastic shell of revolution. *Journal of Applied Mathematics and Mechanics, 35*(3), 397–404.

Khattak, A. R., Garvey, S. D., & Popov, A. A. (2006). Repeated resonances in folded-back beam structures. *Journal of Sound and Vibration, 290*, 309–320.

Khromatoff, V. E. (1972). Properties of spectra of thin circular cylindrical shells oscillating near momentless stress state. *Mechanics of Solids*, Isssue 2, 103–108 (in Russian).

Khromatoff, V. E. (1973). Empirical distribution of natural vibrational frequencies of shells of negative Gaussian curvature. In *Proceedings of the Moscow Power Engineering Institute* (Vol. 164, pp. 43–47) (in Russian).

Khromatoff, V. E. (1975). Wide-band random vibrations of thin shallow shells and their connection to problems of distribution of natural frequencies. In *Proceedings of the 10th All-Union Conference on Theory of Shells and Plates* (pp. 392–402). Tbilisi: "Mezniereba" Publishers (in Russian).

Khromatoff, V. E. (1976). Random wide-band vibrations of shallow orthotropic shells. *Proceedings of the Moscow Power Engineering Institute* (Vol. 280, pp. 100—105, in Russian).

Khromatoff, V. E. (1977). Density of characteristic frequencies of thin orthotropic shells. *International Applied Mechanics, 13*(7), 666–669.

Khromatoff, V. E. (1982). Properties of spectra of natural vibrations of shallow sandwich shells. *Mechanics of Solids, 17*(2), 113–119.

Kim, M., Moon, J., & Wickert, J. A. (2000). Spaital modulation of repeated vibration modes in rotationally periodic structures. *Journal of Vibration and Acoustics, 122*(1), 62–68.

Kim, N. E., & Griffin, J. H. (2003). System identification in higher modal density regions of bladed disks. In *Proceedings of the 8th National Turbine Engine High Cycle Fatigue Conference* (pp. 1–68).

Kindlmann, G., Weinstein, D., Lee, A., Toga, A., & Thompson, P. (2004). Visualization of anatomic covariance tensor fields. In *Engineering in Medicine and Biology Society, IEMBS'04 26th Annual International Conference of the IEEE* (Vol. 1, pp. 1842–1845). San Francisco, CA: IEEE Press.

King, W. W. (1975). Comments on "Bolotin's method applied to the buckling and lateral vibration of stressed plates", *AIAA Journal, 13*(11), 1543–1544.

King, W. W., & Lin, C. C. (1974). Applications of Bolotin's method to vibrations of plates. *AIAA Journal, 1293*, 399–401.

Kiran Kumar, P., Subrahmanyam, J., & Rama Krishna, D. (2016). Experimental investigation of aluminum thin shells subjected to random vibrations. *International Journal on Mechanical Engineering and Robotics, 4*(4), 16–23.

Kitinpornchai, S., Yang, J., & Liew, K. M. (2006). Random vibration of the functionally graded laminates on thermal environments. *Computer Methods in Applied Mechanics and Engineering, 195*(9), 1075–1095.

Kloeden, P. E., & Platen, E. (1992). *Numerical solution of stochastic differential equations*. Berlin: Springer.

Knowles, J. K. (1968). On the dynamic response of a beam to random excitations. *Journal of Applied Mechanics, 35*, 1–6.

Knowles, J. K. (1970). A note on the response of a beam to a randomly moving force. *Journal of Applied Mechanics, 37*(4), 1192–1194.
Kogiso, N., Shao S-W., & Murotsu, Y. (1998). Effect of correlation on relaiblity based design of composite plate subject to buckling. *AIAA Journal, 36*(9), 1706–1713.
Koh, C., & Kelly, J. M. (1990). Application of fractional derivatives to seismic analysis of base-isolated models. *Earthquake Engineering and Structural Dynamics, 21*, 649–664.
Koliopulos, P. K. (1988). Quasi-static and dynamic response statics of linear SDOF system under morison-type wave forces. *Engineering Structures, 10*, 24–36.
Korenev, B. G. (1968). Vibrations of tower structures and certain methods of damping their vibrations. In *Proceedings, Conference on Wind Effects on Buildings and Structures* (Vol. 2, pp. 105–129). Toronto: University of Toronto Press.
Korenev, B. G., & Reznikov, L. M. (1969). Oscillations of structures with dynamic absorbers under steady-state random excitations. *Stroitel'naya Mekhanika i Raschet Sooruzheniy*, (4), 33–39 (in Russian).
Korenev, B. G., & Reznikov, L. M. (1993). *Dynamic vibration absorbers: Theory and technical applications*. New York: Wiley.
Koreshkova, V. S., & Khromatoff, V. E. (2009). On the influence of a transverse magnetic field on the vibration spectra of shallow shells. *Mechanics of Solids, 44*(4), 632–638, 2009.
Kornev, V. M. (1972). Approximate solution of problems on the stability and vibration of elastic shells under condensation of eigenvalues. *Mechanics of Solids, 7*, 119–129.
Kornev, V. M., & Markin, A. B. (1975). Density of eigenmotions of elastic shells under intense dynamic loading. *Journal of Applied Mechanics and Technical Physics*, (5), 173–178 (in Rusian).
Kornev, V. M., & Markin, A. B. (1975). Natural motion density of elastic shells under intensive dynamic loading. *Journal of Applied Mechnics and Technical Physics, 16*(5), 821–825.
Kosco, B. (2015). Statistical independence. In J. Brockman (Ed.), *This idea must die: Scientific theories that are blocking progress* (pp. 538–540). New York: Harper Perennial.
Kostiychenko, A. G., & Sargsyan, I. S. (1979). *Distribution of eigenvalues (Self-Adjoint Ordinary Differential Equations)*. Moscow: "Nauka" Publishing House (in Russian).
Kotulski, Z., & Sobzcyk, K. (1981). Linear systems and normality. *Journal of Statistical Physics, 24*(2), 359–383.
Kotulski, Z., & Sobzcyk, K. (1984). Characteristic functionals of randomly excited physical systems. *Physics A: Statistical Mechanics and Its Application, 123*(1), 261–278.
Kotulski, Z., & Sobzcyk, K. (1987). Effects of paramter uncertainty on the response of vibrator systems to random excitation. *Journal of Sound and Vibration, 119*(1), 159–172.
Kougioumtzoglou, I. A., Fragkoulis, V. C., Pantelous, A. A., & Pirrotta, A., Random vibration of linear and nonlinear structural systems with singular matrices: A frequency domain approach. *Journal of Sound and Vibration, 404*, 84–101.
Kouskoulas, V. J., & Hurty, W. C. (1970). Probabilsitic description of the response of linear structures to random inputs. *Journal of Applied Mechanics, 37*(9), 617–622.
Koval, L. R. (1980). On sound transimission into a stiffened cylindrical shell with rings and stringers treated as discrete elements. *Journal of Sound and Vibration, 71*(4), 511–521.
Kovalenko, I. N., Kuznetsov, N. Yu., & Shurenkov, V. M. (1983). *Random process: A handbook*. Kiev: "Naukova Dumka" Publishers (in Russian).
Kozin, F., & Bogdanoff, J. L. (1960). On the statistical analysis of the motion of some single two-dimensional linear vehicles moving on a random track. *International Journal of Mechanical Sciences, 2*, 168–178.
Kozin, F., & Bogdanoff, J. L. (1963). Comment on: 'The Behavior of Linear Systems With Random Parametric Excitation', by T. K. Caughey and, J. K. Dienes. *Journal of Matehmatical Physics, 92*, 336–337.
Krée, P. (1990). Methods of distributions in the modelling of random vibrations. *Structural Safety, 8*(1–4), 131–141.
Krée, P., & Soize, C. (1986). *Mathematics of random phenomena: Random vibrations of mechanical structures*. Dordrecht: D. Reidel Publishing Company.

Krée, P., & Soize, C. (2012). *Mathematics of random phenomena: Random vibrations of mechanical structures*. Berlin: Springer.
Krenk, S. (1979). Nonstationary narrow-band response and first-passage probability. *Journal of Applied Mechanics*, *46*(4), 919–924.
Krenk, S., Madsen, H. O., & Madsen, P. H. (1983). Stationary and transient response envelopes. *Journal of Engineering Mechanics, 109*(1), 263–277.
Krupien, V. L. (2015). R. V. Fields in equipped multidimensional vibroimpulsive 2D system. *Journal of Machinery Manufracture & R, 44*(6), 501–507.
Krysl, P., & Belytschko, T. (1995). Analysis of thin plates by element—Free Galerkin method. *Computational Mechanics, 17*(1–2), 26–35.
Krysl, P., & Belytschko, T. (1996). Analysis of Thin shells by the element-free Galerkin method. *International Journal of Solids and Structures, 33*(20–22), 3057–3080.
Kubota, Y. (1986). The accuracy of asymptotioc modal analysis (1st Report: Beam Under Point Random Forces). *Bulletin of JSME, 29*(254), 2612–2617.
Kubota, Y., & Dowell, E. H. (1986). Experimental investigation of Asymptotic modal analysis for a rectangular plate. *Journal of Sound and Vibration, 106*(2), 203–216.
Kudriavtsev, E. P. (1864). Application of asymptotic mechod for insevtiagting eigenfrequencies of elastic rectangular plates. *Strength Analysis, 10*, 352–362 (in Russian).
Kuifu, C., & Qunying, J. (1999). On the repeated natural frequencies for torsional vibration of shafts. *Applied Mathematics and Mechanics, 20*(11), 1279–1285.
Kukla, S., & Owczarek, A. (2005). Stochastic vibration of a Bernoulli-Euler beam under random excitation. *Scientific Research of the Institute of Mathematics and Computer Science, 4*(1), 71–78.
Kulkarni, A. M., Banerjee, J. R., & Sinha, P. K., Response of randomly excited orthotropic sandwich plates. *Journal of Sound and Vibration, 41*, 197–205.
Kulterbaev Kh. P., & Islamova, O. B. (2005). Vibration of membranes due to random excitations of different types. *Matematicheskoe Modelirovanie i Kraevye Zadachi (Mathematical Modelling and Boundary-Value Problems),* Part 1, 179–186.
Kulterbaev, Kh. P., & Pshenichkina, V. A. (2006). *Random processes and vibrations of civil engineering structures*. Volgograd: Volgograd State University (in Russian).
Kumar, S., & Srivastava, N. S. (1975). Investigation of noise due to structural vibration using a con-correlation technique. *The Journal of Acoustical Society of America, 57*(4), 769–772.
Kumaraian, M. L., Rebbagondla, J., Mathew, T. V., & Natarajan, S. (2018). Stochastic vibration analysis of functionally graded plates with material randomness using cell-based smoothed discrete shear gap method. *International Journal of Structural Stability and Dynamics*, article 1950037.
Kummer, M. (1978). On resonant classical hamiltonians with two equal frequencies. *Communications in Mathematical Physics, 58*(1), 85–112.
Kunieda, H. (1969). Modal densities of spherical shells, *NASA TM-X-72968*, 1969 (see also *The Shock and Vibration Bulletin, 40*, Part 3, 69–78).
Kurihara, M., & Shimogo, T. (1978). Vibration of an elastic beam subjected to discrete moving loads. *Journal of Mechanical Design, 100*(3), 514–519.
Laderéze, P., Puel, G., Deraemaeker, A., & Romeuf, T. (2006). Validation of structural dynamics models containing uncertainties. *Computer Methods in Applied Mechanics and Engineering, 195*(4–6), 373–393.
Ladygina Ye. V., & Manevitch, A. I. (1993). Free oscillations of a non-linear cubic system with two degrees of freedom and close natural frequencies. *Journal of Applied Mathematics and* Mechanics, *57*(2), 257–266.
Lakis, A. A., & Païdoussis, M. P. (1972). Prediction of the response of a cylindrical shell to arbitrary or boundary-layer-induced random pressure fields. *Journal of Sound and Vibration, 25*(1), 1–27.
Lal, A., & Singh, B. N. (2010). Stochastic free vibration of laminated composite plates in thermal environments. *Journal of Thermoplastic Composite Materials*, *23*(1), 57–77.
Lal, A., Singh, B. N., & Kumar, R. (2007). Natural frequency of laminated composite plate resting on elastic foundation with unceratin system properties. *Structural Engineering and Mechanics*, *27*(2), 199–222.

Lalanne, C. (1999). *Mechanical vibration and shock, random vibration* (Volume III). New York: Taylor and Francis.
Lallement, G., & Kosanek, J. (1993). Parametric correction of self-adjoint finite element models in the presence of multiple eigenvalues. In *Second International Conference on Modern Practice in Stress and Vibration Analysis* (pp. 593–603). IOP.
Lambert, R. F. (1959). *Response of bars (With internal and boundary damping) to transient and random excitation.* WADS Technical Report 58–569, Wright Air Development Center.
Lambert, R. G. (1976a). Fatigue analysis of multi-degree-of-fredom systems under random vibrations. Report SI-683, General Electric.
Lambert, R. G. (1976b). Analysis of fatigue under random vibration. *The Shock and Vibration Bulletin, 46*, Part 3, 55–72.
Lampard, D. G. (1954). Generalization of the Wiener-Khinchin theorem to nonstationary processes. *Journal of Applied Physics, 25*(6), 802–803.
Lancaster, H. O. (1972). Development of the notion of statistical dependence. *Mathematical Chronicle, 2*, 1–16.
Lang, G. F. (1983). Modal density—A limiting factor in analysis. *Sound and Vibration, 17*(3), 20–22.
Langevin, P. (1908). Sur la théorie du mouvement brownien. *Comptes Rendus de l'Académie des Sciences (Paris), 146*, 530–533 (in French).
Langley, R. S. (1986a). Structural response to non-stationary non-white stochastic ground motions. *Earthquake Engineering and Structural Dynamics, 14*, 909–924.
Langley, R. S. (1986b). Comparison of calculation methods for the response of linear oscillation to non-stationary broad band excitation. *Engineering Structures, 8*, 148–158.
Langley, R. S. (1989). A general derivation of the statistical energy analysis equations for coupled dynamic systems. *Journal of Sound and Vibration, 135*, 499–508.
Langley, R. S. (1994). The modal density and mode count of thin cylinders and curved panels. *Journal of Sound and Vibration, 169*, 43–53.
Langley, R. S. (1994). On the modal density and energy flow characteristics of periodic structures. *Journal of Sound and Vibration, 172*(4), 491–511.
Langley, R. S. (1996a). *A survey of analysis methods for dynamic response*. Prepared for the Engineering Science Data Unit. Nov ESD4 Data Item 97033.
Langley, R. S. (1996b). The modal density of anisotropic structural components. *The Journal of Acoustical Society of America, 99*(6), 3481–3487.
Langley, R. S. (1999). Some perspectives on wave-mode duality in SEA. In F. J.Fahy & W. G. Price (Eds.), *IUTAM Symposium on Statistical Energy Analsis* (pp. 1–12). Dordrecht: Springer.
Langley, R. S. (2000). SEA: Current and future research needs. In *Proceedings of the First International AutoSEA Users Conference* (pp. 126–135). San Diego, CA, .
Langley, R. S., & Bremner, P. (1999). A hybrid method for the vibration analysis of complex structural-acoustic systems. *The Journal of the Acoustical Society of America, 142*(4), 1657–1671.
Langley, R. S., & Brown, A. W. M. (2004). The ensemble statistics of the energy of a random system subjected to harmonic excitation. *Journal of sound and Vibration, 275*, 823–846.
Langley, R. S., Cicirello, A., & Deckers, E. (2018). The effect of generalised force correlations on the response statistics of a harmonically driven random system. *Journal of Sound and Vibration, 40*, 456–466.
Langley, R. S., & Cotoni, V. (2004). Response variance prediction in the statistical energy analysis of built-up system. *The Journal of the Acoustical Society of America, 115*, 706–718.
Langley, R. S., Legault, J., Woodhouse, J., & Reynders, E. (2013). On the applicability of the lognormal distribution in random dynamical systems. *Journal of Sound and Vibraition, 332*, 3289–3302.
Lassiter, L. W., Hess, R. W., & Hubbard, H. H. (1957). Experimental study of the response of simple panels to intense acoustic loading. *Journal of Aerospace Sciences, 24*, 19–24.

Lazer, A. C., & McKenna, P. J. (1986). Critical point theory and boundary value problems with nonlinearities crossing multiple eigenvalues II. *Communications in Partial Differential Equations, 11*(15), 1653–1676.
Lazza, A. (2010). An experimental method for accessing the modal density in a reverberation chamber. In *Asia-Pacific International Symposium on Electromagnetic Compatibility* (pp. 578–581). IEEE Press.
Lee, B. H. K. (1983). *Aeroelastic response of an aircraft wing to random loads*. Aeronautical Report LR-613, National Reserch Council, Canada, .
Lee, I. W., & Jung, G. H. (1997). An efficient algebraic method for computation of natural frequency and mode shape sensitivities—Part II. Multiple natural frequencies. *Computers and Structures, 62*(3), 437–443.
Lee, I. W., & Wang, G. H. (1997). An efficient algebraic method for the computation of natural frequency and mode shape sensitivities part 2: Multiple natural frequencies. *Computers and Structures, 62*(3), 437–443.
Lee, I. W., Jung, G. H., & Lee, J. W. (1996). Numerical method for sensitivity analysis of eigensystems with non-repeated and repeated eigenvalues. *Journal of Sound and Vibration, 195*(1), 17–32.
Lee, J. E., & Fassios, S. D. (1993). On the problem of experimental modal analysis based on multiple-excitation multiple-response data, Part 1: Dispersion analysis of continuous-time structural systems. *Journal of Sound and Vibration, 161*(1), 33–56.
Lee, M. C., & Penzien, J. (1983). Stochastic analysis of structures and piping systems subjected to stastionary multiple support excitations. *Earthquake Engineeing and Structural Dynamics, 11*(1), 91–100.
Lee, S. S. (1976). *Lanes of intensified response in structures excited by wide-band random excitation*. Ph.D. Thesis, Department of Mechanical Engineering, Massachusetts Institute of Technology, Cambridge, MA.
Lee T.-N., & Swanson, W. L. (1969). Computer programs for prediction of structural vibrations due to fluctuating pressure enviornment. *Shock and Vibration Bulletin, 40*(3), 57–68.
Lee, W. K., & Cho, D. S. (2000). Damping effect of a randomly excited autoparametric system. *Journal of Sound and Vibration*, 236 (1), 23–31.
Lee, Y. Y., Lam, K. C., Yuen, K. K., Lam, H., & Yao, J. (2003). Active vibration control of an aircraft cabin panel using piezoelectric sensors and actuators. *International Journal of Structural Stability and Dynamics, 3*(1), 131–141.
Lee, Y. Y., Mei, C., & Haung, J. K. (1998). Control of random vibrations of composite plates with piezoelectric actuators. In *Adaptive structures and material systems* (pp. 109–127). New York: ASME Press.
Leehey, P., & Davies, H. G. (1975). The direct and reverberant response of strings and membranes to convecting, random pressure fields. *Journal of Sound and Vibration, 38*(2), 163–184.
Leger, P., & Wilson, E. L. (1988). Modal summation methods for structural dynamic computations. *Earthquake Engineering and Structural Dynamics, 16*, 23–27.
Lei, Z., & Qiu, C. (2000). Neumann dynamic stochastic finite element method of vibration for structures with stochastic parameters to random excitation. *Computers and Structures, 77*(6), 651–657.
Leibowitz, R. C. (1975). Vibroacoustic response of turbulence excited thin rectangular finite plates in heavy and light fluid media. *Journal of Sound and Vibration, 40*, 441–495.
Leissa, A. W. (1969). *Vibration of plates*, NASA SP-160. Washington, D.C.: National Aeronautics and Space Administration.
Leissa, A. W. (1993). *Vibration of shells*. Washington, DC: The Acoustical Society of America.
Lemons, D. S., & Gythiel, A. (1997). Paul Langevin's 1908 paper "on the theory of brownian motion" [Sur la théorie du mouvement brownien. *Comptes Rendus de l'Académie des Sciences (Paris)* 146, 530–533 (1908)], *American Journal of Physics, 65*(11):1079–1081.
Lennox, W. C., & Fraser, D. A. (1974). On the first-passage distribution for the envelope of a nonstationary narrow-band stochastic process. *Journal of Applied Mechanics, 41*(3):793–797.

Leontief, W. (1982). Academic economics, letter to editor. *Science, 217*(4555), 104.
Leung, A. Y. T. (2001). Dynamic stiffness for structures with distributed deterministic or random loads. *Journal of Sound and Vibration, 242*(3), 377–395.
Lewinski, T. (1987). On refined plate model based on kinematic assumptions. *Ingenieur-Archiv, 57*, 133–146.
Lindberg, G. M., & Olson, M. D. (1967). *Vibration modes and random response of a multibay panel system using finite elements*. Report No. AD0669294, Ottawa, Ontario, Canada.
Li, A., & Dowell, E. H. (2004). Asymptotic modal analysis of dynamical systems: The effect of modal cross-correlations. *Journal of Sound and Vibration, 276*(1–2), 65–80.
Li, C. F., & Adhikari, S. (2009). Response statistics of linear stochastic systems: A joint diagnostic approach. In *50th Structures, Structural Dynamics and Materials Conference, AIAA Pape* (pp. 2009–2253). Palm Springs, CA.
Li, G., Rabiz, H., Yelvington, P. E., Oluwole, O. O., Bacon, F., Kolb, C. E., & Schoendorf, J. (2010). Global Sensitivity analysis for systems with independent and/or correlated inputs. *Journal of Physcical Chemistry, 114*(19), 6022–6032.
Li, J. (2016). Probability density evolution of method: Background, significance and recent developments. *Probabilistic Engineering Mechanics, 44*, 111–116.
Li, J., & Chen, J. B. (2004). Probability density evolution of method for dynamic response analysis of structure with uncertain parameters. *Computational Mechanics, 34*(5), 400–409.
Li, J., & Chen, J.-B. (2005). Dynamic response and reliability analysis of structures with uncertain parameters. *International Journal for Numerical Methods in Engineeing, 62* (2), 289–315.
Li, J., & Chen, J. B. (2008). The principle of preservation of probability and the generalized density evolution equation. *Structural Safety, 30*, 65–77.
Li, J., & Chen, J. (2009). *Stochastic dynamics of structures*. Singapore: Wiley (Asia).
Li, J., & Jin, X.-D. (2005). Response of flexure-torsion coupled composite thin-walled beams with closed cross-sections to random loads. *Mechanics Research Communications, 32*(1), 25–41.
Li, J., Hua, H.-X., Shen, R.-Y., & Jin, X.-D. (2004). Stochastic vibration of axially loaded monosymmetric Timoshenko thin-walled beam. *Journal of Sound and Vibration, 274*, 915–938.
Li, J., & Liao, S.-T. (2001). Response analysis of stochastic parameter structures under non-stationary excitation. *Computational Mechanics, 27*, 610–668.
Li, J., Ma, Z., Wang, Z., & Narita, P. (2016). Random vibration control of laminated composite plates with piezoelectric fiber reinforced composites. *Acta Mechanica Solida Sinica, 29*(3), 316–327.
Li, J., Shi, C., Kong, X., Li, X., & Wu, W. (2014). Stochastic response of an axially loaded composite beam exhibiting bending-torsion coupling. *Archive of Applied Mechanics, 84*(1), 109–122.
Li, J., Stewart, M. G., Masia, M. J., & Lawrence, S. J. (2016). Spatial correlation of material properties and structural strength of masonry in horizontal bending. *Journal of Structural Engineering, 142*(110), article 04016112.
Li, L., Hu, Y., & Wang, X. (2012). A parallel way for computing eigenvector sensitivity of asymmetric damped systems with distinct and repeated eigenvalues. *Mechanical System Signal Procedure, 30*, 61–77.
Li, L., Hu, Y., Wang, X., & Ling, L. (2013). Eigensensitivity analysis of damped systems with distinct and repeated eigenvalues. *Finite Elements in Analysis and Design, 72*, 21–34.
Li, Q. S., Zhang, Y. H., Wu, J. R., & Lin, J. H. (2004). Seismic random vibration analysis of tall buildings. *Engineering Strucutres, 26*(12), 1767–1778.
Li, X. J. (2014). Analysis of random response of a simply supported beam subjected to a distributed load. *Applied Mechanics and Materials, 518*, 120–125.
Li, Y.-B., Mulani, S.-B., Kapania, R. K., Fei, Q.-G., & Wu, S.-Q. (2017). Non-stationary random vibration analysis of structures under multiple correlated normal random excitations. *Journal* of *Sound* and *Vibration, 400*, 481–507.
Li, Y.-Y., Zhang, Y.-H., & Kennedy, D. (2017). Random vibration analysis of axially compressed cylindrical shells under turbulent boundary layer in a symplectic system. *Journal of Sound and Vibration, 406*, 161–180.

Liang, J. W., Chauduri, S. R., & Shinozuka, M. (2007). Simulation of nonstationary stochastic processes by spectral representation. *Journal of Engineering Mechanics, 133*, 616–627.

Liang, Z., & Lee, G. C. (2015). *Random vibration: Mechanical, structural, and earthquake engineering applications*. Boca Raton: CRC Press.

Lidsky, V. B., & Tovstik, P. E. (1984). Spectra in theory of shells. *Advances in Mechanics, 7*(2), 26–54 (in Russian).

Lychev, S. A., & Sidorov, Yu. V. (2001). Nonstationary vibrations of sandwich spherical shells with coincident spectra. *Izvestiya VUZOV, Civil Enginering* (4), 31–37 (in Rusian).

Lifshits, M. A. (1995). *Gaussian random functions*. Berlin: Springer.

Liepmann, H. W. (1952). On the application of statistical concepts to the bulleting problem. Journal *of the Aeronautical Sciences, 19*, 793–800.

Liepmann, H. W. (1955). Extension of the statistical approach to buffeting and gust response of wings of finite span. *Journal of Aerospace Scuiences, 22*(3), 197–203.

Lim, K. B., Juang, J., & Ghaemmaghami, P. (1983). Eigenvector derivatives of repeated eigenvalues using singular value decomposition. *Journal of Guidance Control and Dynamics, 12*(2), 282–283.

Lin, B. C., Tadjbakhsh, I. G., Papageorgiou, A., & Ahmadi, G. (1989). Response of base-isolated buildings to random excitations with a Clough-Penzien model. *Earthquake Engineering and Structural Dynamics, 18*, 49–62.

Lin, J. H. (1992). A fast CQC algorithm of PSD matrices for random seismic responses. *Computers and Structures, 44*(3), 683–687.

Lin, J., & Williams, F. W. (1997). Accurate high-speed computation of non-stationary random structural response. *Engineering Structures, 19*(7), 586–593.

Lin, J., Zhang, W., & Li, J. (1994). Structural responses to arbitrarily coherent stationary random excitations. *Computers and Structures, 50*(5), 629–633.

Lin, J. H., Zhang, W. S., & Williams, F. W. (1994). Pseudo-excitation algorithm for nonstationary random seismic responses. *Engineering Strucutres, 16*(4), 270–276.

Lin, J.-H., Shen, W.-P., & Williams, F. W. (1997). Accurate high speed computation of non-stationary random structral response. *Engineering Structres, 19*(7), 586–593.

Lin, J. H., Zhao, Y., & Zhang, Y. H. (2003). Accurate and highly efficient algorithms for structural stationary/non-stationary random responses. *Computer Methods in Applied Mechanics and Engineering, 191*, 103–111.

Lin, J. H., Zhang, Y. H., & Zhao, Y. (2011). Pseudo-excitation method and some recent developments. *Procedia Engineering, 14*, 2438–2453.

Lin, J. H., Song, G. Z., Sun, Y., & Williams, F. W. (1995). Non-stationary random seismic responses of non-uniform beams. *Soil Dynamics and Earthquakes Engineering, 14*, 301–306.

Lin, J. H., Zhang, Y. H., & Zhao, Y. H. (2014). Seismic random response analysis. In W. F. Chen & L. Duan (Eds.), *Bridge engineering handbook: Construction and maintanence* (pp. 133–162). Boca Raton: CRC Press.

Lin, J. T., & Alphenaar, B. (2010). Enhancement of energy harvested from a random vibration source by magnetic coupling of a piezoelectric cantilever. *Journal of Intelligent Material Systems and Structures, 21*(13), 1337–1341.

Lin, O. R., Liu, Z.-X., & Wang, Z.-L. (2001). Cylindrical panel interior noise control using a pair of piezoelectric actuator and sensor. *Journal of Sound and Vibration, 246*(3), 525–541.

Lin, S.-C. (2001). The probabilistic approach for rotating Timoshenko beams. *International Journal of Solids and Structures, 38*(40), 7197–7213.

Lin, Y., Zhao, Y., & Zheng, Y. (2001). Accurate and highly efficient algorithms for structural stationary/non-stationary random responses. *Computer Methods in Applied Mechanics and Engineering, 191*(1–2), 103–111.

Lin, Y. J. (1985). Random vibration of elastic beams and circular plates resting on viscoelastic foundations. *Journal of Vibration, Acoustic Stress, and Reliability in Design, 107*(2), 180–187.

Lin, Y. K. (1962). Stresses in continuous skin-stringer panels under random loading. *Journal of Aerospace Science, 29*, 67–75.

Lin, Y. K. (1963). Nonstationary response of continuous structures to random loading. *The Journal of the Acoustical Society of America, 35*, 222–227.
Lin, Y. K. (1964a). Random Vibration of a Myklestad beam. *AIAA Journal, 2*(8), 1448–1451.
Lin, Y. K. (1964b). On nonstationary shot noise. *The Journal of the Acoustical society of America, 36*, 82–84.
Lin, Y. K. (1965a). Non-stationary excitation in linear systems treated as sequences of random pulses. *The Journal of the Acoustical Society of America, 38*, 453–460.
Lin, Y. K. (1965b) Response of a multi-spanned beam and panel systems under noise excitation, Air Force Systems Command. Report AFML-TR-64–348, Part 1, Wright-Patterson Air Force Base, Ohaio.
Lin, Y. K. (1965c) Transfer matrix representation of flexible airplanes in gust response study. *Journal of Aircraft, 2*(2), 116–121.
Lin, Y. K. (1976). Random vibration of periodic and almost periodic structures. *Mechanics Today, 3*, 93–124.
Lin, Y. K., & Cai, G. Q. (1995). *Probabilistic structural dynamics: Advanced theory and applications*. New York: McGraw-Hill.
Lin, Y. K., & Yang, Y. (1987). Evolutionary Kanai-Tajini earthquake model. *Journal of Engineering Mechanics, 113*, 1119–1137.
Lin, Y. K., Kozin, F., Wen, Y. K., Casciati, F., Schuëller, G. I., Der Kiureghian, A., … Vanmarcke, E. H. (1986). Methods of stochastic structural dynamics. *Structural Safety, 3*(3), 167–194.
Lin, Y. K., & Maekawa, S. (1977). Decomposition of turbulence forcing field and structural response. *AIAA Journal, 15*(5), 609–610.
Lin, Y. K., Maekawa, S., Nijim, H., & Maestrello, L. (1977). Response of periodic beam to supersonic boundary-layer pressure fluctuation. In B. L. Clarkson (Ed.), *Stochastic problems in dynamics* (pp. 468–485). London: Pitman Publishers.
Lin, Y. K., & Shih, T. Y. (1982). Vertical seismic load effect on building response. *Journal of Engineering Mechanics, 107*, 331–343.
Lin, J., Zhang, W., & Zhang, Y. (2000). Seismic analysis of long span structures by means of random vibration approach. *Journal of Building Structures* (1).
Lindberg, G. M., & Olson, M. D. (1967). *Vibration modes and random response of a multi-bay panel system using finite elements*. Report LR-492, National Research Council of Canada.
Liu, B., Yue, Y.-J., Qin, Z.-B., & Chen, F. (2015). Random vibration spectrum analysis based on finite element method of high-pressure cylindrical shells. *Applied Mechanics and Materials, 757*, 147–152.
Liu, S. H., Zhao, D.-S., Tan, T.-S., & Liu, H.-D. (2004). The application of ANSYS in random vibration analysis of steel-truss structure. *Machinary*.
Liu, X. L. (2000). Derivation of formulas for peturbation analysis with modes of close eigenvalues. *Structural Engineering and Mechanics, 10*(5), 427–440.
Liu, Z., Liu, Z., & Xu, Y. (2019). Probability density evolution analysis of a shear-wall structure under stochastic ground motions by shaking table test. *Soil Dynamics and Earthquake Engineering, 122*, 53–66.
Lizarev, A. D. (1967). About low natural frequencies of axisymmetric vibrations of nonshallow spherical shells. *Mechanics of Solids* (3) (in Russian).
Lofrano, E., Paolone, A., & Vasta, M. (2016). Identification of uncertain vibrating beams through a pertubation approach. *Journal of Risk and Uncertainty in Engineering Systems, Part A: Civil Engineering, 2*(2), article C4015006.
Loeb, G. E., White, M. W., & Merzenich, M. M. (1983). Spatial cross-correlation. *Biological Cybernetics, 47*(3), 149–163.
Loh, C. H., Hung, T. Y., Chen, S. F., & Hsu, W. T. (2015). Damage detection in bridge structures using vibration data under random travelling vehicle loads. *Journal of Physics: Conference Series, 628*(1), article 012044.

Loha, T., & Dey, S. (2019). Stochastic longitudinal vibration of single walled carbon nanorods—A non local elasticity approach. In *Advances in computational methods in manufacturing* (pp. 1045–1058). Singapore: Springer.

Lomas, N. S., & Hayek, S. I. (1977). Vibration and acoustic radiation of elastically supported rectangular plates. *Journal of Sound and Vibration, 52*(1), 1–25.

Lord Rayleigh. (1919). On the problem of random vibrations and random flights in one, two, and three dimensions. *Philosophical Magazine, 37*, 321–347.

Lotz, R. (1971). *Random vibration of complex structures*. Ph.D. Thesis, Massachusetts Institute of Technology.

Lotz, R., & Crandall, S. H. (1973). Prediction and measurement of the proportionality constant in statistical energy analysis. *The Journal of the Acoustical Society of America, 54*, 516–524.

Lu, F., Gao, Q., Lin, J. H., & Williams, F. (2006). Non-stationary random ground vibration due to loads moving along a railway track. *Journal of Sound and Vibration, 298*, 30–42.

Lu, F., Kennedy, D., Williams, F. W., & Lin, J. H. (2008). Symplectic analysis of vertical random vibration for coupled vehicle–track systems. *Journal of Sound and Vibration, 317*(1), 236–249.

Luan, X. L., Wang, Y., Jin, X. L., & Huang Z. L. (2017). Optimal locations of piezoelectric patch on wideband random point—driven beam for vibration energy harvesting application. *AIME Journal of Vibration and Acoustics, 170*(1), 011011.

Luco, J. E., & Wong, H. L. (1986). Response of a foundation to a spatially random ground motion. *Earthquake Engineering and Structural Dynamics, 14*(6), 889–908.

Lucor, D., Su, C.-H., & Karniakidis, G. E. (2004). Generalized polynomial chaos and random oscillators. *International Journal for Numerical Methods in Engineering, 60*(3), 571–596.

Luo, Y., & Huang, G. (2017). Characterizing dependence of extreme wind pressures. *Journal of Structural Engineering, 143*(4), article 04016208.

Luongo, A. (1993a). Eigensolution sensitivity for nonsymmetric matrices with repeated eigenvalues. *AIAA Journal, 31*(7), 1321–1328.

Luongo, A. (1993b). Eigensolutions sensitivity for nonsymmetric matrices with repeated frequencies. *AIAA Journal, 31*(12), 2384.

Lutes, L. D. (1986). Cumulants of stochastic response for linear system. *Journal of Engineering Mechanics, 112*(10), 1062–1075.

Lutes, L. D., Chen, Y.-T. T., & Tzung, S.-H. (1980). First passage approximations for simple oscillators. *Journal of Engineering Mechanics, 106*(6), 1111–1124.

Lutes, L. D., & Hu, S.-L. J. (1986). Non-normal stochastic response of linear system. *Journal of Applied Mechanics, 112*(2), 127–141.

Lutes, L. D., & Sarkani, S. (2004). *Random vibrations: Analysis of structural and mechanical systems*. Amsterdam: Elsevier.

Lv, W., & Henry, A. (2016). Direct calculation of modal contributions to thermal conductivity via Green-Kubo modal analysis. *New Journal of Physics, 18*, article 013208.

Lyon, R. H. (1956a). Response of strings to random noise fields. *The Journal of the Acoustical Society of America, 28*, 391–398.

Lyon, R. H. (1956b). Propagation of correlation functions in continuous media. *The Journal of the Acoustical Society of America, 28*, 76–79.

Lyon, R. H. (1963a). Noise reduction of rectangular enclosureswith one flexible wall. *The Journal of the Acoustical Society of America, 35*(11), 1791–1797.

Lyon, R. H. (1963b). *Random vibration studies of coupled structures in electronic equipments. Report ASD-TDR-63–205*, Air Force Systems Command, Wright Patterson Air Force Base, Ohio.

Lyon, R. H. (1967). Spatial response concentrations in extended structures. *Journal of Engineering for Industry, 89*, 754–758.

Lyon, R. H. (1975). *Statistical energy analysis of dynamical systems: Theory and applications*. Cambridge, MA: MIT Press.

Lyon, R. H., & Dejong, R. G. (1995). *Theory and application of statistical energy analysis*. Boston: R. H. Lyon Corporation.

Lyon, R. H., & Eichler, E. (1964). Random vibration of connected structures. *Journal of the Acoustical Society of America, 36*(7), 1344–1354.
Lyon, R. H., & Maidanik, G. (1964). Statistical methods in vibration analysis. *AIAA Journal, 2*, 1015–1024.
Loynes, R. M. (1968). On the concept of the spectrum for non-stationary processes. *Journal of the Royal Statistical Society, Series B, 30*(1), 1–30.
Lyrintzis, C. S., & Vaicaitis, R. (1987). Structure-borne noise generation and transmission. *Probabilisitc Engineering and Mechanics, 2*(3), 114–120.
Ma, X., Leng, X. L., Meng, G., & Fang, T. (2004). Evolutionary earthquake responses of uncertain structure with bounded random parameters. *Probabilistic Engineering Mechanics, 19*(3), 239–246.
MacCalley, R. C. Jr. (1957). The evaluation of random noise integral. *25th Shock and Vibration Bulletin,* Part II, 243.
Mace, B. R. (2003). Statistical energy analysis, energy distribution modes and system modes. *Journal of Sound and Vibration, 264*, 391–409.
Maceyko, M. K. (1983). *Mean-square response of an infinite Bernoulli-Euler beam to non-stationary random excitation*. Master's Thesis, San Diego State University.
Maceyko, M. K., & McGhie, R. D. (1987). Mean-square response of an infinite Bernoulli-Euller beam to nonstationary random excitation. *AIAA Journal, 25*(6), 864–870.
Macvean, D. B. (1980). Response of vehicle accelerating over random profile. *Archive of Applied Mechanics, 49*, 375–380.
Madsen, P. H., & Krenk, S. (1982). Stationary and transient response statistics. *Journal of Engineering Mechanics, 108*, 622–635.
Madsen, H. O., Krenk, S., & Lind, N. C. (1986). *Methods of structural safety*. Englewoods Cliffs, NJ: Prentice Hall.
Maestrello, L. (1963). Measurements and analysis of the respnse field of turbulent boundary layer excited panels. *Journal of Sound and Vibration, 2*, 270–292.
Maestrello, L. (1965). Meausrement of noise radiated by boundary layer excited panels. *Journal of Sound and Vibration, 2*, 100–115.
Maestrello, L. (1967). Use of turbalent model to calculate the vibration and radiation responses of a panel, With practical suggestions for reducing sound level. *Journal of Sound and Vibration, 5*(3), 407–420.
Maestrello, L. (1968). Design criterion of panel structure excityed by turbulent boundary layer. *Journal of Aircraft, 5*, 321–328.
Maestrello, L. (1969). Radiation from a panel response to a supersonic turbulent boundary layer. *Journal of Sound and Vibration, 10*, 261–275.
Maestrello, L. (1996). Control of panel response to turbulent boundary-layer and acoustic excitations. *AIAA Journal, 34*(2), 259–264.
Magrab, E. B., & Burroughs, C. (1972). Forced harmonic and random vibrations of concentric cylindrical shells immersed in acoustic fluids. *Journal of the Acoustical Society of America, 52*, 858.
Mahalingam, S., & Bishop, R.E. D. (1974). The response of a system with repeated natural frequencies to force and displacement excitation. *Journal of Sound and Vibration, 36*(2), 285–295.
Mahmoudkhani, S., Haddadpour, H., & Navazi, H. M., Free and forced random vibration analysis of sandwich plates with thick viscoelastic cores. *Journal of Vibration and Control, 19*(14), 2223–2240.
Mahshingam, S. (1974). The response of a system with repeated natural frequencies to force and displacement excitation. *Journal of Sound and Vibration, 36*(2), 285–295.
Maidanik, G. (1961). Use of delta function for the correlations of pressure fields. *The Journal of the Acoustical Society of America, 33*, 1598–1605.
Maidanik, G. (1977). Some elements in statistical energy analysis. *Journal of Sound and Vibration, 52*, 171–191.

Maidanik, G., Becker, K. J., & Maga, L. J. (2006). Replacement of a summation by an integration in structural acoustics. *Journal of Sound and Vibration, 291*, 323–348.

Maidanik, G., & Lyon, R. H. (1961). Response of strings to moving noise fields. *The Journal of the Acoustical Society of America, 33*(11), 1606–1609.

Makeev, V. P., Grinchenko, N. I., & Pavliuk Yu. S. (1984). *Statistical problems of dynamics of elastic structuires*. Moscow: "Nauka" Publishing House (in Russian).

Maksimov, L. S. (1966). About design of the passive sound isolation on action in the form of random process. *Mechanics of Solids*, (3) (in Russian).

Malekzadeh, P., Bahranifard, F., & Zraee, S. (2013). Three-dimensional free vibration analysis of functonally-graded cylindrical panels with cut-out using Chebyshev-Ritz method. *Composite Structures, 105*, 1–13.

Malhotra, A. K., & Penzien, J. (1970). Nondeterministic analysis of offshore structures. *Journal of Engineering Mechanics, 96*, 985–1003.

Malow, B. Chladni figures: Amazing resonance experiment.

Mandal, R. (2014). *Estimation* of *structural respose by stochastic Galerkin method for elastic large deflection analysis of steel plate under compressive load*. Ph.D. Thesis, Department of System Design for Ocean Space, Yokohama National University, Japan.

Manevitch, A. I., & Manevitch, L. I. (2003). Free oscillations in conservative and dissipative symmetric cubic two-degree-of-freedom systems with close natural frequencies. *Meccanica, 38*(3), 335–348.

Manevitch, A. I. (1999). Bifurcations of stationary free oscillations of a cubic two-degree-of-freedom system having close eigenfrequencies. In *Theoretical foundations of civil engineering* (Vol. 7, pp. 133–140). Warsaw: DW PW (in Russian).

Manohar, C. S., & Adhikari, S. (1998a). Statistics of vibration energy flow in randomly parametered trusses. *Journal of Sound and Vibration, 217*(1), 43–74.

Manohar, C. S., & Adhikari, S. (1998b). Dynamic stiffness of randomly parameterized beams. *Probabilistic Engineering Mechanics, 13*(1), 39–51.

Manohar, C. S., & Ibrahim, R. A. (1999). Progess in structural dynamics with stochastic parameter variations: 1987–1998. *Applied Mechanics Reviews, 52*(5), 177–197.

Manohar, C. S., & Iyengar, R. N. (1994). Free vibration analysis of stochastic strings. *Journal of Sound and Vibration, 176*(1), 35–48.

Manohar, C. S., & Shashirekha, R. B. (1996). Harmonic response analysis of stochastic rods using spatial stochastic averaging. In *IUTAM Symposium on Advances in Nonlinear Stochastic Mechanics* (pp. 289–298). Dordrecht: Kluwer.

Manolis, G. D., Juhn, G., Constantinou, M. C., & Reinhorn, A. M. (1990). *Secondary systems in* base-*isolated structures experimental investigation, stochastic response and stochastic sensitivity*. Technical Report No. NCEER-90–0013, State University of New York at Buffalo, Buffalo, NY.

Manolis, G. D., & Bagtzoglou, A. C. (1992). A numerical comparative study of wave propagation in inhomogeneous and random media. *Computational Mechanics, 10*(6), 397–413.

Maori, S. F., & Caffrey, J. P. (2017). Transient response of a SDOF system with an inerter to non-stationary stochastic excitation. *Journal of Applied Mechanics, 84*(4), article JAM-16–1550.

Marchenko, B. G., & Shcherbak, L. N. (1975). *Linear random process and their applications*. Kies: "Naukova Dumka" Publishing House (in Russian).

Mari, D. D., & Kotz, S. (2001). *Correlation and dependence*. Singapore: Word Scientific.

Mark, W. D. (1970). Spectral Analysis of the convolution and filtering of non-stationary stochastic processes. *Sound and Vibration, 11*, 19–63.

Mark, W. D. (1986). Power Spectrum representation for nonstationary random vibration. In I.Elishakoff & R. H. Lyon (Eds.), *Random vibration-status and recent developments* (pp. 211–240). Amsterdam: Elsevier.

Martin, K. F., & Ghlaim, K. H. (1987). Non orthogonality arising from repeated roots. In *Proceedings of the 5th International Modal Analysis Conference* (pp. 1331–1337).

Masri, S. F. (1978). Response of a multidegree-of-freedom system to nonstationary random excitation. *Journal of Applied Mechanics, 45*, 649–656.

Masri, S. F. (1987). Response of decertified plates to transversal and in-plane non-stationary random excitation. *Journal of Sound and Vibration, 114*, 481–494.
Masri, S. F., & Aryafar, A. (1977). Response of beam to stochastic boundary excitation. *Journal of Engineering Mechanics, 103*, 807–822.
Masri, S. F., & Caffrey, J. P. (2017). Transient response of a SDOF system with an inerter to nonstationary stochastic excitation. *Journal of Applied Mechanics, 84*(4), 041005.
Masri, S. F., & Miller, R. K. (1982). Compact probabilistic representation of random processes. *Journal of Applied Mechanics, 49*, 871–876.
Masri, S. F., & Udwadia, F. (1977). Transient response of a shear beam to correlated random boundary excitations. *Journal of Applied Mechanics, 44*, 487–491.
Masur, E. F. (1984). Optimal structural design under multiple eigenvalue constraints. *International Journal of Solids and Structures, 20*, 211–231.
Masur, E. F. (1985). Some additional comments on optimal structural design under multiple eigenvalue constraints. *International Journal of Solids and Structures, 21*, 117–120.
Masur- Sniady, K., & Sniady P. (2001). Dynamic response of a micro-periodic beam under moving load—Deterministic and stochastic approach. *Journal of Theoretical and Applied Mechanics, 39*, 323–338.
Maury, C., Gardonio, P., Elliott, S. J. (2002). A wavenumber approach to modelling the response of a randomly excited panel, Part II: Application to Aircraft panels excited by turbulent boundary layer. *Journal of Sound and Vibration, 252*, 115–139.
Maxit, L. (2016). Simulation of the pressure field beneath a turbulent boundary layer using realizations of uncorrelated wall plane waves. *The Journal of the Acoustical Society of Anmerica, 140*(2), 1268–1285.
Maymon, G. (1975). Modal densities of stiffened, axially loaded cylindrical shells. *Journal of Sound and Vibration, 42*(1), 115–127.
Maymon, G. (1986). Response of aeronautical structures to random acoustic excitation—A stress modes approach. In I. Elishakoff & R. H. Lyon (Eds.), *Random vibration—Status and recent developments* (pp. 267–278). Amsterdam: Elsevier.
Maymon, G. (1994). Random response of indeterministic structures subjected to stationary random excitation: A practical engineering solution. *Journal of Sound and Vibration, 172*(2), 211–229.
Maymon, G. (2005). What about uncertainties in the model?. In *46th AIAA/ASME/ASVE/AHS/ASC Structures, Structural Dynamics, and Materials Conference* (pp. 1–13).
Maymon, G. (2008). *Structural dynamics and probabilistic analysis for engineers*. Amsterdam: Elsevier.
Maymon, G. (2012). On the uncertainties of a structural model. *Structural Longevity, 7*(1), 1–11.
Mazur-Sniady, K., & Sniady, P. (1986). Dynamic reponse of linear structures to random streams of arbitrary impuslses in time and space. *Journal of Sound and Vibration, 110*(1), 59–68.
Mazzoni, D. (2003). An efficient approximation for the vibro-acoustic response of a turbulent boundary layer excited panel. *Journal of Sound and Vibration, 264*, 951–971.
Mead, D. J. (1964). Bond stresses in a randomly vibrating sandwich plate: Single mode theory. *Journal of Sound and Vibration, 1*(3), 258–259.
Mead, D. J., & Malik, A. K. (1976). An approximate method of predicting the response of periodically supported beams subjected to random connected loading. *Journal of Sound and Vibration, 47*(4), 457–471.
Mead, D. J., & Malik, A. K. (1976). Material damping under random excitation. *Journal of Sound and Vibration, 45*(4), 487–494.
Mead, D. J., & Pujara, K. K. (1971). Space-harmonic analysis of periodically supported beams: Response to convected random loading. *Journal of Sound and Vibration, 14*(4), 525–532.
Mead, D. J., & Wilby, E. W. (1966). The random vibrations of a multi-supported heavily damped beam. *Shock and Vibration Bulletin, 35*(3), 45–55.
Mehlhose, S., Von Scheidt, J., & Wunderlich, R. (1999). Random eigenvalue problems for bending vibrations of beams. *Zeitshcrft für angewandte Mathematik and Mechnik, 79*(10), 693–702.

Melzer, H.-J. (1979). Schlanke Bauwerke Unter Zufaalllast, *VDI-Berichte,* Number 355 (in German).
Melzer, H.-J. (1981). *Tragwerkschwingungen unter Zufallslast mit nicht-Gaussschen Wahrscheinlichkeitsverteilung.* Ph.D. Dissertation, Technical University Munich, Federal Republic of Germany (in German).
Melzer, H.-J., & Schuëller, G. I. (1979). Statistical evaluation of the response of slender structures under stochastic dynamic loading. In *3rd International Conference on Apllications of Statistics and Probability,* (ICASP 3), Sidney, Australia (pp.149–170).
Mendal, R. (2014). *Estimation of structural response by stochastic Galerkin method for elastic large deflection analysis of steel plates under compressor load.* Ph.D. Thesis, Department of System Design for Ocean Space, Yokohoma National University, Japan.
Meng, Q.-S. (1985). Nonstationary random vibration of one-degree and multidegree of freedom systems. *Applied Mathematics and Mechanics, 6*(4), 343–352.
Menh, N. C. (1986). On the investigation of stochastic dynamic systems by successive approximation. *ZAMM: Zeitschrift für angewandte Mathematik and Mechanik, 66*(6), 250–251.
Mercer, C. A., & Hammond, J. K. (1968). On the representation of continuous random pressure fields at a finite set of points. *Journal of Sound and Vibration, 7*(1), 49–61.
Mercier, H., Ammann, W. J., Deischl, F., Eisenmann, J., Floegl, I., Hirsch, G. H., et al. (1995). , Nussbaumer, H., Pretlove A.J., Rainer J.H., Saemann E.-U. and Steinbeisser, L., *Vibration problems in structures: practical guidelines.* Basel: Birkhäuser .
Merz, S., Kinns, R. and Kessissoglou, N. (2009). Structural and acoustic responses of a submarine hull due to propeller forces. *Journal of Sound and Vibration, 325*(1–2), 266–286.
Michaelov, G., Sarkani S., & Lutes L.D. (1999). Spectral characteristic of nonstationary random processes—A critical review. *Structural Safety, 21*(3), 223–244.
Mignolet, M. P. & Spanos, P.D., Recursive simulation of stationary multivariate random processes—Part I. *Journal of Applied Mechanics, 54*(3), 674–680.
Miles, J.W. (1954). On structural fatigue under random loading. *Journal of the Aeronautical Sciences, 21*, 753–762.
Miles, R. N., & Bigelow, S.P. (1994). Random vibration of a beam with a stick-slip end condition. *Journal of Sound and Vibration, 169*(4), 445–457.
Miller, D. K., & Hart F.D. (1967). Modal density of thin circular cylinders, *NASA CR-897.*
Mills-Curran, W.C. (1998). Calculation of eigenvector derivatives for structures with repeated eigenvalues. *AIAA Journal, 26*(7), 867–871.
Mills-Curran, W. C. (1990). Comment on "Eigenvector derivatives for structures with repeated eigenvalues." *AIAA Journal, 28*(10), 1846.
Mignolet, M. P., & Spanos, P. D. (1987). Recursive simulation of stationary multivariate random processes—Part 1. *Journal of Applied Mechanics, 54*(3), 674–680.
Mixson, J. S., Farassat, F., Leatherwood, J. D., Prydz, R., & Revell, J. D. (1983). Interior noise considerations for advanced high-speed turboprop aircraft. *Journal of Aircraft, 20*(9), 791–787.
Mixson, J. S., & Roussos, L. A. (1987). Acoustic fatigue: overview of activities at NASA langley. *NASA TM-89143.*
Morrow, C. T., Troesch, B. A., & Spence, H. R. (1961). Random response of two coupled resonators without loading. *The Journal of the Acoustical Society of America, 33*, 446–455.
Morse, P. M., & Ingard, K. U. (1968). *Theoretical acoustics.* New York: McGraw-Hill.
Moshinski, A. I. (2014). Equation of longitudinal vibrations of a beam with coefficients periodically dependent on the coordinate. *Mechanics of Solids, 49*(5), 518–530.
Moskalenko, V. N. (1969). On vibrations of multispan plates. *Strength Analysis, 14* (in Russian).
Moskalenko, V. N., & Muravin, E. N. (1969). Acoustic radiation of a plate undergoing random vibration. In V. V. Bolotin (Ed.), *Proceedings of the Scientific Technical Conference Based on Results of Scientific Investigations in Years 1968–1969, Section of Energetic Machinery, Subsection of Dynamics and Strength of Machines* (pp. 26–35). Moscow: Power Engineering Institute Press (in Russian).

Moskalenko, V. N., & Sharii, N. V. (1970). Analysis of stiffened shells subjected to random exploitation loads. In *Proceedings of the Moscow Power Engineering Institute, Dynamics and Strength of Machines* (Vol. 74, pp. 29–36). Moscow: Power Engineering Institute Press (in Russian).

Mould, S. Random couscous snaps into beautiful patterns. Retrieved July 8, 2019, from https://www.youtube.com/watch?v=CR_XL192wXw.

Mukherjee, A., & Mukhopadhyay, M. (1988). Response of stiffened plate structures under stochastic excitation. *Computer Methods in Applied Mechanics and Engineering, 71*(3), 273–292.

Mukherjee, S., Gopalakrishnan, S., & Ganguli, R. (2018). Stochastic time domain spectral element analysis of beam structures. *Acta Mechanica*, in press.

Mundkur, G., Bhat, R. B., & Neriya, S. (1994). Vibration of plates With cut-outs using boundary characteristic orthogonal polynomial functions in the Rayleigh-Ritz method. *Journal of Sound and Vibration, 176*(1), 136–144.

Muravin, E. L. (1972). Radiation of infinite plate on supports into an acoustic layer. In M. P. Markovets, Yu. M. Pavlov, & A. B. Matveev (Eds.), *Proceedings of the Moscow Power Engineering Institute, Heat Energetics and Energy Machinery Engineering* (Vol. 104, pp. 75–78) (in Russian).

Muscolino, G. (1986a). Stochastic seismic analysis for non-classically damped linear systems. *Revue roumaine des sciences techniques. Série de mécanique appliquée, 31*(2), 171–177.

Muscolino, G. (1986b). Stochastic analysis of linear structures subjected to multicorrelated filtered noises. *Engineering Structures, 8*, 119–126.

Muscolino, G. (1988a). Nonstationary evelope in random vibration theory. *Journal of Engineering Mechanics, 114*(8), 1396–1413.

Muscolino, G. (1988b). Nonstationary envelope in random vibration theory. *Journal of Engineering Mechanics, 114*(8), 1396–1413.

Muscolino, G. (1990). Analytical evaluation of structural response for stationary multicorrelated input. *Computers and Structures, 34*(2), 319–326.

Muscolino, G. (1991). Non-stationary pre-envelope covariances of non-classically damped systems. *Journal of Sound and Vibration, 149*, 107–123.

Muscolino, G. (1995). Linear systems excited by polynomial forms of non-Gaussian filtered processes. *Probabilistic Engineering Mechanics, 10*, 35–44.

Muscolino, G. (1996). Dynamically modified linear structures: Deterministic and stochastic response. *Journal of Engineering Mechanics, 122*(11), 1004–1051.

Muscolino, G. (2001). Stochastic dynamics for structural engineering problems: A review. In *Civil and structural engineering computing* (pp. 287–318). Saxe-Colurg Publication.

Muscolino, G. (2014). Stochastic analysis of linear systems. In M. Beer, I. A. Kougioumtzoglou, E. Patelli, & I.S-K. Au (Eds.), *Encyclopedia of earthquake engineering* (pp. 1–28). Berlin: Springer.

Muscolino, G., & Ricciardi, G. (1999). Probablity density function of MDOF structural systems under non-normal delta-correlated inputs. *Computer Methods in Applied Mechanics, 168*(1–4), 121–133.

Muscolino, G., & Sofi, A. (2012). Stochastic analysis of structures with uncertain-but-bounded parameters via improved interval analysis. *Probabilistic Engineering Mechanics, 28*, 152–163.

Muscolino, G., Benfratello, S., & Sidoti, A. (2002). Dynamic analysis of distributed parameter system subjected to a moving oscillator with random mass, velocity and acceleration. *Probabilistic Engineering Mechanics, 17*(1), 63–72.

Muscolino, G., Santoro, R., & Sofi, A. (2014). Explicit sensitivities of the response of discretized structures under stationary random processes. *Probabilistic Engineering Mechanics, 35*, 82–95.

Nachum, S., & Altus, E. (2007). Natural frequencies and mode shapes of deterministic and stochastic non-homogeneous rods and beams. *Journal of Sound and Vibration, 302*(4–5), 903–924.

Nadirashvili, N. S. (1988). Multiple eigenvalues of the laplace operator. *Sbornik Mathematics, 61*(1), 225–238.

Naess, A., & Moan, T. (2013). *Stochastic dynamics of marine structures*. Cambridge, U.K.: Cambridge University Press.

Nagem, R. (1991). Random vibration of point-driven beams. *Journal of Sound and Vibration, 151*, 168–174.
Nakagiri, S., Takabatake, H., & Tani, S. (1987). Uncertain eigenvalue analysis of composite laminated plates by the stochastic finite elelment method. *Journal for Engineering for Industry, 109*, 9–12.
Namachchivaya, N. S. (2016). Random dynamical systems: Addressing uncertainty, nonlinearity and predictability. *Meccanica, 51*, 2975–2995.
Naprstek, J., & Frýba, L. (1994). Interaction of a long beam on stochastic foundation with a moving random load. In N. S. Ferguson, H. F. Wolfe, & C.Mei (Eds.), *Structural Dynamics, Recent Advances: Proceedings of the 5th International Conference* (pp. 681–698), ISVR, Southampton.
Narayanan, S., (1975). *Structural optimization in random vibration enironment*. Ph.D. Thesis, Indian Institute of Technology Kanpur.
Narayanan, S., & Nigam, N. C. (1978). Optimum structural design in random vibration environments. *Engineering Optimization, 3*, 97–108
Nascimento, N. (1986). Stochastische Schwingungen punktweise erregter Saiten und Balken. *ZAMM: Zeitschrift für angewandte Mathematik und Mechanik, 63*, T76–T78 (in German).
Nascimento, N. (1984). Der Fokussierungeffect bei stochastish erregten Balken. *ZAMM: Zeitschrift für angewandte Mathematik und Mechanik, 64*, T75–T77 (in German).
Nascimento, N. (1984). *Stochastische Schwingungen eindimensionaler Kontinuierlicher mechanischer Systeme*. Ph.D. Thesis, Technical University Darmstadt (in German).
Nash, W. A. (1965). Response of an elastic plate to a distributed random pressure characterized by a separable cross correlations. In *Developments in mechanics* (pp. 364–375). Oxford: Pergamon Press.
Nash, W. A., & Kanematsu, H. (1960). Random vibrations of shallow shells. *Bulletin of the International Association for Shell and Spatial Structures, 51*, 5–18.
Nataraja, R., & John, D. Y. (1977). Random response of shell structures due to wind. *Journal of Wind Engineering and Industrial Aerodynamics*, *2*(1), 79–97.
Nayfeh, A. H. (1983). Parametrically excited multidegree-of-freedom systems with repeated frequencies. *Journal of Sound and Vibration, 88*(2), 145–150.
Neimark, J. I., & Landa, P. S. (1992). *Stochastic and chaotic oscillations*. Berlin: Springer.
Nelson, H. M. (1972). The modal density for flexural vibration of thick plates and bars. *Journal of Sound and Vibration, 25*(2), 255–261.
Nelson, P. A., & Elliott, S. J. (1992). *Active control of sound*. London: Academic Press.
Nelson, P. A., & Elliott, S. J. (1996). *Active control of vibration*. London: Academic Press.
Nemat-Nasser, S. (1966). On the response of continuous media to random excitations. *International Journal of Solids and Structures, 2*, 371–384.
Newland, D. E. (1966). Calculation of power flow between coupled oscillators. *Journal of Sound and Vibration, 3*, 262–276.
Newland, D. E. (1986). General linear theory of vehicle response to random road roughness. In I. Elishakoff & R. H. Lyon (Eds.), *Random vibration—Status and recent developments* (pp. 303–326). Amsterdam: Elsevier.
Newland, D. E. (1975). *Random vibrations and spectral analysis*. Longmans.
Newland, D. E. (2005). *An introduction to random vibrations, spectral and wavelet analysis*. Mineola, NY: Dover.
Newmark, N. M., & Rosenblueth, E. (1971). *Fundamentals of earthquake engineering*. Englewood Cliffs, N.J.: Prentice Hall.
Newsom, C. D., & Fuller, J. R. (1965). A theoretical approach for determining the response of complex structure to arbitarary random noise fields. Boeing, *Document D6–6897*.
Newsom, C. D., Fuller, J. R., & Sherrer, R. E. (1967). A finite element approach for the analysis of randomly excited complex elastic structures. In *AIAA/ASME 8th Structures, Structural Dynamics and Materials Conference* (pp. 1–10). New York: AIDA Press.
Nielsen, S., & Zhang, Z. (2017). *Stochastic dynamics*. Aarhus Universitetsforlag.

Nieves, F. J., Gascon, F., & Bayon, A. (2002). A Multiple frequency in the two lowest axisymmetric vibration modes of a short cylinder. *Jouranl of Sound and Vibration*, *251*(4), 741–749.
Nigam, N. C. (1972). Structural optimization in random vibration environment. *AIAA Journal*, *10*(4), 551–553.
Nigam, N.C. (1983). *Introduction to random vibrations*. Cambridge, MA: MIT Press.
Nigam, N. C., & Narayanan, S. (1994). *Applications of random vibrations*. Berlin: Springer.
Nikolaenko, N. A. (1967). *Probabilistic methods of dynamical analysis of mechanical engineering structures*. Moscow:"Mashinostroenie" Publishing House (in Russian).
Nikolaenko, N. A., & Nazarov Yu. P. (1988). *Dynamics and seismic resistance of structures*. Mioscow: "Stroyiizdat" Publishing House (in Russian).
Nikolaenko, N. A., & Ulyanov, S. V. (1977). *Statistical dynamics of engineering structures*. Moscow: "Maschinostroenie" Publishing House (in Russian).
Nikolaidis, E., Ruiz, R. H., & Maglaras, G. (1959). Evolution of methods for reliability assesment when Random variables are correlated. In *36th Structures, Structural Dynamics and Materials Conference, AIAA Paper*, 95–1312, CP (p. 1346).
Noh, Y-J., Choi, K. K., & Du, L. (2017). Reliability-based design optimization of problems with correlated input variable using a Gaussian copula. *Structural and Multidisciplinary Optimization*, in press.
Noonan, W. E. (1967). Vibration methods for multiple random excitation. *Shock and Vibration Bulletin*, *37*, Part 2, 89–97.
Norouzi, M. (2014). Random vibration Monte Carlo simulation using multiple harmonic function schemes. *International Journal of Vehicle Noise and Vibration*, *10*(3), 214–225.
Norris, S. (2009). Modal density and modal configuration: Multimodal actions. In *The Routledge Handbook of Multimodal Analysis* (pp. 78–90). Oxon: Routledge.
Norton, M. P., & Karczub, D. G. (2003). *Fundamentals of noise and vibration analysis for engineers*. Cambridge, U.K: Cambridge University Press.
Novak, M., & Mitwalley, H. (1990). Random response of offshore towers with pile-soil-pile intersection. *Journal of Offshore Mechanics and Arctic Engineering*, *112*(1), 35–41.
Ntotsios, E., Thompson, D., & Hussein, M. (2017). The effect of track load correlation on ground-borne vibration from railways. *Journal of Sound and Vibration*, *402*, 142–163.
Ochs, J. B., & Snowdon, J. C. (1975). Transmissibility across simply supported thin plates, I. Rectagular and square plates with and without damping layers. *The Journal of the Acoustical Society of America*, *58*4), 832–840.
Ohayon, R., & Soize, C. (1998). *Structural acoustics and vibration: Mechanical models, variational formulations and discretization*. San Diego: Academic Press.
Ohayon, R., & Soize, C. (2014). *Advanced vibroacoustics: Reduced-order models and uncertainty quantification*. New York: Cambridge University Press.
Ohta, M., Yamaguchi, S., & Uchino, E. (1984). A practical prediction method for the stochastic insulation effect of a sound garrier with arbitrary random noise excitation. *Applied Acoustics*, *17*, 291–301.
Ojalvo, I. U. (1986). *Gradients for large structural models with repeated frequencies*. SAE Technical Paper No. 861789.
Ojalvo, I. U. (1987). Modal sensitivity for structural systems with repeated frequencies. *Sensitivity Analysis in Engineering*, 197–213.
Ojalvo, I. U. (1988). Efficient computation of modal sensitivities for systems with repeated frequencies. *AIAA Journal*, *26*(3), 361–366.
Okafor, F. O., & Udeh, O. T. (2015). Direct method of analysis of an isotropic rectangular characteristic orthogonal polynomials. *Nigerian Journal of Technology*, *34*(2), 232–239.
Olson, M. D. (1970a). *Free vibrations and random response of an integrally stiffened panel*. Report LR-544, National Research Council of Canada.
Olson, M. D. (1970b). A numerical approach to random response problems of an integrally stiffened panel. In *Proceedings of the Conference on Current Developments in Sonic Fatigue*, Southampton University (pp. I.1–I.14).

Olson, M. D. (1972). A consistent finite element method for random response problems. *Computers and Structures*, 2, 163–180.
Olson, M. D., & Lindberg, G. M. (1970). *Free vibrations and random response of an integrally-stiffened panel*. Report No. AD0721517, Natural Aeronautical Establishment, Ottawa, Ontario, Canada.
Olson, M. D., & Lindberg, G. M. (1971). Jet noise excitation of an integrally stiffened panel. *Journal of Aircraft*, *8*(11), 847–855.
Oniashvili, O. D. (1957). *Certain dynamic problems of the theory of shells*, translated and published by M. D. Friedman Inc.
Orabi, I. I., & Ahmadi, G. (1988). Response of structures subjected to horizontal-vertical random earthquake excitations. *Soil Dynamics and Earthquake Engineering*, *7*(1), 9–14.
Orisamolu, T. R. (1985). *Response of discretized plates to parametric and non-parametric non-stationary random excitations*, Master's Theses, Department of Mechanical Engineering, The University of Calgary, Alberta, Canada.
Ornstein, L. S. (1927). Zur Theorie der Brownschen Bewegung für system, worin mechrere temperaturen Vorkommen. *Zeitschrift für Physik*, *41*, 848–856 (in German).
Orris, R. M., & Petyt, M. (1975). Random response of periodic structures by a finite element technique. *Journal of Sound and Vibration, 43*, 1–8.
Osadebe, N. N., Nwokike, V. C., & Oguaghamba, O. A. (2016). Analysis of thin rectangular plate using multi-degrees of freedom Taylor Maclaurin's series in Galerkin's variational method. *Nigerian Journal of Technology, 35*(3), 503–509.
Osadebe, N. N., Atama, C. M., & Oguaghamba, O. A. (2017). Dynamics of orthotropic plates using characteristic orthogonal polynomial-Galerkin method. *Nigerian Journal of Technology, 36*(1), 50–56.
Osorin, V. I. (1960). Analysis of probabilistic spectrum of vibration of linear elastic system. *Proceedings of the Leningrad Polytechnic Instituite*, (210) (in Russian).
Ostoja-Starzhewski, M., & Woods, A. N. (2003). Spectral finite elements for vibrating rods and beams with random field properties. *Journal of Sound and Vibration, 268*(4), 779–797.
Ostrovska, S. (2002). Uncorrelatedness and corellatedness of powers of random variable. *Archiv der Mathematik, 79*(2), 141–146.
Ovchinnikov, I. N. (2006). Heaviest regime of the wide-band random vibration. *Mathematical and Computer Modelling in Machinery and Systems,* 38–54 (in Russian).
Ovseevich, A. I. (2003). Stability of inverted pendulum during fast random vibrations of point of mount. *PMM: Journal of Applied Mathematics and Mechanics, 70* (5), 844–851 (in Russian).
Pace, I. S., & Barnett, S. (1972). Comparison of numerical methods for solving Lyapunov matrix equations. *International Journal of Control, 15*(5), 907–915.
Paez, T. L. (1986). Random vibration of modal frequencies—Experiments and analysis. *Shock and Vibration Bulletin, 56*, 37–49.
Paez, T. L. (1999). *Random vibrations: Assessment of the state of the art*. Report No. SAND99–0469C, Sandria National Laboratories, Livermore, CA.
Paez, T. L. (2006). The history of random vibrations through 1958. *Mechanical Systems and Signal Processing, 20*, 1783–1818.
Page, C. H. (1952). Instantaneous power spectra. *Journal of Applied Physics, 23*(1), 103–106.
Paipetio, S. A. (1984). A Note on the approximate determination of equal eigenfrequencies in elastic systems. *Acta Acustica (United With Acustica), 56*(1), 53–76.
Palashchenko, N. N. (1967). Spectral analysis of vibrations of dynamical systems. *Civil Engineering and Architecture*, (11), 52–56 (in Russian).
Palmeri, A., Ricciardelli, F., Muscolino, G., & De Luca, A. (2004). Random vibration of systems with viscoelastic memory. *Journal of Engineering Mechanics*, *130*(9), 1052–1061.
Palmov, V. A. (1962). The thin plate under random loading. *PMM: Journal of Applied Mathematics and Mechanics, 26*(3), 858–864.
Palmov, V. A. (1965). Thin shells acted on by broad band random loads. *PMM: Journal of Applied Mathematics and Mechanics, 29*, 905–913.

Palmov, V. A. (1967). The correlation properties of the field of vibrations of plates under a broadband random field. *Mechanics of Solids, 2*(5), 136–139 (in Russian).
Palmov, V. A. (1979). Integral method for the analysis of vibration of dynamic structures. *Advances in Mechanics, 2*, 3–24 (in Russian).
Palmov, V. A., & Pupyrev, V. A. (1967a). Forced vibrations of a cylindrical shell with an acoustic medium under a broadband and a high frequency load. *Proceedings of the Leningrad Polytechnic Institute, 279*, 68–76 (in Russian).
Palmov, V. A., & Pupyrev, V.A.(1967b). Vibrations and acoustic radiation of plates under the action of a random load. *Acoustics Journal*, 1392 (in Russian).
Palumbo, D. (2012). Determining correlation and coherence length in turbulent boundary layer flight data. *Journal of Social and Vibration, 331*(16), 3721–3737.
Pandey, M. D., & Ariaratnam, S. T. (1996). Crossing rate analysis of non-Gaussian response of linear systems. *Journal of Engineering Mechanics, 122*(6), 507–511.
Pandey, M. D. (2001). Extreme quantile estimation using order statistics with minimum cross-entropy principle. *Probabilistic Engineering Mechanics, 16*, 31–42.
Pandit, M. K., Singh, B. N., & Sheikh, A. H. (2009). Stochastic free vibration response of soft core sandwich plates using an improved higher-order zigzag theory. *Journal of Aerospace Engineering, 23*(1), 14–23.
Papadimitriou, C., Haralampidis, Y., & Sobczyk, K. (2005). Optimal experimental design in stochastic structural dynamics. *Probabilistic Engineering Mechanics, 20*(1), 67–78.
Papadimitriou, C., Katafygiotis, L. S., & Beck, J. L. (1995). Approximate analysis of response variability of uncertain linear systems. *Probabilistic Engineering Mechanics, 10*, 251–264.
Papadimitriou, C., & Lutes, L. D. (1994). Approximate analysis of higher cumulants for multi-degree-of-freedom random vibration. *Probabilistic Engineering Mechanics, 9*(1), 71–82.
Papadrakakis, M., Stefanou, G., & Papadopoulos, V. (Eds.) (2013). *Computational methods in stochastic dynamics*. Berlin: Springer.
Pappas, M. (1978). Optimal frequency separation of cylindrical shells. *AIAA Journal, 16*(9), 999–1001.
Paramenter, W. W., & Christiansen, R. G. (1973). *Recovery of modal information from a beam undergoing random vibration*. ASME Paper No. 73-WA-DE 10.
Patarabunditkul, S. (2009). *Natural frequency statistics of an uncertain plate*. Bachelor of Engineering Thesis, School of Mechanical and Manufacturing Engineering, The University of New South Wales, Australia.
Paul, M. R., & Cron, M. C. (2004). Stochastic dynamics of nanoscale mechanical oscillator, immersed in a viscous fluid. *Physical Review Letters, 92*, article 235501.
Paulus, M., Dasgupta, A., & Habtour, E. D. (2012). Life estimation modal of a cantilevered beam subjected to complex random vibration. *Fatigue and Fracture of Engineering Materials and Structures, 35*(11), 1058–1070.
Pavlyuk, Yu. S., & Sekulin, V.D. (1975). Analytical estimation of random vibrations of cargo car body. *Vestnik Nsesoiuznogo Nauchno-Issledovatelnogo Institute Zhelezno-Dorozhnogo Transporta*, (5), 35–37 (in Russian).
Pavlyuk Yu. S., Sekulin, V. D., & Reznikov, E.K. (1977). Analytical estimation of random vibrations of sub-axle carriages. *Izvestiya VUZOV, Maschinnostroenie*, 141–145 (in Russian).
Peltier, L. J., & Hambric, S. A. (2007). Estimating turbulent-boundary-layer wall-pressure spectra from CFD RANS solutions. *Journal of Fluids and Structures, 23*(6), 920–937.
Pendered, J. W. (1965). Theoretical investigation into the effects of close natural frequencies in resonance testing. *Journal of Mechanical Engineering Science, 7*(4), 372–379.
Penzien, J., Kaul, M. K., & Berge, B. (1972). Stochastic response of offshore towers to random sea waves and strong motion earthquakes. *Computers and Structures*, 2(5–6), 733–756.
Penzien, J., & Watabe, M. (1975). Simulation of 3-dimensional earthquake ground motions. *Journal of Earthquake Engineering and Structural Dynamics, 3*, 365–373.
Penzien, J., & Watabe, M. (1976). Simulation of 3-dimensional earthquake ground motion. *Bulletin of International Institute of Seismology and Earthquake Engineering, 12*, 103–115.

Pervozvansky, A. A. (1986). High-frequency vibrations in case of considerable dissipation. *Reports of the USSR Academy of Sciences, 288*, 1068–1070.

Pesek, L. (1994). The inverse sensitivity method for parametric correlations of mechanical models with repeated eigenvalues. In H. D. Bui, M. Tanaka, M. Bonnet, H. Maigre, E. Luzzato, & M. Reynier (Eds.), *Inverse problems in engineering mechanics* (pp. 257–262). Rotterdam: Balkema.

Peters, K. A., Schmitz, D., & Wagner, U. (1977). Determination of floor response spectra on the basis of the response spectrum method. *Nuclear Engineering and Design, 44*, 255–262.

Petrov, A. A., & Bazilevskii, S. V. (1979). Influence of cross correlation between generalized coordinates during random vibrations of linear systems. *Stroitelnaya Mekhanika i Raschet Sooruzhenii (Civil Engineering Mechanics and Analysis of Structures)*, (4), 52–56 (in Russian).

Petrov, A. A., & Bazilevskii, S. V. (1977). On accounting spatial correlations of the velocity pulsations when determining dynamic component of wind loading. *Stroitelnaya Mekhanika i Raschet Sooruzhenii (Civil Engineering Mechanics and Analysis of Structures)*, (5), 67–71 (in Russian).

Pertov, V. I., & Finkelstein, R. M. (1966). Orthotropic cylindrical shell in the moving random field of pressures. *Investigations in Elasticity and Plasticity, 5*, 230–237. Leningrad University Press (in Russian).

Petrucci, G., Di Paola, M., & Zuccarello, B. (2000). On the characterisits of dynamic properties of random process by spectral parameters. *Journal of Applied Mechanics, 67*(3), 519–526.

Petrucci, G., & Zuccarello, B. (2004). Fatigue life prediction under wide band random loading. *Fatigue and Fracture of Engineering Materials and Strctures, 27*(12), 1183–1195.

Petyt, M. (1975). Finite element methods for the response of structures to random excitations. In *Proceedings of the World Congress on Finite Element Methods in Structural Mechanics*, Bournemouth, England (pp. 16.1–16.20).

Petyt, M. (1990). *Introduction to finite element vibration analysis*. Cambridge, UK: Cambridge University Press.

Petyt, M., & Lim, S. P. (1978). Finite element analysis of the noise inside a mechanically excited cylinder. *International Journal for Numerical Methods in Engineering, 13*, 109–122.

Pevzner, P., Weller, T., & Berkovits, A. (2000). Further modification of Bolotin method in vibration analysis of rectangular plates. *AIAA Journal, 38*(9), 1725–1729.

Pfaffelmoser, T., & Westermann, R. (2012). Visualization of global correlation structures in uncertain 2D scalar fields. In *Computer graphics forum* (Vol. 31). Wiley Online Library.

Pfaffelmoser, T., & Westermann, R. (2013). Correlation visualization for structural uncertainty analysis. *International Journal for Uncertainty Quantification, 3*(2), 171–186.

Pfalfinger, D. (1981). Probablistic dynamic analysis with ADINA. *Computers and Structures, 13*, 637–646.

Pierson, W. J., & Moskwitz, L. (1964). A proposed spectral form for fully developed wind seas based on the similarity theory of S. A. Kitaigorodskii. *Journal of Geophysics Research, 69*, 5181–5190.

Ping-I Wang, A. (1970). A note on the random responses of a thin cylindrical shell. *Journal of Sound and Vibration, 12*, 393–396.

Pinnola, F. P. (2016). Statistical correlation of fractional oscillator response by complex spectral moments and state variable expansion. *Coummunications in Nonlinear Science and Numerical Simulation, 39*, 343–359.

Pipner, M., & Švejda, J. (1970). The response of the beam to a moving random loads. *Sbornik prací VŠDS a VÚD, 28*, 115–129.

Piszczek, K., & Niziol, J. (1986). *Random vibration of mechanical systems*. London: Ellis Horwood.

Pitarresi, J. M., & Akanda, A. (1993). Random vibration response of a surface mount lead/soldeer joint. *Advances in Electronic Packaging, 4*, 1.

Pitoiset, X., & Preumont, A. (2000). Spectral methods for multiaxial random fatigue analysis of metallic structures. *International Journal of Fatigue, 22*, 541–550.

Pleijel, A. (1955). On the problem of improving Weyl's law for the asymptotic eigenvalue distribution. *Consagno Internationale sulla Equaziono Lineari alle Derivate Partial* (pp. 69–75). Rome: Edizioni Cremonlse.

Pitz, G. F. (1980). Bayesian analysis of random response models. *Psychological Bulletin, 87*(1), 209–212.

Podobnik, B., Grosse, I., Horvatic, D., Ilic, S., Ivanov, P. Ch., & Stanley, H. E. (2009). Quantifying cross-correlations using local and global detrending approaches. *European Physics Journal, B, 71*, 243–250.

Podobnik, B., Horvatic, D., Peterson, A., & Stanley, H. E. (2009). Cross-correlations between volume change and price change. *Proceeding of the National Academy of Sciences of the USA, 106*, 22079–22084.

Politch, J. (1990). Spectroscopic holography: A new method for mode analysis of randomly excited structures. *Journal of the Optical Society of America, 7*(8), 1355–1361.

Poskuryakov, M. N. (1963). About random action on a linearly deformable system. In A. R. Rzhanitsyn (Ed.), *Investigations in analysis of shells and massive strcutres* (pp. 224–230). Moscow: State Publishing House in Civil Engineering, Architecture and Materials (in Russian).

Potapov, V. D. (1999). On the stability of the solutions of linear stochastic integro-differential equations encountered in elasticity and viscoelasticity. *Journal of Applied Mathematics and Mechanics, 63*(5), 785–794.

Potapov, V. D. (2013). Stab of a rod subjected to a random stationary longitudinal force considering transverse shear. *Journal of Machinery Manufracture, 42*(2), 95–101.

Pradlwarter, H. J., & Li W. (1991). On the computation of the stochastic response of highly nonlinear large MDOF-systems modeled by finite elements. *Probabilistic Engineering Mechanics, 6*(2), 109–116.

Pradlwarter, H. J., Schuëller, G. I. (1997). Stochastic structural dynamics—A primer for practical applications. In B. Ayyub, A. Guran & A. Haldar (Eds.), *Uncertainty modeling in vibration, control and fuzzy analysis of structural systems* (pp. 1–27). Singapore: World Scientific.

Pradlwarter, H. J., Schuëller, G. I., Naess, A., Frey, M., Kareem, A., Zhao, J., et al. (1997). Stochastic structural dynamics—A primer for practical applications. In B. Ayyub, A. Guran & A. Haldar (Eds.), *Uncertainty modeling in vibration, control and fuzzy analysis of structural systems*. London: World Scientific.

Prawd, K.S.R. K., & Murthy, P. N. (1970). Complicated nodal patterns on a thin elastic circular plate vibrating in flexure. *The Journal of The Acoustical Society of India, 22*, 19–27.

Prelewicz, D. A. (1972). Response of linear periodically time varying systems to random excitation. *AIAA Journal, 10*(8), 1124–1125.

Prells, U., & Friswell, M. I. (1997). Partial derivatives of repeated eigenvalues and their derivatives. *AIAA Journal, 35*, 1363–1368.

Press, H., & Turkey, J. W. (1956). *Power spectral methods of analysis and application in airplane dynamics*. Monograph 2606, New York: Bell Telephone System.

Pretlove, A. J. (1965). Bond stresses in a randomly vibrating sandwich plates: Multi-modal theory. *Journal of Sound and Vibration, 2*, 1–22.

Preumont, A. (). The generation of non-separable artificail earthequake accelerograms for the design of nuclear power plants. *Nuclear Engineering and Design, 88*, 59–67.

Preumont, A., & Piefort, V. (1994). Predicting random high-cycle fatigue life with finite elements. *Journal of Vibration and Acoustics, 116*, 245–248.

Price, W. G., & Bishop, R.E. D. (1974). *Probabilistic theory of ship dynamics*. London: Chapman & Hall.

Priestley, M. B. (1965). Evolutionary spectra and non-stationary processes. *Journal of the Royal Statistical Society: Series B (Methodological), 27*(2), 204–229.

Priestley, M. B. (1967). Power spectral analysis of non-stationary random process. *Journal of Sound and Vibration, 6*, 86–97.

Proskurykov, M. I. (1963a). On spatial variability of random loading. In A. R. Rzhantsyn (Ed.), *Investigations in analysis of shells of massive structures* (pp. 263–267). Moscow: State Pulhhing House in Civil Engineering, Architecture and Materials (in Russina).

Proskuryakov, M. I. (1963b). On random action on autovibrating system with delayed loads. In A. R. Rzhantsyn (Ed.), *Investigation in analysis of shells of massive structures* (pp. 231–243). Moscow: State Publishing house in Civil Engineering, Architecture and Materials (in Russian).
Pshenichkina, V. A., & Shaposhnikov, N. A. (2014). Vibration of a beam under stochastic forces under moving loads. *Pricaspian Engineering—Civil Engineering Herald,* (2), 42–48 (in Rusian).
Pugachev, V. S. (1962). *Theory of random functions*. Moscow: "Fizmatgiz" Publishing house (in Russian).
Pupyrev, V. A. (1968). Relations of vibratory and and acoustic effects on an elastic plate. *Mechanics of Solids, 3*(1), 55–58.
Putiatin, V. V. (1966). Towards analysis of random excitation action on multidegree system with linear connections. *Mechanics of Solids,* (6) (in Russian).
Putiatin, V. V. (1967). Towards teh probelm on action of vector nonstationary random function on multidegree mechanical system. In *Questions of dynamics and strength* (Vol. 14). Riga: "Zinatne" Publishing House (in Russian).
Putiatin, V. V. (1968). Towards solution of correlation problem for multidegree mechanical system with elastic conenctions. In *Cybernetics and automatic control*. Kiev: Institute of Cybernetics Press (in Russian).
Qiang, Z., Van Dyke, D., Wright, S., & Redmond, M. Dynamic analysis of a building under rocket engine plume acoustic load. In *37th Joint Propulsion Conference and Exhibit*, Paper 2001–3256, Salt Lake City, 8–11 July 2001.
Qiu H-Y., Yang J-M., Butt, S., & Zhong J-H. (2017). Investigation on random vibration of a drillstring. *Journal of Sound and Vibration*, *406*, 74–88.
Qiu, X. J., & Zhi, S-J. (1987). Calculation of structural parameters related with repeated natural frequencies. In *Proceedings of the 5th International Modal Analysis Conference* (pp. 517–523). London: Imperial College of Science and Technology.
Qu, S.-X., Xu, D., & Kang, R. (2013). Analysis of random vibration life of mechanical parts of actuating cylinder based on the finite element. *International Conference on Quality, Reliability, Risk, Maintenance, and Safety Engineering*.
Queck, S. T., Teo, P., & Balendra, T. (1990). Non-stationary structural response with evolutionary spectra using seismological input model. *Earthquake Engineering and Structural Dynamics, 19*, 275–288.
Racicot, R., & Moses, F. (1972). Filtered poisson process for random vibration problems. *Journal of Engineering Mechanics*, *98*, 159–176.
Radin, V. P., & Chirkov, V. P. (1970). Transfer of random vibrations in semi-infinite cylindrical shell from forces acting on the cylinder's. In *Proceedings of the Moscow Power Engineering Institute, Dynamics and Strength of Machines* (pp. 68–74). Moscow: Power Engineering Institute Press in Russian).
Raevsky, S.Ya. (1954). Nonstationary random process in linear systems with constant coefficient. *Proceedings of the Military_Air Academy Names After Zhukovsky*, (495) (in Russian).
Rahman, S. (2007). Stochastic dynamic systems with complex-valued eigensolutions. *International Journal for Numerical Methods in Engineering, 71*(8), 963–986.
Ramachandran, P., & Narayanan, S. (2007). Evaluation of modal density, radiation efficiency and acoustic response of longitudinally stiffened cylindrical shell. *Journal of Sound and Vibration*, *304*(1), 154–174.
Ramu, S. A., & Ganesan, R. (1995). Response and stability of a stochastic beam-column using stochastic FEM. *Computers and Structures, 54*(2), 207–221.
Rand, R., & Vito, R. (1972). Nonlinear vibrations of two-degree-of-freedom systems with repeated linearized natural frequencies. *Journal of Applied Mechanics, 28*, 296–297.
Rangaiah, V. P., & Sadasiva, S. (1984). Note on the prediction of random response of a clamped plate. *Computers and Structures*, *18*(3), 561–563.
Rao, L.V.V. G., & Narayanan, S. (2008). Preview control of random response of a half-car vehicle model traversing rough road. *Journal of Sound and Vibration, 310*(1–2), 352–365.

Rao, S. S. (1981). Reliability-based optimization under random vibration environment. *Computers and Structures, 14*(5–6), 345–355.
Rashidifar, M. A., & Ahmadi, D. (2015). Vibration analysis of randomly oriented carbon nanotube based on FGM beam using Timoshenko theory. *Advances in Mechanical Engineering, 7*(2), article 653950.
Rastehkenari, S. F. (2019). Random vibrations of functionally graded nanobeams based on unified nonlocal strain gradient theory. *Microsystem Technologies, 25*(2), 691–704.
Rastehkenari, S. F., & Ghadiri, M. (2019). Size-dependent random vibration analysis of AFM probe with tip mass considering surface viscoelastic effect. *The European Physical Journal Plus, 134*(11), article 563.
Rattayya, J. V., & Junger, M. C. (1969). Flow excitation of cylindrical shells and associated coincidence effects. *The Journal of the Acoustical Society of America, 36*, 878–884.
Rayleigh, L. (1911). On the calculation of Chladni figures for a square plate. *The London, Edinburg, and Dublin Philosphical Magazine and Journal of Sciences, 22*(128), 225–229.
Reddy, C.V. R., Granesan, N., Rao, B.V. A., & Narayanan, S. (1980). Response of plates With unconstrained layer damping treatment to random acoustic excitation, Part 1: Damping and frequency evaluations. *Journal of Sound and Vibration, 69*(1), 35–43.
Reddy, C.V. R., Ganesan, N., & Rao, B.V. A. (1980). Response of unconstrained layer panels to random excitation at a point. *Journal of Sound and Vibration, 73*, 419–427.
Reddy, C.V. R., Ganesan, N., & Rao, B.V. A. (1981). Response of clamped sandwich panels with visco-elastic core under random acoustic excitation. *Journal of Sound and Vibration, 75*(4), 481–494.
Rees, T. (2010). Chladni plates: The first step towards visualizing sound. Retrieved January 7, 2019, from http://www.hps.cam.ac.uk/whipple/explore/acoustics/ernstchladni/chladniplates/.
Rehak, M., DiMaggio, F., Benaroya, H., & Elishakoff, I. (1987). Random vibrations with MACSYMA. *Computer Methods in Applied Mechanics and Engineering, 61*, 61–70.
Reis, R., & Ravigururajan, T. S. (1982). On minimizing the mean square deflection of thin elastic beams. In G. R. Karr (Ed.), *Developments in theoretical and applied mechanics* (Vol. 21, pp. 125–136). The University of Alabama, Huntsville, AL.
Remington, P. J., & Manning, J. E. (1975). Comparison of SEA power flow prediction with an 'exact' calculation. *The Journal of the Acoustical Society of America, 57*, 374–379.
Ren, G., Zhang, Z.-C., & Cheng, J.-G. (1997). Multiple eigenvalues arising from a class of repetitive substructures. *AIAA Journal, 35*(2), 355–361.
Renauld, M., Elfrink, R., Yambunathan, M., de Nosijer, C., Wang, Z., Rovers, M., et al. (2012). Optimum power and efficiency of piezoelectric vibration harvesters with sinusoidal and random vibrations. *Journal of Micromechanics and Microengineering, 22*, article 105030.
Renger, A. (1979). Equation for the probability density of vibratory systems subjected to simultaneous continuous and discrete stochastic excitations. *ZAMM: Zeitschrift für angewandte Mathematick and Mechanik, 59*(1), 1–13 (in German).
Renji, K. (2000). Experimental modal densities of honeycomb sandwich panels at high frequencies. *Journal of Sound and Vibration, 237*(1), 67–79.
Renji, K. (2004). On the number of modes required for statistical energy analysis-based calculations. *Journal of Sound and Vibration, 269*, 1128–1132.
Renji, K. (2020). Model density and critical frequency of composite panels considering transverse shear, deformation and rotary inertia. *Journal of Vibration and control, 26*, 1–11 (In press).
Renji, K., Nair, P. S., & Narayanan, S. (1996). Modal density of composite honeycomb sandwich panels. *Journal of Sound and Vibration, 195*(5), 687–699.
Renji, K., Nair, P. S., & Narayanan, S. (1997). Critical and coincidence frequencies of flat panels. *Journal of Sound and Vibration, 205*, 19–32.
Rennison, D. C., & Bull, M. K. (1977). On the modal density and damping of cylindrical pipes. *Journal of Sound and Vibration, 54*(1), 39–53.
Rezaee, M., Yam, G. F., & Fathi, R. (2015). Development of "Modal Analysis Free Vibration Response Only" method for randomly excited systems. *Acta Mechanica, 226*(12), 4031–4042.

Rezaeian, S., & Der Kiureghian, A. (2008). Stochastic ground motion with separable temporal and spectral nonstationarities. *Earthquake Engineering and Structural Dynamics, 37*, 1565–1584.

Rezaeian, S., & Der Kiureghian, A. (2010). Simulation of synthetic ground motions for specified earthquake and site characteristics. *Earthquake Engineering and Structural Dynamics, 39*, 1155–1180.

Reznikov, L. M. (1975). Calculating linear multimass systems subjected to a multidimensional stationary random process. *International Applied Mechanics, 11*(2), 189–193.

Reznikov, L. M. (1976). Calculating coordinate variances of multimass systems in transient conditions for stationary random disturbances. *International Applied Mechanics, 12*(5), 512–516.

Reznikov, L. M. (1979). Nonstationary random vibrations of multiple-mass systems. *International Applied Mechanics, 15*(7), 633–638.

Reznikov, L. M. (1984). Statistical characteristics of mechanical systems' oscillations under broad-band random excitations. *Mechnics of Solids*, (4), 35–37 (in Russian).

Riaz, B. (1985). *Response of linear structures to nonstationary seismic excitation.* Ph.D. Thesis, Cornell University.

Ribaulin, A. G., & Sidorenko, A. S. (2013). Stress state of Aviation product with discrete welded connections undergoing random kinematic loading. *Herald of MAI (Moscow Aviation Institute), 80*(1), 183–193 (in Russian).

Ribeiro, J.L. D., & Blessmann, J. (1992). Probability distribution of forces and pressure on a square cylinder. *Journal of Wind Energy and Industrial Aerodynamics, 41–44*, 813–824.

Ribner, H. S. (1956). Spectral theory of buffering and gust response: Unification and extension. *Journal of Aeronautical Sciences, 23*(12), 1075–1077.

Ribner, H. S. (1978). Two point correlations of jet noise. *Journal of Sound and Vibration, 56*(1), 1–19.

Ricciardi, G. (1994). Random vibration of beam under moving loads. *Journal of Engineering Mechanics, 120*(11), 2361–2380.

Rice, S. O. Matheamtical analysis of random noise. *Bell Systems Technical Journal, 23*, 282–332, 1944; *24*, 44–156, 1945.

Richards, E. J., & Mead, D. J. (Eds.) (1968). *Noise and acosutic fatigue in aeronautics*. New York: Wiley.

Richter, M., Von Scheidt, J., & Starkloff, H.-J. (2001). Moment functions for solutions of random boundary value problems. *ZAMM: Zeitschrift für angewandte Mathematick and Mechanik, 81*, 5641–5642.

Riemer, M. (1981). Fokussierungseffekte in Saiten bei stochastischen Punkterregung, *Diploma Work,* Institute of Technical Mechanics, University of Karlsruhe, Federal Republic of Germany (in German).

Riemer, M., & Wedig, W. (1977). Bauwerke unter Wellenlasten—Fokussierungselfbekte be Poissonerregungen, *VDI-Berichte 419*, (pp. 366–389). Düsseldorf: VDI Verlag (in German).

Riera, J. D., & Bucher, C. G. (1989). Discussion to "Explicit Solutions for the Response of Simple Systems Subjected to Nonstationary Random Excitation" by W. D. Iwan and, Z. K. Hou, *Structural Safety, 6*, 77–86.

Ritto, T. G., & Fabro, A. T. (2019). Investigation of random matrix applications on structural dynamics using a tensor decomposition. *Journal of the Brazilian Society of Mechanical Sciences and Engineering*, 41(8), article 352.

Robert, G., & Sabot, J. (1984). Use of random forces to simulate the vibroacoustic response of a plate excited by a hydrodynamic turbulent boundary layer. In *Proceedings of the ASME Winter Meeting: Symposium on Flow-Induced Vibrations* (Vol. 5, pp. 53–61).

Roberts, J. B. (1965a). The response of linear vibratory systems to random impulses. *Journal of Sound and Vibration, 2*(4), 375–390.

Roberts, J. B. (1965b). On the harmonic analysis of evolutionary random vibration. *Journal of sound and Vibration, 2*(3), 336–352.

Roberts, J. B. (1966). The response of a simple oscillator to band-limited white noise. *Journal of Sound and Vibration, 3*(2), 115–126.

Roberts, J. B., Structural Fatigue Under Non-Stationary Random Loading, *Journal of Mechanical Engineering Science*, 8(4):392–405, 1966.

Roberts, J. B. (1971a). The covariance response of linear systems to locally stationary random excitation. *Journal of Sound and Vibration, 17*(3), 299–307.

Roberts, J. B. (1971b). The covariance response of linear systems to non-stationary random excitation. *Journal of Sound and Vibration, 14*, 385–400.

Roberts, J. B. (1972a). An investigation of broad band random vibration simulation. *Journal of Engineering for Industry, 94*(1), 7–14.

Roberts, J. B. (1972b). System response to random impulses. *Journal of Sound and Vibration, 24*, 23–34.

Roberts, J. B. (1973). Distribution of the response of linear systems of poisson distributed random pulses. *Journal of Sound and Vibration, 28*(1), 93–103.

Roberts, J. B. (1975). System response to random impulses. *Journal of Sound and Vibration, 24*, 23–34.

Roberts, J. B., First-Passage Time for the Envelope of a Randomly Excited Linear oscillator, *Journal of Sound and Vibration*, 46(1), 1–14, 1976.

Roberts, J. B., & Spanos, P. D. (2003). *Random vibration and statistical linearization*. Mineola, N.Y.: Dover.

Roberts, J., & Stiillo, D. (1991). Random vibration analysis of a printed wiring board with electronic components. *Journal of the IES, 34*(1), 25–31.

Robin, O., Franco, F., De Rosa, S., Ciappi, E., & Berry, A. (2018). Scaling laws for flat plate vibroacoustic response induced by deterministic and random excitations. In *INTER-NOISE and NOISE-CON Congress and Conference Proceedings*, Institute of Noise Control Engineering (Vol. 257, No. 1, pp. 319–330).

Robson, J. D. (1963). *An introduction to random vibration*. Edinburgh University Press.

Robson, J. D. (1966). The random vibration response of a system having many degrees of freedom. *Aeronautical Quarterly, 17*, 21–30.

Robson, J. D. (1979). Road surface description and vehicle response. *International Journal of Vehicle Design, 1*(1), 25–35.

Robson, J. D. (1981). The description of random vibration. *International Journal of Vehicle Design*, 2(3), 255–275.

Robson, J. D. (1986). Note on the symmetries of higher-order correlation functions and spectral densities. In I. Elishakoff & R. H. Lyon (Eds.), *Random vibration—Status and recent developments* (pp. 341–348). Amsterdam: Elsevier.

Robson, J. D., Dodds, C. J., Macvean, D. B., & Paling, V. R. (1971). *Random vibrations*. Vienna: Springer.

Robson, J. D., & Roberts, J. W. (1965). A theoretical basis for the practical simulation of random motions. *Journal of Mechanical Engineering Science*, *7*, 246–251.

de Rochambeau, M., Ichchou, M., & Troclet, B. (2011). Modeling of random aerodynamic loads applied on fluid-structure coupled systems using rain-on-the-roof equivalent excitation. *Engineering Computations*, *28*(4), 472–491.

Rodgers, J., & Nicewander, W. (1988). Thirteen ways to look at the correlation coefficient. *American Statistician, 42*(1), 59–66.

Roibas, E., Chimero, M., Lopéz-Diez, J., & Simón, F. (2013). A mode count procedure for mid-frequency analysis of complex vibro—Acosutic systems. *Aerospace Science and Technology, 29*(1), 165–174.

Roitenberg, L. Ia. (1965). On the motion of gyrocompass with random displacements of its point of support. *PMM: Journal of Applied Mathematics and Mechanics, 29*(1), 179–186.

Romanov, Yu. I. (1965). Correlation of higher of vibration modes in anti-seismic analyses. *Stroitelnaya Mekhanika i Raschet Sooruzhenii (Civil Engineering Mechanics and Analysis of Structures)*, (1) (in Russian).

Romanovsky, Yu. M. (1961). *Vibration of an elastic airplane subjected to atmospheric turbulence*. Ph.D. Dissertation, Moscow State University (in Russian).

Roozen, N. B. (1992). *Quiet by design: Numerical acousto-elastic analysis of aircraft structures.* Ph.D Thesis, Technical University Eindhoven, The Netherlands.

Rosenblueth, E. (1976). Tall buildings under fire-component earthquake. *Journal of Structural Divison, 102*(2), 453–459.

Rosin, M. F. (1981). *Statistical dynamics and theory of efficiency of authomatic control systems.* Moscow: "Mashinostroenie" Publishing House (in Russian).

Ross, D. (1987). *Mechanics of underwater noise.* Loas Altos:Ca: Peninsula Publishing.

Rosset, J. M., Whitman, R. B., & Dobry, R. (1973). Modal analysis for structures with foundation interaction. *Journal of Structural Division, 99*, 399–416.

Rossi, R. E., Laura. P.A. A., & Mayrizi, M. J. (1992). Numerical experiments on the effect of the value of the shear coefficient upon take natural frequencies of a Timoshenko beam. *Journal of Sound and Vibration, 154*(2), 374–379.

Rossing, T. D., & Fletcher, N. H. (2004). *Principles of vibration and sound.* In T. D. Rossing & N. H. Fletcher (Eds.), (2nd edition). New York: Springer.

Rossing, T. D., & Fletcher, N. H. (2004). Two-dimensional systems: Membranes and plates. In T. D. Rossing & N. H. Fletcher (Eds.), *Principles of vibration and sound* (2nd edition) (pp. 65–94). New York: Springer.

Rouillard, V., & Sek, M. A. (2010). Synthesizing nonstationary, non-Gaussian random vibrations. *Packaging Technology and Science, 23*(8), 423–439.

Roussos, L. A., Powell, C. A., Grosveld, F. W., & Koval, L. R. (1984). Noise transmission characteristics of advanced composite structural materials. *Journal of Aircraft, 21*(7), 528–536.

Rozanov, Yu. A. (1971). *Random proess: A short course.* Moscow: "Nauka" Publishing House (in Russian).

Rozanov Yu. A. (1990). *Stationary random process.* Moscow: "Nauka" Publishing House (in Russian).

Rudder, F. F., Jr. (1972). Study of effects of design details on structural response to acoustic excitation. *NASA CR-1959.*

Rump, S. M. (2001). Computational error bounds for multiple or nearly multiple eigenvalues. *Linear Algebra and Its Applications, 324*(1), 209–226.

Runkle, C. J., & Hart, F. D. (1970). The radiation resistance of cylindrical shells exhibiting axisymmetric mode shapes. *Shock and Vibration Bulletin, 40*, Part 3, 23–29.

Sabetta, F., & Pugliese, A. (1996). Estimation of response spectra and simulation of nonstationary earthquake ground motions. *Bulletin of the Seismological Society of America, 86*(2), 337–352.

Sablik, M. J., Beissner, R. E., Silvus, H. S., & Miller, M. L. (1985). S.E. A., Statistical energy analysis, structural resonances and beam networks. *The Journal of the Acoustical Society of America, 77*(3), 1038–1045.

Sabra, K. G., Winkel, E. S., Bourgoyne, D. A., Elbing, B. R., Ceccio, S. L., Perlin, M., et al. (2007). Using Cross-correlations of turbulent flow-induced ambient vibrations to estimate the structural impulse response, application to structural health monitoring. *The Journal of the Acoustical Society of America, 121*(4), 1987–1995.

Sacgin, A., & Kara M. (2019). The effects of material uncertainty on low-to high frequency vibrations of thin plates using statistical moment based approach. *Materials Research Express, 6*(6), 066528.

Sackman, J. L., & Kelly, J. M. (1979). Seismic analysis of internal equipment and components in structures. *Enginering Structures, 1*(4), 179–190.

Sadasiva, S., Elango, K., & Sethuraman, V. (1975). Prediction of the response of clamped panel to jet noise. In *Proceedings of the Congress of the Indian Society of Theoretical and Applied Mechanics.*

Saeedi, K., & Bhat, R. (2011). Clustered natural frequencies in multi-span beams with restrained characteristic functions. *Shock and Vibration, 18*, 697–707.

Sagirow, P. (1970). *Stochastic methods in the dynamics of satellites.* Vienna: Springer.

Saiful Islam, M., Ellingwood, B., & Corotis, R. B. (1990). Dynamic response of tall buildings to stochastic wind load. *Journal of Structural Engineering, 116*(11), 2982–3002.

Saiful Islam, M., Ellingwood, B., & Corotis, R. B. (1992). Wind-induced response of structurally asymmetric high-rise buildings. *Journal of Structural Engineering, 118*(1), 207–222.
Sakata, M., & Kimura, K. (1979). The use of moment equations for calculating the mean square response of linear system to nonststionary random excitation. *Journal of Sound and Vibration, 67*, 383–393.
Salim, B., Kim, K. B., & Yang, J. N. (1986). Random vibration of rotating machines under earthquake excitations. *Journal of Engineering Mechanics, 12*(6), 550–565.
Salim, S., Iyengar, N.G. R., & Yadav, D. (1994). Free vibration of composite plates with random material characteristics. In N. S. Ferguson, H. F. Wolfe, & C. Mei, (Eds.), *Structural Dynamics, Recent Advances: Proceedings of the 5th International Conference* (pp. 814–823). ISVR, Southampton.
Samali, B., Kim, K. B., & Yang, J. N. (1986). Random vibration of rotating machines under earthquake excitation. *Journal of Engineering Mechanics, 112*(6), 550–565.
Samorodnitsky, G., & Taqqu, M. S. (1994). *Stable non-Gaussian random process. Stochastic models with infinite variance.* London: Chapman and Hall.
Samuels, J. C., & Eringen, A. C. (1958). Response of a simply supported timoshenko beam to a purely random Gaussian process. *Journal of Applied Mechanics, 25*, 496–500.
Sandberg, G., & Göranson, P. (1988). A Symmetric finite element formulation for acoustic fluid-structure interaction analysis. *Journal of Sound Vibration, 123*, 507–515.
Sandi, H. (1986). Random vibrations in some structural engineering problems. In I. Elishakoff & R. H. Lyon (Eds.), *Random vibration—Status and recent developments* (pp. 349–378). Amsterdam: Elsevier.
Sandrararajan, C., & Reddy, D. V. (1973). Response of recangular orthotropic plates to random excitation. *Earthquake Engineering and Structural Dynamics, 2*, 161–170.
Sankararaman, S., & Mahaderan, S. (2013). Separating the contributions of variability and parameter uncertainty in probability distribution. *Reliability Engineering and System Safety, 112*, 187–199.
Sankin Yu. N. (2008). Random vibrations of a complex electromechanical system with weak damping. *Transactions of Ulianovsk State Technical University, 3*, 20–28 (in Russian).
Sano, M., & Fujimori, Y. (1977). *Response of a thin plate structures to high intensity noise: Analysis and experiments.* Technical Report of National Aerospace Laboratoty, TR-506 (in Japanese).
Saragoni, G. R., & Hart, G. C. (1974). Simulation of artificial earthquakes. *Earthquake Engineering and Structural Dynamics, 2*, 249–267.
Sarkar, A., & Ghanem, R. (2002). Mid-frequency structural dynamics with parameter uncertainty. *Computer Methods in Applied Mechanics and Engineering, 194*(47–48), 5499–5513.
Sarkar, A., & Ghanem, R. (2003). A substructure approach for the midfrequency vibration of stochastic systems. *The Journal of Acousical Society of America, 113*(4), 1922–1934.
Sarkar, K., Gangali, R., Ghosh, D., & Elishakoff, I. (2016). Closed-form solutions and uncertainty quantification for gravity-loaded beams. *Meccanica, 51*(6), 1465–1479.
Saunders, H. (1972). Response of a multilayer plate to random excitation. *Shock and Vibration Bulletin, 42*, Part 5, 27–32.
Sato, R., & Lapwood, E. R. (1977). The asymptotic distribution of torsional eigenfrequencies of a spherical shell. *Journal of Physics of the Earth, 25*(3), 257–282.
Savin, E. (2017). Transient vibrational power flows in slender random structures: Theorectical modeling and numerical simulations. *Probabilistic Engineering Mechanics, 28*, 194–205.
Schenk, C. A., Pradlwarter, H. J., & Schuëller, G. I. (2004). On the dynamic stochastic response of FE models. *Probabilistic Engineering Mechanics, 19*(1–2), 161–170.
Scherrer, R. J., & Schuëller, G. I. (1982). On the correlation between XYZ—Components of earthquake acceleration. In *Proceedings of the 7th ECEE Conference*, Athens (Vol.2, pp. 169–177).
Schiegel, L. B. (1963). Covariand matrix approximation. *AIAA Journal, 1*(11), 2672–2673.
Schiehlen, W. O. (1974). Kovarianzanalyse von Zufallsschwingungen, *VDI-Bericht 221,* Duesseldorf (in German).

Schiehlen, W. O. (1986). Random vehicle vibrations. In I. Elishakoff & R. H. Lyon (Eds.), *Random vibration—Status and recent developments* (pp. 379–388). Amsterdam: Elsevier.
Schilling, R. L., & Partzsch, L. (2014). *Brownian motion: An introduction to stochastic processes*. (2nd edition). Bertin: Walter de Gruyter.
Schock, R. W., & Tuell, L. P. (1984). An investigation into the probabilistic combination of quasi-static and random accelerations. In *A Collection of Technical Papers, AIAA/ASME/ASCE/AHS 25th Structures, Structural Dynamics and Materials Conference,* Part I (pp. 193–199).
Schuëller, G. I. (1981). *Einführung in die Sicherheit und Zuverlässigkeit von Tragwerke*, Berlin: Verlag von Wilhelm Ernst & John (in German).
Schuëller, G. I. (Ed.) (1997). A state of the art report on computational stochastic mechanics. *Probabilistic Engineering Mechanics, 12*(4), 197–321.
Schuëller, G. I., & Pradlwarter, H. J. (2006). Computational stochastic structural analysis (GOSSAN)—A software tool. *Structural Safety, 28*(1–2), 68–82.
Schuëller, G. I., Pradlwarter, H. J., & Schenk, C. A. (2003). Non-stationary response of large linear FE models under stochastic loading. *Computers and Structures, 81*(8–11), 937–947.
Schuëller, G. I., & Shinozuka, M. (Eds.) (1987). *Stochastic methods in structural dynamics*. Boston: Martinus Nijhoff.
Schimansky-Geier, L., & Poeschel T. (2014). *Stochastic dynamics*. Berlin: Springer.
Seçgin, A., & Kara, M. (2019). Stochastic vibration analyses of laminated composite plates via a statistical moments-based methodology. *Journal of Vibration Engineering & Technologies, 7*(1), 73–82.
Segalman, D. J., Fulcher, C. W., Reese, G. M., & Field, R. V. (2000). An efficient method for calculation RMS Von Mises stress in a random vibration environment. *Journal of Sound and Vibration, 230*, 393–410.
Serbin, H. (1962). Response of lightly damped structures to random pressure field. *The Shock and Vibration Bulletin, 31*, Part 3, 55–79.
SenGupta, G. (1975). Current developments in interior noise and sonic fatigue research. *Shock and Vibration Digest, 7*(10), 3–20.
SenGupta, G. (1978a). Reduction of cabin noise and vibration by intrinsic structural tuning. *AIAA Journal, 16*(6), 545–546.
SenGupta, G. (1978b). Low frequency cabin noise reduction based on the intrinsic structural tuning concepts. *NASA CR-145262*.
SenCupta, G. (1979). Reduction of cabin noise during cruise conditions by stringer and frame damping. *AIAA Journal, 17*(3), 229–236.
Sepahvand, K., Marburg, S., & Hardtke, H. J. (2012). Stochastic free vibration of orthotropic plates using generalized polynomial chaos expansion. *Journal of Sound and Vibration, 331*(1), 167–179.
Seyranian, A. P. (1987). Multiple eigenvalues in optimization problems. *PMM: Journal of Applied Mathematics and Mechanics, 51*, 349–352.
Seyranian, A. P. (1993). Sensitivity analysis of multiple eigenvalues. *Journal of Structural Mechanics, 21*(2), 261–284.
Seyranian, A. P., Lund, E., & Olhoff, N. (1994). Multiple eigenvalues in structural optimization problems. *Structural Optimization, 8*(4), 207–227.
Shah, A., & Gasparini, D. A. (1983). *Random vibration of continuous beams*, Research Report 83–2, Case Western Reserve University, Cleveland, Ohio.
Shaker, A., Abdelrahman, W., Tawfik, M., & Sadek, E. (2008a). Stochastic finite element analysis of the free vibration of functionally graded material plates. *Computational Mechanics, 41*(5), 707–714.
Shaker, A., Abdelrahman, W. G., Tawfik, M., & Sadek, E. (2008b). Stochastic finite element analysis of the free vibration of laminated composite plates. *Computational Mechanics, 41*(4), 493–501.
Sharan, A. M., & Reddy, V. R. (1986). Design of machine-tools: The random response calculations of machine tools subjected to stochastic excitations using dynamic condensation technique. *Journal of Mechanisms, Transmissions, and Automation in Design, 108*(2), 145–151.

Sharan, A. M., Sankar, S., & Sankar, T. S. (1982). A new approach for the calculation of response spectral density of a linear stationary random mutidegree of freedom system. *Journal of Sound and Vibration, 83*(4), 513–519.
Shen, F., Zheng, M., Shi, D. F., & Xu F. (2003). Using the cross-correlation technique to extract modal parameters on response-only data. *Journal of Sound and Vibration, 259*(5), 1163–1179.
Shen, L., Ostoja-Starzewski, M., & Porcu, E. (2015). Responses of first-order dynamical systems to Matérn, Cauchy, and Dagum excitations. *Mathematics and Mechanics of Complex Systems, 3*(1), 27–41.
Shih, T. Y. (1980). *Random vibrations of structures under parametric and non-parametric earthquake loads*. Ph.D. Thesis, University of Illinois at Urbana-Champaign.
Shih, T. Y., & Chen, Y. C. (1984). Stochastic earthquake response of tapered column. *Journal of Engineering Mechanics, 110*(8), 1185–1210.
Shihab, S., & Preumont, A. (1989). Non-stationary random vibrations of linear multi-degree-of-freedom system. *Journal of Sound and Vibration, 132*(3), 457–471.
Shimogo, T. (1958). On the vibration of the beam under time-space random loading. *Proceedings of the 8th Japan National Congress for Applied Mechanics* (pp. 389–392).
Shimogo, T. (1960). On the beam under random loading. *Bulletin of the Japan Society of Mechanical Engineers, 3*, 60–65.
Shinozuka, M. (1964). *Random vibration of a beam column*, Report TR-10, Columbia University, Institute for the Study of Fatigue and Reliability.
Shinozuka, M. (1970). Random processes with evolutionary power. *Journal of Engineering Mechanics, 96*, 543–545.
Shinozuka, M. (1971). Simulation of multivariate and multidimensional random process. *The Journal of the Acoustical Society of America, 49*(1), 357–367.
Shinozuka, M. (1972). Monte Carlo solution of structural dynamics. *Computers and Structures, 2*, 855—874.
Shinozuka, M., & Deodatis, G. (1988). Response variability of stochastic finite element systems. *Journal of Engineering Mechanics, 118*(3), 499–519.
Shinozuka, M., & Deodatis, G. (1991). Simulation of stochastic processes by spectral representation. *Applied Mechanics Reviews, 44*(4), 191–204.
Shinozuka, M., & Deodatis, G. (1996). Simulation of multi-dimensional gaussian stochastic fields by spectral representation. *Applied Mechanics Reviews, 49*(1), 29–53.
Shinozuka, M., Kako, T., & Tsurui, A. (1986). Random vibration analysis in finite element formulation. In I. Elishakoff & R. H. Lyon (Eds.), *Random vibration—Status and recent developments* (pp. 415–450). Amsterdam: Elsevier.
Shinozuka, M., & Astill, C. J. (1972). Random eigenvalue problems in structural analysis. *AIAA Journal, 10*(4), 456–462.
Shinozuka, M., & Deodatis, G. (1991). Simulation of stochastic processes by spectral representation. *Applied Mechanics Reviews, 44*(4), 191–204.
Shinozuka, M., & Deodatis G. (1996). Simulation of multi-dimensional Gaussian stochastic fields by spectral representation. *Applied Mechanics Reviews, 49*, 29–53.
Shinozuka, M., & Henry, L. (1965). Random vibration of a beam column. *Journal of Engineering Mechanics, 91*, 123–143.
Shinozuka, M., & Jan, C.-M. (1972). Digital simulation of random processes and its applications. *Journal of Sound and Vibration, 25*(1), 111–128.
Shinozuka, M., & Sato, Y. (1967). Simulation of nonstationary random process. *Journal of Engineering Mechanics, 93*, 11–40.
Shinozuka, M., Kako, T., & Tsurui, A. (1986). Random vibration analysis in finite element foundation. In I. Elishakoff & R. H. Lyon (Eds.), *Random vibration—Status and recent developments* (pp. 415–425). Amsterdam: Elsevier.
Shinozuka, M., & Yang, J. N. (1969). Numerical Fourier transform for random vibration. *Journal of Enginering Mechanics, 95*, 731–746.

Shinozuka, M. Yang, J. N. (1971). Peak structural response to non-stationary random excitations. *Journal of Sound and Vibration*, *16*(4), 505–517.
Shires, D. (2011). On the time compression (test acceleration) of broadband random vibration tests. *Packaging Technology and Science, 24*(2), 75–87.
Shoenberger, R. W. (1976). Comparison of the subjective intensity of sinusoidal, multifrequency, and random whole-body vibration. *Aviation, Space, and Environmental Medicine*.
Shmakov, V. P. (1969). Some problems of axisymmetric vibrations of a spherical shell. In *Isskedovaniya no Teorii Sooruzhenii (Investigations into Theory of Structures)*, *17*, 1969 (in Russian).
Shorter, P. (1998). *Combining finite elements and statistical energy analysis*. Ph.D. Thesis, University of Auckland.
Shorter, P. J. (2000). On the use of the sturm sequence to evaluate modal density. *The Journal of the Acoustical Society of America, 108*(5), 2557.
Si, L. T., Zhao, Y., Zhang, Y. H., & Ouyang, H. (2016). Random vibration of an elastic half-space subjected to a moving stochastic load. *Computers and Structures*, *168*, 92–105.
Sieniawska, R., & Sniady, P. (1990a). First passage problem of the beam under a random stream of moving forces. *Journal of Sound and Vibration, 136*(2), 177–185.
Sieniawska, R., & Sniady, P. (1990b). Life expectancy of highway bridges due to traffic laod. *Journal of Sound and Vibration, 140*(1), 31–38.
Sibani, P., & Jensen, H. J. (2013). *Stochastic dynamics of complex systems: From glasses to evolution*. London: Imperial Collge Press.
Silaev, A. (1972). *Spectral theory of suspension and road vehicles*. Moscow: "Mashinostroenie" Publishing House (in Russian).
Silaev, A. A. (1972). *Spectral theory of vehicle cushioning*. Moscow: "Mashinostroenie" Pubishing House (in Russian).
Simiu, E., & Scanlan, R. H. (1978). I *Wind Effects on structures: An introduction to wind engineering*. New York: Wiley.
Singh, B. N., Yadav, D., & Iyengar, N.G. R. (2001). Free Vibration of laminated spherical panels with random material properties. *Journal of Sound and Vibration*, *244*, 55–64.
Singh, B. N., Vyas, N., & Das, P. (2009). Stochastic free vibration of Smart random plates. *Journal of Structural Engineering and Mechanics*, *31*(5), 481–506.
Singh, B. N., Yadav, D., & Iyengar, N.G. R. (2002). Free vibration of composite cylindrical shells with random material properties. *Computers and Structures*, *58*(4), 435–442.
Singh, M. P. (1975). Generation of seismic floor spectra. *Journal of Engineering Mechanics Division*, *101*(5), 593–607.
Singh, M. P. (1984). Structural response under multicomponent earthquakes. *Journal of Engineering Mechanics, 110*(5), 761–775.
Singh, M. P. (1980). Seismic response by SRSS for non-proportional damping. *Journal of Engineering Mechanics, 106*(6), 1405–1419.
Singh, M. P. (1988). Seismic design of secondary structures. *Probabilistic Engineering Mechanics*, *3*, 151–158.
Singh, M. P., & Abdelnaser, A. S. (1991a). Random vibrations of Timoshenko beams with generalized boundary conditions. In Y. K. Lin & I. Elishakoff (Eds.), *Stochastic structural dynamics 1: New theoretical developments* (pp. 173–192). Berlin: Springer.
Singh, M. P., & Abdelnaser, A. S. (1991b). Random vibration of externally damped viscoelastic Timoshenko beams with general boundary conditions. *Journal of Applied Mechanics, 60*, 149–156.
Singh, M. P., & Abdelnaser, A. S. (1992). Random response of symmetric cross-ply composite beams with arbitrary boundary conditions. *AIAA Journal*, *30*(4), 1081–1088.
Singh, M. P., & Chu, S. L. (1970). Stochastic considerations in seismic analysis of structures. *Journal of Earthquake Engineering and Structural Dynamics*, *4*, 245–307.
Singh, M. P., & Ghafory-Ashtiany, M. (1984). Structural response under multicomponent earthquakes. *Journal of Engineering Mechanics*, *110*, 761–775.

Singh, M. P., & Ghafory-Ashtiany, M. (1986). Modal time history analysis of non-classically damped structures for seismic motions. *Earthquake Engineering and Structural Dynamics, 14*(1), 133–146.
Singh, M. P., Khdeir, A. A., Maldonaldo, G. O., & Reddy, J. N. (1988). Random response of antisymmetric angle-ply laminated plates. *Structural Safety, 6*, 115–127.
Singh, M. P., Khdeir, A. A., Maldonado, G. O., & Reddy, J. N. (1989). Random response of symmetric angle-ply laminated plates. *Structural Safety, 16*(2–4), 115–127.
Singh, M. P., & Maldonaldo, G. O. (1991). An improved response spectrum method for calcualting seismic design response, Part I: Classically damped structures. *Earthquake Engineering and Structural Dynamics, 20*, 621–635.
Singh, M. P., & Mehta, K. B. (1983). Seismic design response by an alternative SRSS rule. *Earthquake Engineering and Structural Dynamics, 11*, 711–721.
Singh, R., & Lee, C. (1993). Frequency response of linear systems with parameter uncertainties. *Journal of Sound and Vibration, 168*(1), 71–92.
Singh, R., & Nair, S. S. (1987). Proportional damping approximations for systems with repeated roots. In *Proceedings of 5th International Modal Analysis Conference* (pp. 1022–1337). Imperial College of Science and Technology, London.
Singh, M. P., & Wen, Y. K. (1977). Nonstationary seismic response of light equipment. *Journal of Engineering Mechanics, 103*, 1035–1048.
Skalmierski, B., & Czogala, Z. (1969). On certain problems of non-stationary vibration of shells under random loadings. *Rozprawy Inzynierskie, 17*(2), 219–228 (in Polish).
Skalmierski, B., & Tylikowski, A. (1972). *Proclsy Stochastycznl w Dynamice (The Stochastic Proceeses in Dynamics)*. Warszawa: PWN (in Polish).
Skalmierski, B., & Tylikowski, A. (1982). *Stochastic processes in dynamics*. Martinus Nijhoff Publishers.
Skorokhod, A. V. (1979). *Random process in problems of mathematics physics*. Kies, Institute of Mathematics, Ukranian Academy of Science (in Russian).
Skudrzyk, E. (1980). The mean value method of predicting the dynamic response of complex vibration. *The Journal of the Acoustical Society of America, 67*(4), 1105–1135.
Slavič, J., Knez, L., & Boltežar, M. (2017). The importance of harmonic versus random excitation for a human finger. *International Journal of Mechanical Sciences, 131*, 507–515.
Smeby, W., & Der Kiureghain, A. (1985). Modal combination rules for multicomponent earthquake excitation. *Earthquake Engineering and Structural Dynamics, 13*(1), 1–12.
Smilansky, U., & Stöckmann, H. J. (Eds.), *Nodal patterns in physics and mathematics: From Chlandni's seminal work to modern applications a historic scientific perspective*. Berlin: Springer.
Smith, A. E., Ryan, P. B., & Evans, J. S. (1992). The effect of neglecting correlations when propagating uncertainty and estimating the population distribution of risk. *Risk Analysis, 12*(4), 467–474.
Smith, P. W. Jr. (1981). Spatial correlations of random response in point-excited waveguides. *The Journal of the Acoustical Society of America, 69*, 1337–1342.
Smith, P. W. Jr. (1980a). Private communication, 29 May 1980.
Smith, P. W. Jr. (1980b). Private communication, 19 September 1980.
Smith, P. W. Jr. (1980c). Private communication, 20 October 1980.
Smith, P. W. Jr., & Lyon, R. H. (1965). *Sound and structural vibration*. NASA CR-160.
Smith, S. T., & Bradford, M. A. (1999). Numerical convergence of simple and orthogonal polynomials for the unilateral plate buckling problem using the Rayleigh-Ritz method. *International Journal for Numerical Methods in Engineering, 44*(11), 1685–1707.
Smiliansky, U., & Stöckmann, H. J. (2007). Preface. *European Physical, Special Topics, 145*, 1–2.
Sniady, P. (1984a). Dynamic response of linear structures to a stream of random impulses in a "Space-Time" system. *Journal of Sound and Vibration, 95*, 41–47.
Sniady, P. (1984b). Vibration of a beam due to a random stream of moving forces with random velocity. *Journal of Sound and Vibration, 97*(1), 23–33.

Sniady, P. (1989). Dynamic response of linear structures to a random stream of pulses. *Journal of Sound and Vibration, 131*, 91–102.
Sniady, P., Adamowski, R. Kougut, G., & Zielichowski-Haber, W. (2008). Spectral stochastic anaylsis of structures with uncertain parameters. *Probabilistic Engineering Mechanics, 23*, 76–83.
Sniady, P., Podwórna, M., & Idzikowski, R. (2019). Stochastic vibrations of the Euler–Bernoulli beam based on various versions of the gradient nonlocal elasticity theory. *Probabilistic Engineering Mechanics,* in press.
Sniady, P., Sieniawska, R., & Zukowski, S. (1993). A train of pulses in load modelling. *Structural Safety, 13*(1–2), 29–43.
Snieder, R., Wapenaar, K., & Wegler, U. (2007). Unified Green's function retrieved by cross-correlation; connection with energy principles. *Physical Review E, 75*(3), article 036103.
Snyder, D. I. (1975). Random point processes. New York: Wiley.
Sobczyk, K. (1972). Free vibration of elastic plte with random properties—The eigenvalue problem. *Journal of Sound and Vibration, 22*(1), 33–39.
Sobczyk, K. (1970). Random vibrations in homogeneous elastic systems. *Proceedings of Vibration Problems, 4*(11), 369–384.
Sobczyk, K. (1973). *Metody Dynamiki Statystycznej (Methiods of Statistical Dynamics)*. Warszawa: PWN (in Polish).
Sobczyk, K. (1985). *Stochastic wave propagation*. Amsterdam: Elsevier.
Sobczyk, K. (1990). *Stochastic differential equations: With applications to physics and engineering*. Berlin: Springer.
Sobczyk, K. (2001). Information dynamics: Premises, challenges and results. *Mechanical Systems and Signal Processing, 15*(3), 475–498.
Sobczyk, K. (2006). Stochastic dynamics and reliability of degrading systems. *Bulletin of the Polish Academy of Sciences, Technical Sciences, 54*(1), 125–137.
Sobczyk, K., Macvean, D. B., & Robson, J. D. (1977). Response of profile-imposed excitation with randomly varying traversal velocity. *Journal of Sound and Vibration, 52*(1), 37–49.
Sobczyk, K., & Treebicki, J. (1999). Stochastic response of degrading elastic systems. In *IUTAM Symposium on Rheology of Bodies With Deflects* (pp.99–108). Berlin: Springer.
Sobczyk, K., Wedrychowia, S., & Spencer, B. F. Jr. (1996). Dynamics of arctural systems with spatial randomness. *International Journal of Solids and Structures, 33*(11), 1651–1669.
Soedel, W., & Dhar, M. (1978). Difficulties in finding modes experimentally when several contribute to resonance. *Journal of Sound and Vibration, 58*(1), 27–38.
Soize, C. (1994). *The Fokker-Planck equation for stochastic dynamical systems and its explicit steady state solution*. New Jersey: World Scientific.
Soize, C. (1986). Probabilistic structural modeling in linear dynamic analysis of complex mechanical systems, I—Theoretical elements. *La Recherche Aérospatiale (English Edition), 5*, 23–48.
Soize, C. (1998). Reduced models in the medium frequency range for general dissipative structural dynamics systems. *European Journal of Mechanics, A: Solids, 17*(4), 657–685.
Soize, C. (2005). Random matrix theory for modeling random uncertainties in computational mechanics. *Computer Methods in Applied Mechanics and Engineering, 194*(1216), 1333–1366.
Soize, C. (2010). Generalized Probabilistic approach of uncertainties in computational dynamics using random matrices and polynomial chaos decompositions. *International Journal for Numerical Methods in Engineering, 81*(8), 939–970.
Soize, C., & Batou, A. (2011). Stochatic reduced-order model in low-frequency dynamics in presence of numerous local elastic modes. *Journal of Applied Mechanics, 78*(6), article 061003.
Sokolov, A. A. (1990). Vibration diagnostics of slow periodic parameter variations in a system with random external excitation. *Mechanics of Solids, 25*(1), 46–52.
Sokolskii, A. G. (1974). On the stability of an autonomous Hamiltonian system with two degrees of freedom in the case of equal frequencies. *PMM: Journal of Applied Mathematics and Mechanics, 38*(5), 791–799.
Solodov, A. V., & Solodov, A. A. (1988). *Statistical dynamics of systems with point process*. Moscow: "Nauka" Publishing House (in Russian).

Solodovnikov, V. V. (1960). *Statistical dynamics of linear systems of automatic control*. Moscow: Fizmatgiz ", Publishing House (in Russian).
Solomos, G. P., & Spanos, P. D. (1984). Oscillator response to nonstationary excitation. *Journal of Applied Mechanics, 51*, 907–912.
Solomos, G. P., & Spanos, P.-T. D. (1984). Oscillator response to nonstationary excitation. *Journal of Applied Mechanics, 51*(4), 907–912.
Soong, T. T. (1973). *Random differential equations in science and engineering*. New York: Academic Press.
Soong, T. T. (1989). Stochastic structural dynamics research vs. practice. *Structural Safety, 6*(2–4), 129–134.
Soong, T. T., & Bogdanoff, J. L. (1963). On the natural frequenciez of a disordered linear chain of N degrees of freedom. *International Journal of Mechanical Sciences, 5*, 237–265.
Soong, T. T., & Costantinou, M. (2014). *Passive and active structural vibration control in civil engineering*. Berlin: Springer.
Soovere, J. (1984). *Dynamic response of acoustically excited stiffened composite honeycomb panels*. Ph.D. Thesis, University of Southampton.
Soovere, J. (1986). Random vibration analysis of stiffened honeycomb panels. *Journal of Aircraft, 23*(6), 537–544.
Soovere, J., & Clarkson, B. L. (1967). Frequency response functions from cross-correlation: Bartlett weighting function. *AIAA Journal, 5*, 601–603.
Souf, M. B., Ichchou, M. N., Bareille, O., & Haddar, M. (2013). On the dynamics of uncertain coupled structures Through a wave based method in mid-and high-frequency ranges. *Computational Mechanics, 52*(4), 849–860.
Sounders, K. D. (1958). A power-spectrum equation for stationary random gusts, including a simple example. *Journal of the Aeronautical Sciences, 25*(5), 295–3012.
Soyluk, K. (1995). Comparison of random vibration methods for multi-support seismic excitation analysis of long-span bridges. *Engineering Structures, Computers and Structures, 56*(6), 871–895.
Soyluk, K. (2004). Comparison of random vibration methods for multi-support seismic excitation analysis of long-span bridges. *Engineering Structures, 26*(11), 1573–1583.
Spanos, P. D. (1978a). Non-stationary random vibration of a linear structure. *International Journal of Solids and Structures, 14*(10), 861–867.
Spanos, P. D. (1978b). Energy analysis of structural vibrations under modulated random excitation. *Journal of Structural Mechanics, 6*, 289–302.
Spanos, P. D. (1980). Earthquake energy spectral equations. *Journal of Engineering Mechanics, 106*, 147–159.
Spanos, P. D. (1982). Numerics for common first-passage problems. *Journal of Engineering Mechanics, 108*, 864–881.
Spanos, P. D. (1983a). Digital synthsis of response-design spectrum compatible earthquake records for dynamic analysis. *The Shock and Vibration Digest, 15*(3), 21–30.
Spanos, P. D. (1983b). Spectral moments calculations of linear output system. *Journal of Applied Mechanics, 50*, 901–903.
Spanos, P. D. (1987). An approach to calculating random vibration integrals. *Journal of Applied Mechanics, 54*, 409–413.
Spanos, P. D., & Chen, T. W. (1981). Random response to flow-induced forces. *Journal of Engineering Mechanics, 107*(6), 1173–1190.
Spanos, P. D., & Iwan, W. D. (1978). Computational aspects of random vibration analysis. *Journal of Engineering Mechanics, 104*(6), 1403–1415.
Spanos, P. D., & Ghanem, R. (1991). Boundary element formulation for random vibration problems. *Journal of Engineering Mechanics, 117*(2), 409–423.
Spanos, P. D., & Lutes, L. D. (1980). Probability of response to evolutionary, process. *Journal of Engineering Mechanics, 106*, 213–224.

Spanos, P. D., & Miller, S. M. (1994). Hilbert transform generalization of a classical random vibration integral. *Journal of Applied Mechanics*, *61*(3), 575–581.

Spanos, P. D., & Vargas Loli, L. M. (1985). A statistical approach to generation of design spectrum compatible earthquake time histories. *Soil Dynamics and Earthquake Engineering, 4*(1), 2–8.

Spanos, P. D., & Zeldin, B. A. (1997). Random vibration of systems With frequency dependent parameters in fractional derivatives. *Journal of Engineering Mechanics, 123*(3), 290–292.

Spanos, P. D., & Zeldin, B. A. (1998). Monte Carlo Treatment of random fields: A broad perspective. *Applied Mechanics Reviews, 51*, 219–237.

Spencer, B. F. Jr., & Elishakoff, I. (1988). Reliability of uncertain linear and nonlinear systems. *Journal of Engineering Mechanics*, *114*(1), 135–148.

Spiegel, M. R. (1952). The random vibration of a string. *Quarterly of Applied Mechanics, 10*(10), 25–33.

Spirer, H. F., Spirer, L., & Jable, A. J. (1998). *Misused statistics* (second edition). New York: Marcel Dekher.

Spreemann, D., Hobbmann, D., Folkmer, B., & Manoli, Y. (2008). Numerical optimization approach for resonant electromagnetic vibration transducer designed for random vibration. *Journal of Micromechanics and Microengineering*, 18, article 104001.

Srinivasan, R. S. (1991). Random response of thick laminated rectangular plates. *Journal of Vibration and Acoustics*, *113*(3), 286–291.

Srinivasan, S. K. (1974). *Stochastic point processes and their applications*. London: Griffin.

Srinivasan, S. K., Subramanian, R., & Kumaraswamy, S., Response of linear vibratory systems to non-stationary stochastic impulses. *Journal of Sound and Vibration, 6*, 169–179.

Stadler, W., & Shreeves, R. W. (1970). The transient and steady-state response of an infinite Bernoulli-Euler beam with damping and an elastic foundation. *Quarterly Journal of Mechanics and Applied Mathematics*, *23*, 197–208.

Stafford, P. J., Sgobba, S., & Marano, G. (2009). An energy based envelope function for the stochastic simulation of earthquake accelerograms. *Soil Dynamics and Earthquake Engineering, 29*, 1123–1133.

Stanisic, M. M. (1968). Response of plates to random loads. *The Journal of the Acoustical Society of America, 43*, 1351–1357.

Starkloff, H.-J., Von Scheidt, J., & Wunderlich, R. (2001). Random vibration of systems with weakly correlated random excitation. *ZAMM: Zeitschrift für angewandte Mathematik and Mechanik, 81*, 5649–5650.

Stratonovich, R. L. (1963). *Topics in the theory of random noise* (Vol. 1). New York: Gordon and Breach.

Stearn, S. M. (1970). *Stress distribution in randomly excited structures*. Ph.D. Thesis, University of Southampton, England.

Stearn, S. M. (1971). The concentration of dynamic stress in a plate at a sharp change of section. *Journal of Sound and Vibration*, *15*(3), 193–200.

Stefanou, G., & Papadrakakis, M. (2004). Stochastic finite element analysis of shells with combined random material and geometric properties. *Computer Methods in Applied Mechanics and Engineering*, *193*(1–2), 139–160.

Steinwolf, A. (1994). Analysis and approximation of probability distortion of vehicle random vibrations with consideration of kurtosis and skewness values. In N. S. Ferguson, H. F. Wolfe & C. Mei (Eds.), *Structural Dynamics, Recent Advances: Proceedings of the 5th International Conference* (pp. 785–794). ISVR, Southampton.

Steinwolf, A. (2006). Random vibration testing beyond PSD limitations. *Sound and Vibration*, *40*(9), 12–21.

Steinwolf, A., Ferguson, N. S., & White, R. G. (2000). Variations in steepness of the probability density function of beam random vibration. *European Journal of Mechanics-A/Solids*, *19*(3), 19–341.

Steinwolf, A., & Rizzi, S. A. (2006). Non-Gaussian analysis of turbulent boundary layer fluctuating pressure on aircraft skin panels. *Journal of Aircraft*, *43*(6), 1662–1675.

Stewart, J. K., & Colwell, R. C. (1939). The calculation of Chladni patterns. *The Journal of the Acoustial Society of America, 11*(1), 147–151.
Steinwolf, A., & White, R. G. (1997). Probability Density Functions of acoustically induced strains in experiments with composite plates. *AIAA Journal, 35*(12), 1853–1861.
Strasberg, M. (1958). Response of plates and membranes to pressure fluctuations of a turbulent boundary layer. *The Journal of the Acoustial Society of America, 30*, 680.
Strawderman, W. A. (1969a). Turbulent-flow-excited vibration of a simply supported, recangular flate plate. *The Journal of the Acoustical Society of America, 45*, 177–192.
Strawderman, W. A. (1969b). Turbulent-induced plate vibrations: An evaulation of finite—And infinite-plate models. *The Journal of the Acoustical Society of America, 46*, 1294–1307.
Strawderman, W. A., & Christman, R. A. (1972). Turbulence-induced plate vibrations: Some effects of fluid loading on finite and infinite plates. *The Journal of the Acoustical Society of America, 52*(5B), 1537–1552.
Stumpf, H. J. (1960). *Response of mechanical systems to random excitation*. Ph.D. Thesis, California Institute of Technology.
Su, L., & Ahmadi, G. (1988). Earthquake Response of linear continuos structures by the method of evolutionary spectra. *Engineering Structures, 10*, 47–56.
Su, L., & Ahmadi, G. (1988). Response of frictional base isolation systems to horizontal—Vertical earthquake excitations. *Probabilistic Engineering Mechanics, 3*, 12–21.
Su, L., & Ahmadi G. (1992). Probalistic Responses of base-isolated structures to El Centro 1940 and Mexico City 1985 earthquakes. *Engineering Structures, 14*, 217–230.
Su, L., Ahmadi, G., & Tadjbakhsh, I. G. (1988). Response of base-isolated shear beam structures to random excitations. *Probabilistic Engineering and Mechanics, 5*, 35–46.
Su, L., Orabi, I. I., & Ahmadi, G. (1988). Nonstationary earthquake response of a sliding ridged structure. *International Journal of Engineering Sciences, 26*, 1013–1026.
Sun, J. Q., & Hsu, C. S. (1989). Random vibration of hinged elastic shallow arch. *Journal of Sound and Vibration, 132*(2), 299–315.
Sukhir, E. L., *Probabilistic methods in ship structural mechanics and vibrations*. Nikolaev Shipbuilding Institute (in Russian).
Sun, J.-Q. (2006). *Stochastic dynamics and control*. Amsterdam: Elsevier.
Sundararajan, C., & Reddy, D. V. (1973). Response of rectangular orthotropic plates to random excitation. *Earthquake Engineering and Structural Dynamics, 2*(2), 161–170.
Sun, H. L., Zhang, P. Q., Chen, H. B., & Zhang, K. (2006). Active control of a structure with continuously claosely spaced natural frequencies. *Journal of Sound and Vibration, 294*, 15–22.
Sun, L., & Greenberg, B. S. (2000). Dynamic response of linear systems to moving stochastic sources. *Journal of Sound and Vibration, 229*(4), 957–972.
Sun, T. C. (1979). A finite element method for random differential equations with random coefficients. *SIAM Journal of Numerical Analysis, 16*(6), 1019–1035.
Sun, W. J., & Kareem, A. (1989). Response of MDOF systems to nonstationary random excitation. *Engineering Structures, 11*(2), 83–91.
Sveshnikov, A. A. (1966). *Applied methods of the theory of random functions*. New York: Pergamon Press House.
Sveshnikov, A. A. (1968). *Applied methods of random function theory*. Moscow: "Nauka" Publishing House (in Russian).
Svetlitsky, V. A. (1976). *Random vibrations of mechanical systems*. Moscow: "mashinostroenie" Publising House (in Russian).
Svetlitsky, V. A. (1986). *Dynamics of flying vehicle launch*. Moscow: Nauka" Publishing House (in Russian).
Svetlitsky, V. A. (2003). *Statistical dynamics and reliability theory for mechanical structures*. Berlin: Springer.
Svetlitsky, V. A. (2004a). *Statistical mechanics and theory of reliability*. Moscow: Moscow State Technical University Press (in Russian).

Svetlitsky, V. A. (2004b). *Engineering vibration analysis: Worked problems* (Vol. 1). Berlin: Springer.

Sweers, J. E. (1964). Prediction of response and fatigue life of honeycomb sandwich panels subjected to acoustic excitation. In W. J. Trapp & D. M. Forney (Eds.), *Acoustical fatigue in aerospace structures* (pp. 389–402). Syracuse University Press.

Szechenyi, E. (1971a). Modal densities and radiation efficiencies of unstiffened cylinders using statistical methods. *Journal of Sound and Vibration, 19*(1), 65–81.

Szechenyi, E. (1971b). Sound Transmission through cylinder walls using statistical considerations. *Journal of Sound and Vibration, 19*(1), 83–94.

Szilard, R. (2004). *Theories and applications of plate analysis: Classical, numerical and engineering methods*. New York: Wiley.

Tack, D. H., & Lambert, R. G. (1962). Response of bars and plates to boundary layer turbulence. *Journal of Aerospace Science, 29*, 311–322.

Tahmasebi, P. Hezarkhani, A., & Sahimi, M. (2012). Multiple-point geostatistical modeling based on the cross correlation functions. *Computational Geosciences, 16*(3), 779–797.

Tajimi, H. (1960). A standard method of determining the maximum response of a building structure during an earthquake. *Proceedings of the 2nd Word conference on Earthquake Engineering, 2*, 781–798.

Takahashi, Y. (1979). Two Hundred years of Lichtenberg figures. *Journal of Electrostatics, 6*(1), 1–13.

Takewaki, I. (2000). Soil-structure random response reduction via TMD-VD simultaneous use. *Computer Methods in Applied Mechanics and Engineering, 190*(5–7), 677–690.

Tamboli, J. A., & Joshi, S. G. (1999). Optimum design of a passive suspension system of a vehicle subjected to actual random road excitations. *Journal of Sound and Vibration, 219*(2), 193–205.

Tamura, Y., Kim, Y. C., Kikuchi, H., & Hibi, K. (2014). Correlation and combination of wind force components and responses. *Journal of wind Engineering and Industrial Aerodynamics, 125*, 81–93.

Tan, D. Y. (1994). Discrete analysis method for random vibration of structures subjected to temporally and spatially correlated coloured excitation. *Computers and Structures, 50*, 619–627.

Tan, D. Y., & Yang, Q.-S. (1995). Structural random vibration analysis by using recursive formulas. *Probabilistic Engineering Mechanics, 10*, 11–21.

Tan, D. Y., Yang, Q. S., & Zhao, C. (1992). Discrete analysis method for random vibration of structures subjected to spatially correlated filtered white noises. *Computers and Structures, 43*(6), 1051–1056.

Tang, D. M., & Dowell, E. H. (1986). Random response of beams and plates with slipping at support boundaries. *AIAA Journal, 24*(8), 1354–1361.

Tang, J., & Wang, W. L. (1999). On calculation of sensitivity for non-defective eigenproblems with repeated roots. *Jouranl of Sound and Vibration, 225*(4), 611–631.

Tang, X.-D., & Zuo, L. (2012). Vibration energy harvesting from random force and motion excitations. *Smart Materials and Structures, 21*(7), article 075025.

Tellier, M., & Iwankiewicz, R. (2005). Response of linear dynamic systems to non-erlang renewal impulses: Stochastic equations approach. *Probabilistic Engineering Mechanics, 20*(4), 281–295.

Teytel, M., How Rare are Multiple Eigenvalues?, *Communications in Pure and Applied Mathematics, 7*, 917–934, 1999.

Tezak, E. G., Nayfeh, A. H., & Mook, D.T. (1982). Parametrically excited non-linear multidegree-of-freedom systems with repeated natural frequencies. *Journal of Sound and Vibration, 85*(4), 459–47.

Thesing, T., & Bishop, N. (2016). *Modern methods for random fatigue of automotive parts*. SAE Technical Paper No. 2016–01–0372.

Thomas, D. R., Nelson, P. A., & Elliott, S. J. (1993). Active control of the transmission of sound through a thin cylindrical shell, Part 1: The minimization of vibrational energy. *Journal of Sound and Vibration, 167*, 91–111.

Thomas, J. H. (1968). Random vibrations of thin elastic plates. *ZAMP: Zeitschrift für die angewandte Mathematik und Physik, 19*, 921–926.
Thompson, J. J., & Holy, Z. J. (1975). Random pressure induced vibration of pin-ended cylindrical rods and tubes. *Nuclear Engineering and Design, 33*(2), 370–380.
Thomson, W. T. (1962). Continuous structures excited by correlated random forces. *International Journal of Mechanical Sciences, 4*, 109–114.
Thomson, W. T., & Barton, M. V. (1957). The response of mechanical systems to random excitation. *Journal of Applied Mechanics, 24*, 248–251.
Tickell, R. G., & Holmes, P. (1978). Approaches to fluid loading, probabilistic and deterministic "Analyses". In *Numerical mehtods in offshore engineering*. New York: Wiley.
Tikhonov, V. I., & Tolkachev, A. A. (1965). The effect of non normal fluctuations on linear systems. In P. I. Kuznetsov, R. L. Stratonovich & V. I. Tikhonov (Eds.), *Non-linear transformation of stochastic processes*. New York: Pergamon Press.
To, C.W. S. (1988). Direct integration operation and their stability for random response of multi-degree-of-freedom systems. *Computers and Structures, 30*(4), 865–874.
To, C.W. S. (1982). Nonstationary random response of a MDOF system by the theory of evolutionary spectra. *Journal of Sound and Vibration, 83*, 273–291.
To, C.W. S. (1984). Time-dependent variance and covariance of responses of structures to non-stationary random excitations. *Journal of Sound and Vibration, 93*, 135–156.
To, C.W. S. (1986a). Stochastic central difference method in structural dynamics. *Computers and Structures, 23*(6), 813–818.
To, C.W. S. (1986b). Response statistics of discretized structures to non-stationary random excitation. *Journal of Sound and Vibration, 105*, 217–231.
To, C.W. S. (1992). A stochastic version of the newmark family of algorithms for discretized dynamic systems. *Computers and Structures, 44*(3), 667–673.
To, C.W. S. (2001). On computational stochastic structural dynamics applying finite elements. *Archives of Computational Methods in Engineering, 8*(1), 3–40.
To, C.W. S., *Stochastic structural dynamics:Application of finite element methods*. New York: Wiley.
To, C.W. S., & Chen, T. (2007). Optimal control of random vibration in plate and shell structures with distributed piezoelectric components. *International Journal of Mechanical Sciences, 49*(12), 1389–1398.
To, C.W. S., & Liu, M. L. (1994). Random responses of discretized beams and plates by the stochastic central difference method with time co-ordinate transformation. *Computers and Structures, 53*, 727–738.
To, C.W. S., & Liu, M. L. (2000). Large nonstationary random responses of shell structures with geometrical and material nonlinearities. *Finite Elements in Analysis and Design, 35*, 59–77.
To, C.W. S., & Orisamolu, I. R. (1987). Response of a two-degree-of-freedom system to random disturbances. *Computers and Structures, 25*(3), 311–320.
To, C.W. S., & Orisamolu, I. R. (1987). Response of discretized plates to transversal and in-plane non-stationary random excitations. *Journal of Sound and Vibration, 114*(3), 481–494.
To, C.W. S., & Wang, B. (1993a). Time-dependent response statistics of axisymmetric shell structures. *Journal of Sound and Vibration, 164*(3), 554–564.
To, C.W. S., & Wang, B. (1993b). Response analysis of discretized plates under in-plane and external nonstationary random excitations. In *Proceedings of the 14th ASME Bieenial Conference in Mechanical Vibration and Noise*, DE-*56*, 51–62. New York: ASME Press.
To, C.W. S., & Wang, B. (1996). Nonstationary random response of laminated composite structures by a hybrid strain-based flat triangular shell element. *Finite Elements in Analysis and Design, 23*(1), 23–35.
Tobias, S. A. (1951). A theory of imperfection for the vibrations of elastic bodies of revolution. *Engineering, 172*(4470), 409–410.
Torano, J., Rodrigues, R., Diego, I., Rivas, J. M., & Casal, M. D. (2006). FEM models including randomness and its application to the blasting vibration prediction. *Computers and Geotechnics, 33*(1), 15–28.

Tovstik, P.E. (1972). About the density of vibration frequency of thin shells of revolution. *PMM: Journal of Applied Mathematics and Mechanics, 36*(2), 270–278 (in Russian).
Tovstik, P. E., Tovstik, T. M., & Shekhovtsov, V.A. (2005). Modelling of vibrations of sea stationary platform during random excitation. *Herald of Saint-Pertsburg University, Series 1, Matheamtics, Mechanics, Astronomy,* (4) (in Russian).
Tovstik, P. E., Tovstik, T. M., & Shekhovtsov, V.A. (2005). About the effect of spectral density form on vibration of the sea platform. *Herald of Saint-Pertsburg University, Series 1, Matheamtics, Mechanics, Astronomy,* (4) (in Russian).
Trapp, W. J., & Forney, D.M. (Eds.) (1965). *Acoustical fatigue in aerospace structures*. Syracuse University Press.
Tripathi, V., Sinh, B. N., & Shukla, K.K. (2007). Free vibration of laminated composite plates with random properties. *Composite Structures, 81*, 96–104.
Trubert, M.R.P. (1963). Response of elastic structures to statistically correlated multiple random excitations. *The Journal of the Acoustical Society of America, 35*.
Tsai, D.-M., & Lin, C.-T. (2003). Fast normalized cross correlation for defect detection. *Pattern Recognition Letters, 24*(15), 2625–2631.
Tso, M. H., Yuan, J., Wong, W. O. (2012). Suppression of random vibration in flexible structures using a hybrid vibration absorber. *Journal of Sound and Vibration, 331*, 974–986.
Tubino, F., Carassale, L., & Solari, G. (2003). Seismic response of multisupported structres by proper orthogonal decomposition. *Earthquake Engineering and Structural Dynamics, 32*(11), 1639–1654.
Tung, C.C. (1967). Random response of highway bridges to vehicle loads. *Journal of Engineering Mechanics, 93*, 79–94.
Tung, C.C. (1969a). Response of highway bridges to renewal traffic loads. *Journal of Engineering Mechanics, 95*, 41–57.
Tung, C.C. (1969b). Life expectancy of highway bridges to vehicle loads. *Journal of Engineering Mechanics, 95*(6), 1417–1428.
Tupolev, A.A. (1979). Problems of dynamics and strength during design of supersonic passenger airplanes. *Mashinovedenie,* (1), 3–7 (in Russian).
Tustin, W., & Mercado, R. (1984). *Random vibration in perspective*. Tustin Institute of Technology.
Tylikowski, A. (1969). Vibration of a linear system caused by a chance process with uniformly varying frequency. *Engineering Transactions, 17*, 269–279 (in Polish).
Udwadia, F. E. (1987a). Response of uncertain dynamic systems. I. *Applied Mathematics and Computation, 22*(2–3), 115–150.
Udwadia, F. E. (1987b). Response of uncertain dynamic systems. II. *Applied Mathematics and Computation, 22*(2–3), 151–187.
Uhlenbeck, G. E., & Ornstein, L.S. (1930). On the theory of the Brownian motion, I. *Physical Review, 36*, 823–841.
Uhlenbeck, G. E., & Van Lear, G.A. (1931). The Brownian motion of strings and elastic rods. *Physical Review, 38*, 1583–1598.
Uhlenbeck, G. E., & Wang, M.C. (1945). On the theory of the Brownian motion, II. *Review of Modern Physics, 17*(2), 323–342.
Ullmann, D. (1983). *Ernst Florens Friedrich Chladni*. Leipzig: Teubner (in German).
Ushkalov, V.F. (1986). Random vibrations of railway vehicles. In I. Elishakoff & R.H. Lyon (Eds.), *Random vibration—Status and recent developments* (pp. 475–498). Amsterdam: Elsevier.
Ushkalov, V. F., & Reznikov, L. R. (1982). *Stochastic dynamics of railway vehicles.* Kiev: "Naukova Dumka" Publishers (in Russian).
Vaicaitis, R. (1973). Generalized random forces for rectangular panels. *AIAA Journal, 11*(7), 984–988.
Vaicaitis, R. (1974). Free vibration of beams with random characteistics. *Journal of Sound and Vibration, 35*(1), 13–21.
Vaicaitis, R. (1977). Noise transmission by viscoelastic sandwich panels. *NASA-TND-8516*.

Vaicaitis, R. (1982a). Recent research on noise transmission into aircraft. *Shock and Vibration Digest, 14*(8), 13–18.
Vaicaitis, R. (1982b). Noise transmission into propellant aircraft. *Shock and Vibration Digest, 17*(8), 15–20.
Vaicaitis, R., Grosveld, F. W., & Mixson, J.S. (1985). Noise transmission through aircraft panels. *Journal of Aircraft, 22*(4), 303–310.
Valeev, K.G. (1971). Vibration of cylindrical panle under action of random loading. In M.S. Karnishin (Ed.), *Theory of plates and shells*, pp. 17–22. Moscow: "Nauka" Publishing House (in Russian).
Valeev, K. G., & Kvitka, V.E. (1973). Determination of the stress state of a flat panel in the acoustic field of an exhaust jet. *Soviet Applied Mechanics, 6*(4), 423–427.
Van Flotow, A., & Mercadel, M. (1995). The measurement of noise and vibration transmitted into aircraft cabins. *Sound and Vibration, 29*(1), 16–19.
Van Hoolst, T. (1995). Resonances between two stellar oscillation modes with nearly equal frequencies. *Astronomy and Astrophysics, 295*, 371–392.
Van Lear, G. A., & Uhlenbeck, G. E. (1931). Brownian motion of strings and elastic rods. *Physical Review, 38*, 1583–1598.
Vanmarcke, E.H. (1972). Properties of spectral moments with applications to random vibration. *Journal of Engineering Mechanics, 98*, 425–446.
Vanmarcke, E.H. (1975). On the distribution of first passage time for normal stationary random function. *Journal of Engineering Mechanics, 42*, 215–220.
Vanmarcke, E.K. (1976). Structural response to earthquakes. In C. Lomnitz & E. Rosenblueth (Eds.), *Seismic risk and engineering devisions* (pp. 287–335). Amsterdam: Elsevier.
Vanmarcke, E.H. (1979). Some recent developments in random vibration. *Applied Mechanics Reviews, 32*(10), 1197–1202.
Vanmarcke, E.H. (1983). *Random fields : Analysis and synthesis*. Cambridge, MA: MIT Press.
Vanmarcke, E.H. (1986). Random vibration approach to seismic safety evaluation. In I. Elishakoff & R.H. Lyon (Eds.), *Random vibration—Status and recent developments* (pp. 499–510). Amsterdam: Elsevier.
Vasiliev, D.G. (1984). Improved asymptotic formula of courant for the distribution function of Plate's eigenvalues. *Proceedings of the Seminar on Theory and Numericakl Methods of Analysis of Plates and Shells* (Vol.2, pp. 90–96). Tbilisi: Tbilisi University Press (in Russian).
Vasiliev, D.G. (1986). Two-term asymptotics of spectrum of a thin elastic shell. *Advances of Mathematical Sciences, 41*(4), 182–183 (in Russian).
Vasiliev, D. G. (1987). Resonance phenomena in theory of elasticity and hydroelasticity. In *Extended abstract of the habilitation dissertation,* Institute of Mechanics Problems, Academy of USSR Sciences (in Russian).
Vasiliev, D. G., & Goldenveizer, A. L. (1991). Distribution of free vibration frequencies in two-and three dimensional bodies. In N. Kh.Arutiunian, I. F. Obraztsov, & V. Z.Parton (Eds.), *Mechanics of deformable solids* (pp 227–242).
Vedeneev, V. V. (2013). Effect of damping on flutter of simply supported and clamped panels at low supersonic speeds. *Journal of Fluids and Structures, 40*, 366–372.
Venini, P., & Mariani, C. (1997). Free vibrations of uncertain composite plates via stochastic Rayleigh-Ritz approach. *Computers and Structures, 64*(1–4), 407–423.
Ventzel, E. S., & Krauthammer, T. (2001). *Thin plates and shells:Theory, analysis, and applications*. Boca Raton: Taylor and Francis.
Ventzel, E. S., & Ovcharov, L. A. (2000). *Theory of random processes and its applications*. Moscow: "Vysshaya Shkola" Publishing House (in Russian).
Verechovsky, M. (2008). Simulation of simply cross-correlated random fields by series expansion methods. *Structural Safety, 30*(4), 337–363.
Vessel, K. N., Ram, Y. M., & Pang, S.-S. (2005). Sensitivity of repeated eigenvalues to perturbations. *AIAA Journal, 43*, 582–585.

Vickery, B. J. (1978). A model for the prediction of the response of chimneys to vortex shedding. In *Proceedings, 3rd International Symposium on Design of Industrial Chimeys* (pp. 157–162). Essen: Vulcan-Verlag.
Vielaverde, R. (1988). Rosenblueth's modal combination rule for systems with non-classical dambing. *Earthquake Engineering and Structural Dynamics, 16*, 315–328.
Vijayakumar, K. (1971). A new method for analysis of flexural vibration of rectangular orthotropic plates. *Journal of Aeronautical Society of India, 23*(4), 1974–1994.
Villaverde, R. (1988). Rosenblueth's modal combination rule for systems with non-classical damping. *Earthquake Engineering and Structural Dynamics, 16*(3), 315–328.
Virchis, V. J., & Robson, J. D. (1971). Response of an accelerating vehicle to random road undulation. *Journal of Sound and Vibration, 18*(3), 423–427.
Vitiello, P., De Rosa, S., & Franco, F. (2008). Convected field analysis of flat panels response to turbulent boundary layer induced excitation. *Aerospace Science and Technology, 12*(1), 91–104.
Vlasov, V. Z. (1949). *General theory of shells and its technological applications*. Gostekhizdat, Moscow-Leningrad (in Russian).
Volmir, A. S., & Kildibekov, I. G. (1965). Probabilistic characteristics of behaviour of cylindrical shell under acoustic loading. *Priladnaya Mekhanika (Applied Mechanics), 1*(3), 1–9 (in Russian).
Von Neumann, J., & Wigner, E. (1929). Über das Verhalten von Eigenwerten bei Adiabatischen Prozessen. *Zeitschrift für Physik A, 30*, 467–470 (in German).
Von Scheidt, J. (1983). Random vibrations generated by weakly correlated loads. In *Problems of stochastic analysis and applications* (pp. 173–247). Technical University Zwickau.
Von Scheidt, J., Fellenberg, B., & Wöhrl, U. (1994). *Analyse and simulation Stochastistischer Schwingungosysteme*. Stuttgart: B.G. Teubner (in German).
Von Scheidt, J., & Purkert, W. (1983). *Random eigenvalue problems*. Berlin: Akademic Verlag.
Von Scheidt, J., Stakloff, H.-J., & Wunderlich, R. (2000). Asymptotic expansions of integral functionals of e-Correlated processes. *Journal of Analysis and Applications, 19*(1), 255–268.
Von Scheidt, J., Starkloff, H.-J., & Wunderlich, R. (2002). Random transverse bibrations of a one-sided fixed beam and model reduction. *ZAMM: Zeitschrift für angewandte Mathematik und Mechanik, 82*(11–12), 831–845.
Von Scheidt, J., Starkloff, H.-J., & Wunderlich, R. (2002). Remarks on randomly excited oscillators. *ZAMM: Zeitschrift für angewandte Mathematik und Mechanik, 82*(11–12), 847–859.
Voznanov, A. N., & Postnov, V. A. (1973). On random vibrations of a spherical segment. In *Theory of shells and plates* (pp. 423–425). Moscow: "Nauka" Publishing House (in Russian).
Vu, T. V., Khosravifard, A., Hematiyan, M. R., & Bui, T. Q. (2018). Enhanced meshfree method With new correlation functions for functionally graded plates using a refined inverse sin shear deformation plate theory. *European Journal of Mechanics-A/Solids*.
Waberski, A. (1978a). Vibration statistics of thin plates with complex form. *AIAA Journal, 16*(8), 788–794.
Waberski, A. Stochastic vibration of plate with complex form and clamped on the boundary. *Engineering Transactions, 26*(2) (in Polish).
Waberski, A. (1979). Errata: Vibration statistics of thin plates of complex form. *AIAA Journal, 17*(5), 544.
Waberski, A. (1980). A note on the analysis of non-stationary random vibrations of plates of complex form. *Journal of Sound and Vibration, 70*(3), 453–457.
Waberski, A. (1982). The effect of non-stationary random vibrations of plates of complex form. *Applied Acoustics, 15*, 171–189.
Wade, N.-J. (2005). Sound and sight: Acoustic figures and visual phenomena. *Perception, 34*(10), 1275–1290.
Wagner, H. and Bhat, B. R. (1970). Linear response of an elastic plate to actual random load. *Ingenieur Archiv, 39*(3), 149–185.
Walker, A. M. (1973). Discussion on the papers by Priestley, and Tori and Hammond. *Journal of the Royal Statistical Society, 35*(2) 180–182.

Wallaschek, J. (1987). *Spektraldichtemethode und Integralkovarianzanalyse bei Stochastischen Schwingungen mechanischer Kontinua*. Darmstadt: VDI Verlag (in German).

Wallaschek, J. (1987). The influence of the correlation of the exciting-field process in stochastic beam vibrations. *ZAMM-Zeitschcrift für angewandle Mathematik und Mechanik, 67*(4) (in German).

Wallaschek, J. (1988). Integral covariance analysis for random vibrations of linear continuous mechanical systems. *Dynamics and Stability of Systems, 3*(1–2), 99–107.

Wallaschek, J. (1989). A comparison of different mathematical models for the spatial correlation of distributed excitation forces. In *Structural safety and reliability* (pp. 1371–1374). NewYork: ASCE.

Wan, F. Y. M. (1972). Linear partial differential equations with random forcing. *Studies in Applied Mathematics, 51*(2), 163–178.

Wan, F. Y. M. (1973). A direct method for linear dynamical problems in continuum mechanics with random loads. *Studies in Applied Mathematics, 52*(3), 259–276.

Wan, F. Y. M., & Lakshmikantham, C. (1972). Rotor blade response to random loads: A direct domain approach. In *AIAA Paper 72-169*, San Diego, CA.

Wang, B. (1991). *Non-stationary random response of plates with geometric nonlinearity*. M.E.S. Thesis, University of Western Ontario, Canada.

Wang, C., & Lai, J. C. S. (2005). Discussions on " On the Number of Modes Required for Statistical Energy Analysis-Based Calculations". *Journal of Sound and Vibration, 281*, 475–480.

Wang, C.-Y., & Vaicaitis, R. (1998). Active control of vibrations and noise of double wall cylindrical shells. *Journal of Sound and Vibration, 216*(5), 865–888.

Wang, D. J., Chou, C. S., & Wang, Q. S. (2003). Qualitative Properties of Frequencies and Modes of F169–175.

Wang, D., Jiang, J. S., & Zhang, W. (2004). Characteristics of sensitivity analysis of repeated frequencies. *AIAA Journal, 42*(9), 1939–1943.

Wang, D., Zhang, W. H., & Jiang, J. S. (2005). What are the repeated frequencies? *Journal of Sound and Vibration, 281*, 1186–1194.

Wang, J., Shang, P. J., & Ge, W. J. (2012). Multifractal cross-correlation analysis based on statistical moments. *Fractals, 20*, 271–279.

Wang, J. M., Fleet, D. J., & Hertzmann, A. (2007). Gaussian process dynamical models for human motion. *IEEE Transactions on Pattern Analysis and Machine Intelligence, 30*(2), 283–298.

Wang, H. W., & Deyou, Z. (2001). Study on the modal density of a free damped layered panel. *Applied Acoustics, 1*, 9.

Wang, L., Yang Z.-C., & Waters, T. P. (2010). Structural damage detection using cross correlation functions of vibration response. *Journal of Sound and Vibration, 239*(24), 5070–5086.

Wang, M. C., & Uhlenbeck, G. E. (1945). On the theory of Brownian motion II. *Review of Modern Physics, 17*(2–3), 323–342.

Wang, R.-T., & Lin, T.-Y. (1998). Random vibration of multi-span Timoshenks beam due to a moving load. *Journal of Sound and Vibration, 213*(1), 127–138.

Wang, W. L. (1966). A family of modal methods for comuting eigenvector derivatives with repeated roots. *Acta Mechanica Sinica, 12*(2), 154–168.

Wang, X.-G., & Sun, J. Q. (2003). Random fatigue of a higher order sandwich beam with parameter uncertainties. *Journal of Sound and Vibration, 260*, 349–356.

Wang, Y. T., & Foschi, R. O. (1997). Random field stiffness projective and reliability of laminated beams. *Structural Safety, 11*(3–4), 191–202.

Wang, Z. N. (1984). *Modal analysis of random vibrations and time domain identification of modal parameters*. Doctoral Thesis, Northwestern Polytechnic University, Xian, People's Republic of China (in Chinese).

Wapenaar, K. (2004). Retrieving the elastodynamic Green's function of an arbitrary inhomogeneous medium by cross-correlation. *Physical Review Letters, 93*(25), article 254301.

Warburton, G. B. (1975). Comment on "Applications of Bolotin's Method to Vibrations of Plates" By W.W. King and, C.C. Lin. *AIAA Journal, 13*(6), 839.

Warburton, G. B. (1964). *The dynamical behaviour of structures*. London: Pergamon Press (Second edition, Elsevier, 2014).

Warner, J. V., & Bernhard, R. J. (1990). Digital control of local sound fields in an aircraft passenger compartment. *AIAA Journal, 28*(2), 284–289.

Weaver, R. L. (1989). Spectral statistics in elastodynamics. *The Journal of the Acoustical society of America, 85*, 1005–1013.

Weaver, R. L. (1997). Multiple-scattering theory for mean response in a plate with sprung masses. *The Journal of the Acoustical Society of America, 101*(6), 3466–3474.

Weaver, W., Jr., & Johnston, P. R. (1987). *Structural dynamics by finite elements*. Englewood Cliffs: Prentice-Hall.

Wedig, W. (1974). *Zufallschwingungen*, VDI-Berichte Nr.221 (pp. 85–97) (in German).

Wedig, W. (1977). Moments and probability densities of parametrically excited systems and continuous systems. In *VII. Internationale Konferenz über nichtlineare Schwingungen, Berlin. II* (Vol. 2, pp. 469–492).

Wedig, W. (1979). Zufallschwingungen von querangeströmten Saiten. *Ingenieur-Archiv, 45*, 325–335 (in German).

Wedig, W. (1980a). Stochastic vibrations of slim masts induced by turbulent flows- modal and integral covariance methods in hybrid systems. In *Recent advances in structural dynamics* (pp. 131–146).

Wedig, W. (1980b). Stationäre Zufallschwingungen von Balken – eine neue Methode zur Kovarianzanabyse. *ZAMM: Zeitschrift für angewandte Mathematik und Mechanik, 60*, T89–T91 (in German).

Wedig, W. (1980c). Modale und integrale Kovarianzanalyse hybrider Systeme. *ZAMM: Journal of Applied Mathematics and Mechanics/Zeitschrift für angewandte Mathematik und Mechanik, 60*(8), 277–291 (in German).

Wedig, W. (1982). Stochastic systems with distributed parameters. In *Identification of Vibrating Structures* (pp. 493–510). Vienna: Springer.

Wedig, W. (1983). Covariance analysis of distributed systems under stochastic point forces. In H. W. Knobloch & H. W. Schmidt (Eds.), *Equidiff: Lecture notes in mathematics* (pp. 641–660). Berlin: Springer.

Wedig, W. (1984). A critical review of methods in stochastic structural dynamics. *Nuclear Engineering and Design, 79*, 281–287.

Wedig, W. (1987). Stochastische Schwingungen—Simulation, Schaetzung, und Stabilitaet. *Zeitschrift für angewandte Mathematik und Mechanik, 67*(4) T34–T42 (in German).

Wedig, W. (2003). Dynamics of half-car models under stochastic road excitations. In *ASME 2003 International Mechanical Engineering Congress and Exposition* (pp. 269–275). American Society of Mechanical Engineers.

Wedig, W., & Riemer, M. (1981). Bauwerke unter Wellenlast – Tokussierungseffekte bei Poissonerregung. In *VDI – Berichte* (pp. 201–207) (in German).

Wedig, W., & Riemer, M. (1982). Bettungs – und Dämpfungsanalyse homogen verteiller Systeme. *VDI-Berichte, 456*, 201–207 (in German).

Wedig, W. (2012). Digital simulation of road–vehicle systems. *Probabilistic Engineering Mechanics, 27*(1), 82–87.

Wedig, W. (2014). Multi-time-scale dynamics of road–vehicle systems. Probabilistic *Engineering* Mechanics, *37*, 180–184.

Weeks, E., & Cest, T. (1980). Complex stress response and reliability analysis of a composite elastic-viscoelastic missile configuration using finite elements. *Mechanics Research Communications, 7*(2), 59–63.

Wei, F. S., & Zhang, D. W. (1996). Eigenvector derivatives with repeated eigenvalues using generalized inverse technique. *AIAA Journal, 34*(10), 2206–2209.

Wei, G., Nong, Z., & Jinchen, J. (2009). A new method for random vibration analysis of stochastic truss structures. *Finite Element Analysis and Design, 45*(3), 190–199.

Weidenhammer, F. (1962). Das Schwingungsfundament unter Zufallserregung. *Ingeneur-Archiv, 31*(6), 433–443 (in German).
Wen, Y. K. (1990). *Structural load modeling and combination for performance and safety evaluation*. Oxford: Elsevier.
Wen, Y. K., & Gu P. (2004). Description and simulation of nonstationary processes based on Hilbert spectra. *Journal of Engineering Mechanics, 130*, 942–951.
Weyl, H. (1915). Das asymptotische Verteilungsgestz der Egenschwingungen eines beliebig gestalteten elastischen Körpers. *Rendiconti del Circolo Matematico di Palermo, 30*, 1–50 (in German).
Weyl, H. (1950). Ramifications, old and new of eigenvalue problem. *Bulletin of American Mathematical Society, 56*, 115–139.
Whitehead, D. S. (1962). *The analysis of blade vibration due to random excitation*. Reports and Memoranda No. 3253, Aeronautical Research Council, London: Her Majesty, Stationery Office.
White, P. H. (1969). Cross correlation in structural systems: Dispersion and nondispersion waves. *The Journal of the Acoustical Society of America, 45*(5), 1118–1128.
White, P. M. (1960). Sound transmission through a finite, closed, cylindrical shell. *The Journal of the Acoustical Society of America, 40*, 1124–1130.
White, R. G. (1978). A Comparison of the statistical properties of the response of aluminum alloy and CFRP plates to an acoustic excitation. *Composites*, 251–258.
White, P. H. (1969). Cross-correlation in structural systems. *The Journal of the Acoustical Society of America, 45*, 1118–1128.
White, R. G. (1978). A comparison of some statistical properties of the responses of aluminum alloy and CFRP plates to acoustic excitation. *Composites, 9*, 251–258.
White, R. G., & Teh, C. E. (1981). Dynamic behavior of isotropic plates under combined acoustic excitation and static in-plane compression. *Journal of Sound and Vibration, 75*(4), 527–547.
White, R. G. (1990). Developments in acoustic fatigue design process for composite aircraft structures. *Composite Structures, 16*, 171–192.
Whitney, C. (1990). *Random processes in physical systems*. New York: Wiley.
Wiener, N. (1930). Generalized harmonic analysis. *Acta Mathematica, 55*, 117–258.
Wijker, J. J. (2009). *Randomvibrations in spacecraft structural design: Theory and applications*. Berlin: Springer.
Wijker, Y. (2018). *Miles' equation in random vibrations: Theory and applications in spacecraft structures design*. Berlin: Springer.
Wilby, J. F. (1967). *The response of panels to turbulent boundary-layer excitation*. Ph.D. Thesis, University of Southmpton.
Wilby, J. F. (1989). Noise transmission into propeller-driven airplanes. *The Shock and Vibration Digest, 21*(6), 3–9.
Wilby, J. F., & Gloyna, F. L. (1972). Vibration measurements of an airplane fuselage structure. i: turbulant forcendary layer excitation. *Journal of Sound and Vibration, 23*(4), 467–486.
Wilkinson, J. H. (1968a). Almost diagonal matrices with multiple or close eigenvalues. *Linear Algebra and Its Applications, 1*(1), 1–12.
Wilkinson, J. P. D. (1968b) Modal densities of certain shallow structural elements. *The Journal of the Acoustical Society of America, 43*(2), 245–251.
Williams, F. W., & Banerjee, J. R. (1984). Accurately computed modal densities for panels and cylinders, including corrugations and stiffnesses. *Journal of Sound and Vibration, 93*(4), 481–488.
Wilms, E. V. (1997). The motion of a damped n-degree of freedom system with a single natural "frequency". *Journal of Sound and Vibration, 204*(4), 679–680.
Wilms, E. V., & Cohen, H. (1989). A 3-N degree of freedom system with two natural frequencies. *Mechanics Research Communications, 16*(1), 59–63.
Wilson, J. R., & Hunt, R. C. (2000). Problems with correlated data. *IEEE Transactions on Reliability, 49*(2), 127–130.
Wirsching, P. H., & Campbell, G. W. (1974). Minimal structural response under random excitation using the vibration absorber. *Earthquake Engineering and Structural Dynamics, 2*(4), 303–312.

Witt, M. (1987). Dynamic response and reliability of laminated plates to random loading. In I. Elishakoff & H. Irretier (Eds.), *Refined dynamical theories of beams, plates and shells and their applications* (pp. 274–285). Berlin: Springer.

Witt, M., & Sobczyk, K. (1980). Dynamic response of laminated plates to random loading. *International Journal of Solids and Structures, 16*, 231–238.

Wittig, L. E. (1971). *Random vibration of point driven strings and plates*. Ph.D. Thesis, Department of Mechanical Engineering, Massachusetts Institute of Technology, MA.

Wojtkiewica, S. F., Johnson, E. A., Bergman, L. A., Grigoriu, M., & Spencer, B. F. (1999). Response of stochastic dynamical systems driven by additive Gaussian and Poisson white noise: Solution of forward Kolmogorov equation by a spectral finite difference method. *Computer Methods in Applied Mechanics, 168* (1–4), 73–89.

Wood, A. D., & Zaman, F. D. (1977). Free vibration of randomly inhomogeneous plates. *Journal of Sound and Vibration, 52*(4), 543–552.

Woodhouse, J. (1981). An approach to the theoretical background of statistical energy analysis applied to structural vibration. *The Journal of the Acoustical Society of America, 69*, 1695–1709.

Woodhouse, J. (1981). An introduction to statistical energy analysis of structural vibration. *Applied Acoustics, 14*, 455–469.

Wright, M., & Weaver, R. (Eds.). (2010). *New directions in linear acoustics and vibration: Quantum chaos, random matrix theory, and complexity*. Cambridge, UK: Cambridge University Press.

Wu, Y., Zhang, J. S., Chen, B., & Shen, S. Z. (2007). Identification of the dominant vibration modes of single-layer reticulated shells under wind action. *International Journal of Space Structures, 22*(2), 123–132.

Wunderlich, R. (1999). *Systems of random differential equations and model reduction*. Habilitation Thesis, Techanical University Chemnitz, Federal Republic of Germany.

Xi, D. C., Chen, Q. H., & Cai, G. Q. (1986). Control effects of damped sandwich beam on random vibration. *Journal of Vibration, Acoustics, Stress, and Reliability in Design, 108*(1),65–68.

Xiao, W., Fang, F., & Liu, J. (2017). Analysis of axial compressive loaded beam under random support excitation. *Journal of Sound and Vibration, 410*, 378–388; Cross-Correlation Analysis, *Chaos, Solitons and Fractals, 96*, 59–69, 2017.

Xie, G., Thompson, D. J., & Jones, C. J. C. (2002). *The influence of boundary conditions on the mode count and modal density of two-dimensional system*. ISVR Technical Memorandum No. 894, Institute of Sound and Vibration Research, University of Southampton.

Xie, G., Thompson, D. J., & Jones, C. J. C. (2004). Mode count and modal density of structural systems: relationships with boundary conditions. *Journal of Sound and Vibration, 274*(3), 621–651.

Xiong, H., & Shang, P. Y. (2016). Weighted multifractal cross-correlation analysis based on Shannon entropy. *Communication in Nonlinear Science and Numerical Simulation, 30*, 268–283.

Xu, L., Chen, J., & Li, J. (2012). Probability density evolution analysis of engineering structures via cubature points. *Computational Mechanics, 50*(1), 135–156.

Xu, L., & Zhai, W.-M. (2017). Stochastic analysis model for vehicle-track coupled systems subject to earthquakes and track random irregularities. *Journal of Sound and Vibration, 407*, 209–225.

Xu, K. (1992). *Dynamic analysis and control of structures with closely spaced natural frequencies*. Ph.D. Thesis, Northwestern University.

Xu, K., & Igusa, T. (1992a). Dynamic characteristics of multiple substructures with closely spaced frequencies. *Earthquake Engineering and Structural Dynamics, 21*(12), 1059–1070.

Xu, K., & Igusa, T. (1992b). Nonstatonary response of structures with closely spaced frequencies. *Journal of Engineering Mechanics, 118*(7), 1387–1405.

Xu, K., Warnitchai, P., & Igusa, T. (1993). Control of a structure with two closely spaced frequencies. *Journal of Guidance, Control, and Dynamics, 16*(6), 1177–1178.

Xu, X. F., & Stefanou, G. (2012). Convolved orthogonal expansions for uncertainty propagation: Application to random vibration problems. *International Journal for Uncertainty Quantification, 2*(4).

Xu, Y., Qian, Y., Chen, J., & Song, G. (2015). Stochastic dynamic characteristic of FGM beams with random material properties. *Composite Structures, 133*, 585–594.
Xu, Y. L., Zhang, W. S., Ko, J. M., & Lin, J. H. (1999). Pseudo-excitation method for vibration analysis of wind-excited structures. *Journal of Wind Engineering and Industrial Aerodynamics, 83*, 443–454.
Xu, Z., & Wu, B. (2008). Derivatives of complex eigenvectors with distinct and repeated eigenvalues. *International Journal of Numerical Methods in Engineering, 75*, 945–963.
Xue, S. D., Cao, Z., & Wang, X. S. (2002). Random vibration study of structures under multi-component seismic excitations. *Advances in Structural Engineering, 5*(3), 185–192.
Yadav, D., Kamle, S., & Talukdar, S. (1997). Transverse vibration of multimass loaded variable section beams under random excitation. *Journal of Sound and Vibration, 207*, 151–173.
Yadav, D., & Nigam, N. C. (1978). Ground induced non-stationary response of vehicles. *Journal of Sound and Vibration, 61*(1), 117–126.
Yadav, D., & Verma, N. (1998). Free vibration of composite circular cylindrical shells with random material properties, part 1: General theory. *Computers and Structures, 41*(3–4), 331–338.
Yadav, D., & Verma, N. (2001). Free vibration of composite circular cylindrical shells with random material properties, part II: Applications. *Computers and Structures, 51*(4), 371–380.
Yaglom, A. M. (2004). *An introduction to the theory of stationary random processes*. Courier Corporation.
Yamada, T., & Takemiya, H. (1968). Studies on the responses of multi-degrees of freedom systems subjected to random excitation with applications to the tower and pier systems of long suspension bridges. *Memoirs of Faculty of Engineering, Kyoto University, 30*, 371–396.
Yamada, Y., Iemura, H., Kawano, K., & Venkataramana, K. (1989). Seismic response of offshore structures in random seas. *Earthquake Engineering and Structural Dynamics, 18*(7), 965–981.
Yamamoto, Y., & Baker, J. W. (2013). Stochastic model for earthquake ground motion using wavelet packets. *Bulletin of Seismological Society of America, 103*, 3044–3056.
Yamazaki, T., & Kuroda, K. (2008). Verification of modal density description of analytical SEA parameters for structure-borne sound by using FEM. *Transactions of the Japan*, Society of Mechanical Engineers, Series C, *73*(726), 1963–1970.
Yan, H., Parrett, A., & Nack, W. (2000). Statistical energy analysis by finite elements for middle frequency vibration. *Finite Elements in Analysis and Design*, 35(4), 297–304.
Yanev, P. I. (1970). *Response of simple inelastic systems to random excitation*. Thesis, Massachusetts Institute of Technology.
Yang, C., Xiu, D., & Kirby, M. (2013). Visualization of covariance and cross-covariance fields. *International Journal for Uncertainty Quantification, 3*(1), 25–38.
Yang, D., Chen, G., & Zhou, J. (2017). Exact solutions of fully nonstationary random vibration for rectangular Kirchhoff plates using discrete analytical method. *International Journal of Structural Stability and Dynamics, 17*(10), article 1750126.
Yang, D.-X., Chen, G.-H., & Zhou, J.-L. (2017). Exact solutions of fully nonstationary random vibration for rectangular Kirchhoff plates using discrete analytical methods. *International Journal of Structural Stability and Dynamics*, article 1750126.
Yang, D.-X., & Zhou, J.-L. (2015). A stochastic model and synthesis for near-fault impulsive ground motions. *Earthquake Engineering and Structural Dynamics, 44*, 243–264.
Yang, I. M. (1970). *Stationary random response of multidegree-of-freedom systems*. Ph.D. Thesis, California Institute of Technology, Pasadena, CA.
Yang, I. M., & Iwan, W. D. (1972). Calculation of correlation matrices for linear systems subjected to nonwhite excitation. *Journal of Applied Mechanics, 39*(6), 559–562.
Yang, J., Liew, K. M., & Kitipornchai, S. (2005). Stochastic analysis of compositionally graded plates with system randomness under static loading. *International Journal of Mechanical Sciences, 47*(10), 1519–1541.
Yang, J. N. (1972). Nonstationary envelope process and first excursion probability. *Journal of Structural Mechanics, 1*(2), 231–248.

Yang, J. N. (1973). First-excursion probability in non-stationary random vibration. *Journal of Sound and Vibration, 27*(2), 165–182.

Yang, J. N., & Heller, R. A. (1975). Random vibration of compliant wall. *Journal of Engineering Mechanics, 102*, 1041–1057.

Yang, J. N., & Liu, S. C. (1981). Distribution of maximum and statistical response spectra. *Journal of Engineering Mechanics, 107*(6), 1089–1102.

Yang, J. N., & Shinozuka, M. (1970). First passage time problem. *The Journal of the Acoustical Society of America, 47*, 393–394.

Yang, T. Y. (1986). *Random vibration of structures*. New York: Wiley.

Yang, T. Y., & Kapania, R. K. (1984). Finite element random response analysis of cooling tower. *Journal of Engineering Mechanics, 1101*(4), 589–609.

Yang, Z. Yu, Z., & Sun, H. (2007). On the cross correlation function amplitude vector and its application to structural damage detection. *Mechanical Systems and Signal Processing, 21*(7), 2918–2932.

Yang, Z. C., Yu, Z. F., & Sun, H. (2007). On the cross-correlation function amplitude vector and its application to structural damage detection. *Mechanical Systems and Signal Processing, 21*(7), 2918–2932.

Yao, G., & Gao, X. (2011). Controllability of regular systems and defective systems with repeated eigenvalues. *Journal of Vibration and Control, 17*(10), 1574–1581.

Yaroslavskii, V. Yu. (1974). Accounting for influence of correlation of bending-torsional modes of buildings for analysis on seismic excitations. *Referativnyi Sbornik TSINIS, Seismic Constructions, 5*, Moscow (in Russian).

Yasui, H., Marukawa, H., Katagiri, J., Tamura, Y., & Watanabe, N. (1999). Study of wind-induced response of long-span structure. *Journal of Wind Energy and Industrial Aerodynamics, 83*, 277–288.

Yaules, M., Dangupta, A., & Habtown, E. D. (2012). Life estimation model of a cantilever beam subjected to complex random vibration. *Fatigue and Fracture of Engineering Materials and Structures, 35*(1), 1058–1070.

Yiang, Z. Q., & Zhou, W. X. (2011). Multifractal detrending moving-average cross-correlation analysis. *Physical Review E, 8*, article 016106.

Yin, X. F., Wang, L., Kong, B., Song, G., & Liu, Y. (2018). Probability analysis of the vibration of bridges with rough surface under stochastic traffic. *International Journal of Structural Stability and Dynamics, 18*(9), article 1850108.

Ying, Z. G., Feng, J., Ni, Y. Q., & Zhu, W. Q. (2011). Elastic potential response analysis of a piezoelectric shell under random micro-vibration excitations. *Smart Materials and Structures, 20*, article 105029.

Yousafzai, A. H., & Ahmadi, G. (1982). Deterministic and stochastic earthquake response analysis of the containment shell of a nuclear power plant. *Nuclear Engineering Design, 72*, 309–320.

Yu, D., Al-Yafawi, A., Nguyen, T. T., Park, S., & Chung, S. (2011). High-cycle fatigue life prediction for Pb-free BGA under random vibration loading. *Microelectronics Reliability, 51*(3), 649–656.

Yu, H. J., Yu, K., Li, X., & Zhao, R. (2015). Modal density of sandwich panels based on an improved ordinary sandwich panel theory. *Composite Structures, 131*, 927–938.

Yu, L. (2010). *Fatigue reliabilityof ship structures*. Glasgow, UK: Glasgow University Press.

Yu, R., Wang, L., Liu, H., & Zhou, X. (2005). The CCQC3 method for multi-component seismic responses of non-classically damped irregular structures. In *Structures Congress 2005: Metropolis and Beyond* (pp. 1–12).

Yu, R.-F., Zhou, X.-T. and Tuan, M.Q. (2012). Dynamic response analysis of generally damped linear system with repeated eigenvalues. *Structural Engineering and Mechanics*, 42(4), 449–469.

Yu, R.-F., Zhou, X.-Y., Abduwahit, A. (2016). Complex complete quadratic combination method for damped system with repeated eigenvalues. *Earthquake Engineering and Engineering Vibration, 15*, 537–550.

Zang, Z., & Chen, S. (1991). The standard deviations of the eigen solutions for random MDOF systems. *Computers and Structures, 39*, 603–607.

Zaretsky, M. V., & Sidorenko, A. S. (2012). Dynamics of the aviation product structure during kinematic random excitation. In *Proceedings of MAI (Moscow Aviation Institute),* No. 58 (in Russian).

Zaretsky, M. V., & Sidorenko, A. S. (2013). Durability index estimate of aviation product under random forces. In *Proceedings of MAI (Moscow Aviation Institute),* No. 70 (pp. 1–28) (in Russian).

Zembaty, Z. (1987). On the reliability of tower-shaped structures under seismic excitation. *Earthquake Engineering and Structural Dynamics, 15,* 761–775.

Zembaty, Z. (1988). A note on non-stationary stochastic response and strong motion duration. *Earthquake Engineering and Structural Dynamics, 16,* 1189–1200.

Zembaty, Z., & Krenk, S. (1993). Spatial seismic excitations and response spectra. *Journal of Engineering Mechanics, 119*(12), 2449–2460.

Zelnik, D. S. (1973). On multiplicity of frequencies of cylindrical shells. In *Theory of shells and plates* (pp. 572–575). Moscow: "Nauka" Publishing House (in Russian).

Zeman, J. L., & Bogdanoff, J. L. (1969). A comment on complex structural response to random vibrations. *AIAA Journal, 7,* 1225–1231.

Zerva, A. (1993). Effect of spatial variability and propagation of seismic ground motions on the response of multiple supported structures. *Probabilistic Engineering Mechanics, 6,* 212–221.

Zhang, M., & Schmidt, R. (2014). Sensitivity analysis of an auto-correlation_function_based damage index and its application in structural damage detection. *Journal of Sound and Vibration, 333,* 7352–7363.

Zhang, H.-Y., Bai, C.-Q., & Mas, Y.-J. (2015). Stochastic finite element modeling and response analysis of rotor systems with random properties under random loads. *Journal of Mechanical Science and Technology, 29*(8), 3083–3090.

Zhang, L.-J., Lee, C.-M., & Wang, Y. S. (2002). A study on nonstationary random vibration of a vehicle in time and frequency domains. *International Journal of Automotive* Technology, *3*(3), 101–109.

Zhang, T. S., & Fang, T. (2000). Random response of linear viscoelastic systems under random excitation. In J. H. Zhang & X. N. Zhang (Eds.), *Proceedings of International Conference on Advanced Problems in Vibration Theory and Applications, ICAPV 2000* (pp. 580–584). Beijing: Science Press.

Zhang, Z., Jiang, C., Ruan, X. X., & Guan, F. J. (2017). A novel evidence theory model dealing with correlated variables and the corresponding structural reliability analysis method. *Structural and Multidisciplinary Optimization,* 1–16.

Zhang, Z. F., & Chen, S. H. (1991). The Standard deviations of the eigensolutions for random MDOF systems. *Computers and Structures, 39*(6), 603–607.

Zhang, Y., & Der Kiureghian, A. (1994). First-excursion probability of uncertain structures. *Probabilistic Engineering Mechanics*, 9, 135–143.

Zhang, Y. H., Li Q. S., Lin, J. H., & Williams, F. W. (2009). Random vibration analysis of long-span structures subjected to spatially varying ground motions. *Soil Dynamics and Earthquake Engineering, 29*(4), 620–629.

Zhang, Y. Q., & Wang, W. L. (1995). Eigenvector derivatives of generalized nondefective eigenvector with repeated eigenvalues. *Journal for Gas Turbines and Power*, 117(1), 207–212.

Zhang, Y.-W., Lin, J. H., Zhao, Y., Howson, W. P., & Williams, F. W. (2010). Symplectic random vibration analysis of a vehicle moving on an infinitely long periodic track. *Journal of Sound and Vibration, 329*(21), 4440–4454.

Zhai, G.-F., Chen, Y.-H., & Ren, W.-B. (2007). Random vibration analysis of switching apparation based on Monte Carlo method. *Journal of Zhejiang University-Science A, 8*(3), 422–427.

Zhao, H., & Grigoriu, M. (2016). Reliability of linear systems subjected to wind loads. *Journal of Risk and Uncertainty in Enginering Systems, Part A: Civil Engineering, 2*(2), article C4015003.

Zhao, J., Miao, X., Li, F., & Li, H. (2018). Probabilistic diagnostic algorithm-based damage detection for plates with non-uniform sections using the improved weight function. *Journal of Vibration Engineering & Technologies, 6*(3), 249–260.

Zhao, J. X., Kong, X. R., Wang, S. N., & Luo, W. B. (2007). Experimental identification of modal density parameters of aluminum honeycomb sandwich panel. *Journal of Harbin Institute of Technology, 5*, 33.

Zhao, Y., Deng, Z., & Han, Y. (2019). Dynamic response analysis of structure with hybrid random and interval uncertainties. *Chaos, Solitons & Fractals*, article 109495.

Zharov, A. M. (1964). Reaction of a structure to nonstationary seismic action. *Structural Mechnics and Analysis of Structures* (6), 37–41 (in Russian).

Zheleznov, N. A. (1959). Some questions of spectral correlation theory of nonstationary signals. *Radiotechnics and Electronics* (3), 359–373 (in Russian).

Zheng, H., Cai, C., Pau, G. S. H., & Liu, G. R. (2005). Minimizing vibration response of cylindrical shells through layout optimization of passive constrained damping treatments. *Journal of Sound and Vibration, 279*, 739–756.

Zheng, S. (1998). Combined method for calculating eigenvector derivatives with repeated eigenvectors. *AIAA Journal, 26*(3), 428–431.

Zheng, Z.-B., & Dai, H.-Z. (2017). Simulation of Multi-dimensional random fields by Kahrunen-Loève expansion. *Computer Methods in Applied Mechanics and Engineering, 324*, 221–247.

Zhinzher, N. I. (1994). Dynamic edge effects in aeroelastic vibrations of plates. *Mechanics of Solids, 19*(5), 168–172.

Zhinzher, N. I. & Belostotsky A. M. (1976). Asymptotic distribution of eigenfrequencies of shells of shells of rotation. *Proceedings of the Moscow Power Engineering Institute* (Vol. 280, pp. 13—18, in Russian).

Zhinzher, N. I., & Khromatoff, V. E. (1971). Application of an asymptotic method to the investigation of the vibration spectra of orthotropic circular cylindrical shells. *Mechanics of Solids* (6), 95–101 (in Russian).

Zhinzher, N. I., & Khromatoff, V. E. (1973). Frequency density of free vibrations of orthotropic circular cylindrical shells. In *Theory of shells and plates* (pp. 464–468). Moscow: "Nauka" Publishing House (in Russian).

Zhon, Z.-B., & Lus, J. (1998). Responses of a new active vibration isolation system to random vibrations. *Chinese Physics Letters, 15*(6), 429–436.

Zhong, W. X. (1996). Review of a "High-Efficency Algorithm Series for Structural Random Response". *Progress in Natural Sciences, 6*, 257–268.

Zhongsheng, L. (1994). The concept of principal mode shapes and measures of contrllability and observability for mode shapes with repeated frequencies. *Acta Mechanica Sinica, 4*.

Zhou, B., Guo, W., Han, X., & Tan, S. K. (2016). Random seismic response analysis of jacket structure with Timoshenko's beam theory. *Ships and Offshore Structures, 11*(4), 438–444.

Zhou, Y. X., Zhang, Y. M., & Yao, G. (2017). Stochastic forced vibration analysis of a tapered beam with performance deterioration. *Acta Mechanica, 228*, 1393–1406.

Zhu, D., Yi, X., Wang, Y., & Sabra, K. (2010). Structural damage detection through cross-correlation analysis of mobile sensing data. In *World Conference on Structural Control and Monitoring*.

Zhu, M., Yang, Y., Guest, J. K., & Shields, M. D. (2017). Topology optimization for linear stationary stochastic dynamics applications to frame structures. *Structural Safety, 67*, 116–131.

Zhu, W.-Q. (1986). Wide band random vibration of structures. In I. Elishakoff & R. H. Lyon (Eds.), *Random vibration—Status and recent developments* (pp. 525–536). Amsterdam: Elsevier.

Zhu, W. Q. (1992). *Random vibration*. Beijing: Science Publishing (in Chinese).

Zhu, W. Q., & Cai, G. Q. (2011). Random vibration of viscoelastic system under broad-band excitations. *International Journal of Non-Linear Mechanics, 46*(5), 720–726.

Zhu, W. Q., & Ying, L. (1986). Wide band random vibration of rectangular plates with commensurable aspect ratio. In *Proceedings of the International Conference on Vibration Problems in Engineering* (pp. 890–895). Xi'an Joaitung University.

Zibdeh, H. S. (1995). Stochastic vibration of an elastic beam due to random moving loads and deterministic axial forces. *Engineering Structures, 17*, 530–535.

Zibdeh, H. S., & Abu-Hilal, M. (2003). Stochastic vibration of laminated coated beam transversed by a random moving load. *Engineering Structures, 25*(3), 397–404.

Zibdeh, H. S., & Juma, S. H. (1999). Dynamic response of a rotating beam subjected to a random moving load. *Journal of Sound and Vibration, 223*(5), 741–758.

Zibdeh, H. S., & Rackwitz, R. (1995). Response moments of an elastic beam subjected to Poissonian moving loads. *Journal of Sound and Vibration, 188*, 479–495.

Zibdeh, H. S., & Rackwitz, R. (1996). Moving loads on beams with general boundary conditions. *Journal of Sound and Vibration, 195*(1), 85–102.

Ziegler, F. (1987). Random vibrations: A spectral method for linear and nonlinear structures. *Probabilistic Engineering Mechanics, 2*(2), 92–99.

Ziegler, F., Irschik, H., & Heuer, R. (1986). Nonstationary response of polygonally shaped membranes to random excitation. In I. Elishakoff & R. H. Lyon (Eds.), *Random vibration—Status and recent developments* (pp. 555–565). Amsterdam: Elsevier.

Ziola, S. M., & Gorman, M. R. (1991). Source location in thin plates using cross-correlation. *The Journal of the Acoustical Society of America, 90*(5), 2551–2556.

Zhou, Q., He, D., & Zhao, Y. (2019). Stationary random vibration of a viscoelastic timoshenko cantilever beam under diverse random processes. *Proceedings of the Institution of Mechanical Engineers, Part C: Journal of Mechanical Engineering Science* (in press).

Zubaydi, A., Haddara, M. R., & Swamidas, A. S. J. (2000). On the use of the Autocorrelaion function to identity the damage in the side shell of a ship's hull. *Marine Structures, 13*(6), 537–551.

Zuccarello, B., & Adragna, N. F. (2007). A novel frequency domain method for predicting fatigue crack growth under wide band random loading. *International Journal of Fatigue*, *29*(6), 1065–1079.

Zvdokimov, M. A., & Kuznetsov, V. A. (2008). Random vibrations in elastic medium. *Transactions of Samara State Technical University* (1), 20–22 (in Russian).

Zverev, A. Ya., & Efimtsov, B. M. (2012). A comparative estimate of acoustic emission of thin-walled structures by a turbulent boundary layers for various representations of the mutual spectrum. *Acoustical Physics, 58*(4), 420–425.

Zwikker, C., & Kosten, C. W. (1949). *Sound absorbing materials*. Amsterdam: Elsevier.

Author Index

I. Elishakoff, *Dramatic Effect of Cross-Correlations in Random Vibrations of Discrete Systems, Beams, Plates, and Shells*,
https://doi.org/10.1007/978-3-030-40394-2

I

L

T

W

Subject Index

I. Elishakoff, *Dramatic Effect of Cross-Correlations in Random Vibrations of Discrete Systems, Beams, Plates, and Shells*,
https://doi.org/10.1007/978-3-030-40394-2

GPSR Compliance

The European Union's (EU) General Product Safety Regulation (GPSR) is a set of rules that requires consumer products to be safe and our obligations to ensure this.

If you have any concerns about our products, you can contact us on ProductSafety@springernature.com

In case Publisher is established outside the EU, the EU authorized representative is:

Springer Nature Customer Service Center GmbH
Europaplatz 3
69115 Heidelberg, Germany

Batch number: 10756376

Printed by Printforce, the Netherlands